Bioinformatics Algorithms

Bioinformatics Algorithms

Design and Implementation in Python

Miguel Rocha

University of Minho, Braga, Portugal

Pedro G. Ferreira

Ipatimup/i3S, Porto, Portugal

Academic Press is an imprint of Elsevier
125 London Wall, London EC2Y 5AS, United Kingdom
525 B Street, Suite 1650, San Diego, CA 92101-4495, United States
50 Hampshire Street, 5th Floor, Cambridge, MA 02139, United States
The Boulevard, Langford Lane, Kidlington, Oxford OX5 1GB, United Kingdom

Library of Congress Cataloging-in-Publication Data
A catalog record for this book is available from the Library of Congress

British Library Cataloguing-in-Publication Data
A catalogue record for this book is available from the British Library

ISBN: 978-0-12-812520-5

For information on all Academic Press publications
visit our website at https://www.elsevier.com/books-and-journals

Publisher: Mara Conner
Acquisition Editor: Chris Katsaropoulos
Editorial Project Manager: Serena Castelnovo
Production Project Manager: Vijayaraj Purushothaman
Designer: Miles Hitchen

Typeset by VTeX

Contents

CHAPTER 1

Introduction

1.1 Prelude

In the last decades, important advances have been achieved in the biological and biomedical fields, which have been boosted by important advances in experimental technologies. The most known, and arguably most relevant, example comes from the impressive evolution of sequencing technologies in the last 40 years, boosted by the large investment in the Human Genome Project mainly in the 1990's [92,150].

Additionally, other high-throughput technologies for measuring gene expression, protein or compound concentrations in cells, have led to a real revolution in biological and medical research. All these techniques are currently able to generate massive amounts of the so called omics data, that can be used to foster scientific research in the life sciences and promote the development of novel technologies in health care, biotechnology and related areas.

Merely as two examples of the impact of these novel technologies and produced data, we can pinpoint the impressive development in areas such as personalized (or precision) medicine and metabolic engineering efforts within industrial biotechnology.

Precision medicine addresses the growing trend of tailoring treatments to the characteristics of individual (or groups of) patients. This has been made increasingly possible by the availability of genomic, epigenomic, gene expression, and other types of data about specific patients, allowing to determine distinct risk profiles for certain diseases, or to study differentiated effects of treatments correlated to patterns in genomic, epigenomic or gene expression data. These data allow to design specific courses of action based on the patient's profiles, allowing more accurate diagnosis and specific treatment plans. This field is expected to grow significantly in the coming years, as it is confirmed by projects such as the 100,000 Genomes Project launched by the UK Prime Minister David Cameron in 2012 (`https://www.genomicsengland.co.uk/the-100000-genomes-project/`) or the launch of the Precision Medicine Initiative, announced in January 2015 by President Barack Obama, and which has started in February 2016.

Cancer research is an area that largely benefited from the recent advances in molecular assays. Projects such as the *Genomic Data Commons* (`https://gdc.cancer.gov`) or the *International Cancer Genome Consortium* (ICGC, `http://icgc.org/`) are generating comprehensive and multi-dimensional maps of the genomic alterations in cancer cells from hundreds of individuals in dozens of tumor types with a visible scientific, clinical, and societal impact.

Bioinformatics Algorithms. DOI: 10.1016/B978-0-12-812520-5.00001-8

Other current large-scale efforts boosted by the use of high-throughput technologies and led by international consortia are generating data at an unprecedented scale and changing our view of human molecular biology. Of notice are projects such as the 1000 Genomes Project (`www.internationalgenome.org/`) that provides a catalog of human genetic variation across worldwide populations; the Encyclopedia of DNA Elements (ENCODE, `https://www.encodeproject.org/`) has built a map of functional elements in the human genome; the Epigenomics Roadmap (`http://www.roadmapepigenomics.org/`) is characterizing the epigenomic landscapes of primary human tissues and cells or the Genotype-Tissue Expression project (GTEx, `https://www.gtexportal.org/`) which is providing gene expression and quantitative trait loci from more than 50 human tissues.

On the other hand, metabolic engineering is related to the improvement of specific microbes used in industrial biotechnological processes to produce important compounds as bio-fuels, plastics, pharmaceuticals, foods, food ingredients and other added-value compounds. Strategies used to improve host microbes include blocking competing pathways through gene deletion or inactivation, overexpressing relevant genes, introducing heterologous genes or enzyme engineering.

In both cases, the impact of data availability has been tremendous, opening new avenues for scientific advance and technological development. However, this has also raised significant challenges in the management and analysis of such complex and large volumes of data. Biological research has become in many aspects very data-oriented and this has been intricately connected to the ability to handle these huge amounts of data generating novel knowledge, or as Florian Markowetz recently puts it "All biology is computational biology" [108]. Therefore, the value of the sophisticated computational tools that have been developed to address these data processing and analysis has been undeniable.

This book is about Bioinformatics, the field that aims to handle these biological data, using computers, and seeking to unravel novel knowledge from raw data. In the next section, we will discuss further what Bioinformatics is, and the different tasks and scientific disciplines that are involved in the field. To close the chapter, we will overview the content of the remaining of the book to help the reader in the task of better navigating it.

1.2 What is Bioinformatics

Bioinformatics is a multi-disciplinary field at the intersection of Biology, Computer Science, and Statistics. Naturally, its development has followed the technological advances and research trends in Biology and Information Technologies. Thus, although it is still a young field, it is evolving fast and its scope has been successively redefined. For instance, the National Institute of Health (NIH) defines Bioinformatics in a broad way, as the "research, development, or application of computational tools and approaches for expanding the use of biological, medical, biological, behavioral, or health data" [79]. According to this definition, the tasks involved include data acquisition, storage, archival, analysis, and visualization.

Some authors have a more focused definition, which relates Bioinformatics mainly to the study of macromolecules at the cellular level, and emphasize its capability of handling large-scale data [105]. Indeed, since its appearance, the main tasks of Bioinformatics have been related to handling data at a cellular level, and this will also be the focus of this book.

Still in the previous seminal document from the NIH, the related field of Computational Biology is defined as the "development and application of data-analytical and theoretical methods, mathematical modeling, and computational simulation techniques to the study of biological, behavioral, and social systems". Thus, although deeply related, and sometimes used interchangeably by some authors, the first (Bioinformatics) relates to a more technologically oriented view, while the second is more related to the study of natural systems and their modeling. This does not prevent a large overlap of the two fields.

Bioinformatics tackles a large number of research problems. For instance, the *Bioinformatics* (`https://academic.oup.com/bioinformatics`) journal publishes research on application areas that include genome analysis, phylogenetics, genetic, and population analysis, gene expression, structural biology, text mining, image analysis, and ontologies and databases.

The National Center for Biotechnology Information (NCBI, `https://www.ncbi.nlm.nih.gov/Class/MLACourse/Modules/MolBioReview/bioinformatics.html`) unfolds Bioinformatics into three main areas:

- developing new algorithms and statistics to assess relationships within large data sets;
- analyzing and interpreting different types of data (e.g. nucleotide and amino acid sequences, protein domains, and protein structures);
- developing and implementing tools that enable efficient access and management of different types of information.

This book will focus mainly on the first of these areas, covering the main algorithms that have been proposed to address Bioinformatics tasks. The emphasis will be put on algorithms for sequence processing and analysis, considering both nucleotide and amino acid sequences.

1.3 Book's Organization

This book is organized into four logical parts encompassing the major themes addressed in this text, each containing chapters dealing with specific topics.

In the first part, where this chapter is included, we introduce the field of Bioinformatics, providing relevant concepts and definitions. Since this is an interdisciplinary field, we will need to address some fundamental aspects regarding algorithms and the Python programming language (Chapter 2), cover some biological background needed to understand the algorithms put forward in the following parts of the book (Chapter 3).

The second part of this book addresses a number of problems related to sequence analysis, introducing algorithms and proposing illustrative Python functions and programs to solve them. The Bioinformatics tasks addressed will cover topics related with basic sequence processing and analysis tasks, such as the ones involved in transcription and translation (Chapter 4), algorithms for finding patterns in sequences (Chapter 5), pairwise and multiple sequence alignment algorithms (Chapters 6 and 8), searching homologous sequences in databases (Chapter 7), algorithms for phylogenetic analysis from sequences (Chapter 9), biological motif discovery with deterministic and stochastic algorithms (Chapters 10, 11), and finally Hidden Markov Models and their applications in Bioinformatics (Chapter 12).

The third part of the book will focus on more advanced algorithms, based in graphs as data structures, which will allow to handle large-scale sequence analysis tasks, such as the ones typically involved in processing and analyzing next-generation sequencing (NGS) data. This part starts with an introduction to graph data structures and algorithms (Chapter 13), addresses the construction and exploration of biological networks using graphs (Chapter 14), focuses on algorithms to handle NGS data, addressing the tasks of assembling reads into full genomes (in Chapter 15) and matching reads to reference genomes (in Chapter 16).

The book closes with Part IV, where a number of complementary resources to this book are identified (Chapter 17), including interesting books and articles, online courses, and Python related resources, and some final words are put forward.

As a complementary source of information, a website has been developed to complement the book's materials, including code examples and proposed solutions for many of the exercises put forward in the end of each chapter. This website is available in the following URL: `https://github.com/miguelfrocha/BioinformaticsAlgorithmsBook`.

CHAPTER 2

An Introduction to the Python Language

In this chapter, we provide a brief introduction to Python, in its version 3, which will be used as the programming language in this book. We will discuss the different ways of using Python to solve problems, covering basic data structures and functions pre-defined by the language, but also discussing how a programmer can define new functions, modules, and programs/scripts. We will address the basic algorithmic constructs, such as conditional and cyclic instructions, as well as data input/output, files and exception handling. Finally, we will cover the paradigm of object-oriented programming and its implementation in Python using classes and methods, also browsing through some of the main pre-defined classes and their methods.

2.1 Features of the Python Language

Python is an interpreted language that can be run both in script or in interactive mode. It was created in the early 1990s by Guido van Rossum [149], while working at Centrum Wiskunde & Informatica in Amsterdam.

Python has two main versions still in use by the community: 2.x (where the last release was 2.7 in 2010) and 3.x, where new releases have been coming out gradually (last at the time of writing was 3.6 in the end of 2016). In this book, we will use Python 3.x, since it is the most recent and eliminates some quirks of the previous Python 2.x releases, being also the predictable future of the language. Due to some compatibility issues, a number of programmers still use the previous 2.x versions, but this scenario is rapidly changing. Most of the examples in this book will still work in Python 2 and the reader should not face difficulties in switching to that version if that is a requirement for some reason.

As its creator puts it, Python is "a high-level scripting language that allows for interactivity". It combines features from different programming paradigms including imperative, scripting, object-oriented, and functional languages.

We emphasize the following features of the language:

- **Concise and clear syntax.** The syntax not only improves code readability, but also allows an easy-to-write code that increases programming productivity.
- **Code indentation**. Opposed to other languages that typically use explicit markers, such as begin-end blocks or curly braces to define the structure of the program, Python only

Bioinformatics Algorithms. DOI: 10.1016/B978-0-12-812520-5.00002-X

uses the colon symbol ":" and indentation to define blocks of code. This allows for a more concise organization of the code with a well defined hierarchical structure of its blocks.

- **Set of high-level and powerful data types**. Built-in data types include primitive types that store atomic data elements or container types that contain collections of elements (preserving or not the order of their elements). The language offers a flexible and comprehensive set of functions to manage and manipulate data structures designed with built-in data types, which makes it a self-contained language in the majority of the coding situations.
- **Simple, but effective, approach to object-oriented programming**. Data can be represented by objects and the relations between those objects. Classes allow the definition of new objects by capturing their shared structural information and modeling the associated behavior. Python also implements a class inheritance mechanism, where classes can extend the functionality of other classes by inheriting from one or more classes. The development of new classes is, therefore, a straightforward task in Python.
- **Modularity**. Modules are a central aspect of the language. These are pieces of code previously implemented that can be imported to other programs. Once installed, the use of modules is quite simple. This not only improves code conciseness, but also development productivity.

Being an interpreted language means that it does not require previous compilation of the program and all its instructions are executed directly. For this to be possible, it requires a computer program called interpreter that understands the syntax of the programming language and executes directly the instructions defined by the programmer.

In the interactive mode, there is a working environment that allows the programmer to get a more immediate feedback on the execution of each code statement through the use of a shell or command line. This is particularly useful in learning or exploratory situations. If a proper interpreter is installed, typing "python" in the command line of your operating system will start the interactive mode that is indicated by the prompt symbols ">>>". Python 3's interpreter can be easily downloaded and installed from `https://www.python.org/downloads/`.

An extended version of the Python command line is provided by *Jupyter notebooks* (`http://jupyter.org/`), a web application which allows to create and share documents that contain executable Python code, together with explanatory text and other graphical elements in HTML. This allows to test your code similarly to the Python shell, but also to document it.

In the script mode, a file containing all the instructions (a program or script) is provided to the interpreter, which is then executed without further intervention, unless explicitly declared in the code with instructions for data input. Larger blocks of code will be presented preferentially in script mode.

Both modes are also present in many of the popular Integrated Development Environments (IDE), such as *Spyder*, *PyCharm* or *IDLE*. We recommend that the reader becomes familiar with one of these environments, as these are able to provide a working environment where a number of features are available to increase productivity, including tools to support program development and enhanced command lines to support script mode.

One popular alternative to easily setup your working environment is to install one of the platforms that already include a Python distribution, a set of pre-installed packages, a tool to manage the installed packages, a shell, a notebook, and an IDE to write and run your programs. One of such environments is *anaconda* (`https://www.anaconda.com/`), which has free versions for the most used computing platforms. Another alternative is *canopy* from *Enthought* (`https://www.enthought.com/product/canopy`). We invite the user to explore this option, which although not being mandatory, greatly increases productivity, since they easily create one (or several distinct) working environments.

In computer programming, an algorithm is a set of self-contained instructions that describes the flow of information to address a specific problem or task. Data structures define the way data is organized. Computer programs are defined by the interplay of these two elements: data structures and algorithms [155].

Next, we introduce each of the main Python built-in data types and flow control statements. With these elements in hand, the reader will be able to write its own computer programs. Whenever possible, we will use examples inspired by biological concepts, which could be nucleotide or protein sequences or other molecular or cellular concepts.

For illustrative purposes of the coding structure, we will sometimes use pseudo-code syntax. Pseudo-code is a simplified version of a programming language without the specifics of any language. This type of code will be used to convey an algorithmic idea or the structure of code and has no meaning to the Python interpreter.

Also, comments are instructions that are ignored by the code interpreter. They allow the programmer to add explanatory notes throughout the text that may help later to interpret the code. The symbol # denotes a comment and all text to the right of it will be ignored. These will be used throughout the code examples to add explanations within the programs or statements.

The Python language is based on three main types entities which are covered in the following sections:

- **Variables** or **objects** which can be built-in or defined by the programmer. They handle data storage.
- **Functions** are the program elements that are used to define data processing, resembling the concept of mathematical functions, typically taking one or more inputs and possibly returning a result (output).

- **Programs** are developed for the solution of a single or multiple tasks. They consist of a set of instructions defining the information flow. During the execution of a program, functions can be called and the state of variables and objects are altered dynamically.

Within functions and programs, a set of statements are used to describe the data flow, including testing or control structures for conditional and iterative looping. We will start by looking at some of Python's pre-defined variables and functions, and will proceed to look at the algorithmic constructs that allow for the definition of novel functions or programs.

2.2 Variables and Pre-Defined Functions

2.2.1 Variable Types

Variables are entities defined by their names, referring to values of a certain type, which may change their content during the execution of a program. Types can be atomic or define complex data structures that can be implemented through *objects*, instances of specific *classes*. Types can be either pre-defined, i.e. already part of the language, or defined by the programmer.

Pre-defined variable types in Python can be divided into two main groups: primitive types and containers. Primitive types include numerical data, such as integer (`int`) or floating point (`float`) (to represent real numbers). Boolean, a particular case of the integer type, is a logical type (allowing two values, `True` or `False`).

Python has several built-in types that can handle and manage multiple variables or objects at once. These are called *containers* and include the *string*, *list*, *tuple*, *set*, and *dictionary* types. These data types can be further sub-divided according to the way their elements are organized and accessed. Strings, lists, and tuples are sequence types since they have an implicit order of their elements, which can be accessed by an index value.

Sets and dictionaries represent a collection of unordered elements. The set type implements the mathematical concept of sets of elements (any object), where the position or order of the elements is not maintained. Dictionaries are a mapping type, since they rely on a hashing strategy to map keys to the corresponding values, which can be any object.

One important characteristic of some of these data types is that once the variables are created their value cannot be changed. These are called immutable types and include strings, tuples, and sets. An attempt to alter the composition of a variable of one of these types generates an error.

Table 2.1 provides a summary of the different features of Python primitive and container data types. The last column indicates if the container type allows different types of their elements or not.

Table 2.1: Features of Python built-in data types.

Data type	Complexity	Order	Mutable	Indexed	Heterogeneous
int	primitive	-	yes	-	-
float	primitive	-	yes	-	-
complex	primitive	-	yes	-	-
Boolean	primitive	-	yes	-	-
string	container	yes	no	yes	no
list	container	yes	yes	yes	yes
tuple	container	yes	no	yes	yes
set	container	no	no	no	yes
dictionary	container	no	yes	no	yes

In Python, the data type is not defined explicitly and it is assumed during the execution by taking into account the computation context and the values assigned to the variable. This results in a more compact and clear code syntax, but may raise execution errors, for instance when the name of a variable is incorrectly written or when non-operations are performed (e.g. sum of an integer with a string).

2.2.2 Assigning Values to Variables

In Python, the operator = is used to assign a value to a variable name. It is the core operation in any computer program, that allows to define the dynamics of the code. This operator is different from ==, that is used for a comparison between two variables, testing their equality.

Therefore, following the syntax: *varname = value*, the variable named *varname* will hold the corresponding value. The right side of an assignment can also be the result of calling a function or a more complex expression. In that case, the expression is evaluated before the corresponding resulting value is bound to the variable.

When naming variables composed by multiple words, boundaries between them should be introduced. In this book, we will use the underscore convention, where underscore characters are placed between words (e.g. *variable_name*).

We will use the interactive mode (shell) to look in more detail on how to declare variables of the different built-in types and what type of operations are possible on these variables.

Python allows variables to have an undefined value that can be set with the keyword `None`. In some situations, it may be necessary to test if a variable is defined or not before proceeding.

```
>>> x = None
>>> x == None
True
```

If a variable is no longer being used it can be removed by using the del clause.

```
>>> del x
```

2.2.3 Numerical and Logical Variables

Numeric variables can be either integer, floating point (real numbers), or complex numbers. Boolean variables can have a True or False value and are a particular case of the integer type, corresponding to 1 and 0, respectively.

```
# integer
>>> sequence_length = 320
# floating
>>> average_score = 23.145
# Boolean
>>> is_sequence = True
>>> contains_substring = False
```

Multiple variables can also be assigned with the same value in a single line:

```
>>> a = b = c = 1
```

Multiple values can be assigned to different variables in a single instruction, in the order they are declared. In this case, variables and values are separated by commas:

```
>>> a, b, c = 1, 2, 3
```

Assignments using expressions in the right hand side are possible. In this case, the evaluation of the expression follows the arithmetic precedence rules.

```
>>> a = 2*(1+2)
```

Variable values can also be swapped in the same line.

```
>>> a,b,c = c,a,b
```

The following binary arithmetic operators can be applied to two numeric variables:

- + – sum;
- - – difference;
- * – multiplication;

Table 2.2: Mathematical and character functions.

Function	Description
abs(*x*)	absolute value of *x*
round(*x*, *n*)	*x* rounded to a precision of *n* places
pow(*x*, *y*)	*x* raised to power of *y*
ord(*c*)	ASCII numerical code for character *c*
chr(*x*)	ASCII string (with a single character) for numerical code *x*

- **/** – division;
- ** – exponentiation;
- **//** – integer division; and,
- **%** – modulus operator (remainder of the integer division).

All the usual arithmetic priorities apply here as well. Some examples are shown in the following code:

```
>>> x = 5
>>> y = 4
>>> x + y
9
>>> x * y
20
>>> x / y
1
>>> x // y
1
>>> z = 25
>>> z % x
0
>>> z % y
1
>>> x ** y
625
```

Table 2.2 describes examples of mathematical functions and functions to convert between numerical values and characters based on the ASCII code.

Examples of the use of these functions follow:

```
>>> abs(-3)
3
```

```
>>> round(3.4)
3.0
>>> float(2)
2.0
>>> int(3.4)
3
>>> int(4.6)
4
>>> int (-2.3)
-2
>>> 0.00000000000001
1e-14
>>> 2.3e-3
0.0023
>>> chr(97)
'a'
>>> ord('a')
97
```

The package *math* includes a vast number of useful mathematical and scientific functions, including trigonometric functions (**sin**, **cos**, **tan**), square root (**sqrt**), and others as **factorial**, logarithm (**log**) and power function (**exp**), where **exp**(x) returns e^x. By importing this package, these functions become available in the current working session.

With these capacities, the interactive environment of Python becomes a powerful scientific calculator, as shown in the examples below:

```
>>> import math
>>> math.sqrt(4)
2.0
>>> math.sin(0.5)
0.479425538604203
>>> math.log(x)
1.6094379124341003
>>> math.pi
3.141592653589793
>>> math.tan(math.pi)
-1.2246467991473532e-16
>>> math.e
2.718281828459045
```

```
>>> math.exp(1)
2.718281828459045
>>> math.log(math.e)
1.0
```

Notice in the previous examples that the constants pi and e are also available within the package.

When updating a variable x through an arithmetic operation that depends on the current state of x, the assignment operator can be preceded by a mathematical operator, +=, -=, *=, /=, %= or **=. As an example, the two following expressions are equivalent:

```
# equivalent statements
>>> a += 3
>>> a = a+3
```

Given two Boolean variables x and y, the logical operations **and**, **or**, and **not** provide a logical result of True or False, returning respectively the logical conjunction, disjunction, and negation.

2.2.4 Containers

2.2.4.1 Lists

Lists allow the storage and processing of sequences of values of different types. They can be defined by square brackets enclosing a sequence of comma-separated values. The notation [] defines an empty list.

A list with the integer values from 1 to 5 and 7 can be declared as follows:

```
>>> x = [1, 2, 3, 4, 5, 7]
```

Each of the values in a list can be accessed by an index that defines the position of the value within the sequence. Indexes are integer values that range from 0 (first position) to the number of elements on the list minus 1 (last position). To access the third element of the previously defined list, we can use the syntax $x[2]$. Since lists are mutable objects, we can also directly change their values, for instance with $x[0] = -1$, setting the first element to be -1.

By using negative indexes, the elements of the list can be accessed backwards, where $x[-1]$ corresponds to the last element of the list, i.e. 7, $x[-2]$ to the second last element, and so on. Elements can also be removed from lists with the **del** statement.

```
>>> x = [1, 2, 3, 4, 5, 7]
>>> del x[4]
>>> x
[1, 2, 3, 4, 7]
```

The list object can handle heterogeneous data. Thus, as the example below shows, a list may contain data from different types including other lists.

```
>>> y = [1, 2, "A", "B", [4, "C"]]
```

The + operator can be used to concatenate (join) lists together:

```
>>> [1,2,3] + [4,5,6]
[1, 2, 3, 4, 5, 6]
```

Slicing is a powerful mechanism to generate sub-lists, i.e. lists containing selected elements that preserve their order from the original list. The general syntax for slicing is *list_name*[*startslice* : *endslice* : *step*]. Note that a more compact syntax for slicing can be used by omitting some arguments. In the case where it is possible to omit arguments, default values are assumed. Also, the *endslice* is always one position after the last selected element.

Examples of slicing on lists follow below:

```
>>> x
[1, 2, 3, 4, 7]
# elements from index 1 to 2
>>> x[1:3]
[2, 3]
# elements from index 0 to 2
>>> x[:3]
[1, 2, 3]
# elements from index 3 to end of list
>>> x[3:]
[4, 7]
# all elements but the last element
>>> x[:-1]
[1, 2, 3, 4]
# every two elements
>>> x[::2]
[1, 3, 7]
# skipping first and last elements
```

```
>>> x[1:-1]
[2, 3, 4]
# reversing the list
>>> x[::-1]
[7, 4, 3, 2, 1]
```

Python offers a set of several useful functions for list management. One of the most frequent operations to perform on a list is to determine its length. The function **len** returns the number of elements in a list.

Matrices can also be implemented in Python using lists of lists, each representing a row (or a column) of the matrix. As an example, the following code creates a matrix with 3 rows and 3 columns, prints the number of rows and columns, checks the element on the third row and second column, and gets all elements of the last row.

```
>>> m = [[1,2,3],[4,5,6],[7,8,9]]
>>> print("Number of rows:" , len(m))
>>> print("Number of columns:", len(m[0]))
Number of rows: 3
Number of columns: 3
>> m[2][1]
8
>>> m[-1]
[7, 8, 9]
```

2.2.4.2 Strings

Strings are sequences of characters, which can be defined by text enclosed by the characters "..." or '...'. A string can be visualized using the function **print** that requires parentheses to enclose the object to be printed, as shown in the example below.

```
>>> txt = "This is a string"
>>> print(txt)
This is a string

>>> suffix = "as an example"
>>> txt = txt + " " + suffix
>>> print(txt)
This is a string as an example
```

Strings are ordered sequences. Therefore, sub-sequences can be generated through slicing in the same way as with lists.

```
>>> txt[0:4]
'This'
>>> txt[0:4][::-1]
'sihT'
```

Strings are immutable objects. The application of **del** or the attempt to assign new values will generate an error.

```
>>> txt[0] = "t"
Traceback (most recent call last):
  File "<stdin>", line 1, in <module>
TypeError: 'str' object does not support item assignment
```

2.2.4.3 Tuples

Tuples represent a third type of ordered sequences. They can be declared by assigning a sequence of values separated by commas within the container ( ). They share many of the properties of lists with the exception that once created they are immutable. Some examples of their use follow:

```
>>> t = (1, "a", 2, "c", [1,2,3])
>>> t[1]
'a'
>>> t[-2:]
('c', [1, 2, 3])
>>> coords = (10,20)
>>> x,y = coords
>>> coords[1]
20
>>> coords[1] = 25
Traceback (most recent call last):
  File "<stdin>", line 1, in <module>
TypeError: 'tuple' object does not support item assignment
```

2.2.4.4 Sets

Sets are non-ordered collections of immutable objects. They are defined by the syntax **set()**. They are particularly useful for membership testing or removing duplicates from lists, since they directly implement the mathematical concept of a set.

```
>>> set([1,2,3])
{1, 2, 3}
# intersection between sets
>>> set([1,2,3]) & set([1,2,4])
{1, 2}
# union between sets
>>> set([1,2,3]) | set([1,2,4])
{1, 2, 3, 4}
```

Other operators on sets include: - (difference), ^ (symmetric difference), and the mathematical inclusion relations <= (is subset) or >= (is superset).

2.2.4.5 Dictionaries

Dictionaries are unordered containers that provide a mapping association between keys and values. Each key should be unique. Variables of this type are defined by key/value pairs separated by the colon symbol and enclosed by the container **{ }**.

```
# an empty dictionary
>>> translate_numeric_text = {}
>>> translate_numeric_text = {"one":1, "two":2, "three":3,
1:"one", 2:"two", 10:"many"}
>>> translate_numeric_text
{1: 'one', 2: 'two', 10: 'many', 'three': 3, 'two': 2, 'one': 1}
```

A value in a dictionary is accessed by the corresponding key and the access is done with square brackets []:

```
>>> print (translate_numeric_text['one'])
1
>>> translate_numeric_text["one"] + translate_numeric_text["two"]
3
```

Values in a dictionary can be directly altered or deleted:

```
>>> translate_numeric_text["ten"]=10
>>> del translate_numeric_text["three"]
```

2.2.5 Variable Comparison

Depending on their types, variables can be compared in different ways, as it is the case with the pre-defined types we have covered above. There are a number of comparison operators that can be used, all of which require two variables and return a Boolean result:

- < (less than);
- > (greater than);
- == (equal to);
- <= (less than or equal to);
- >= (greater than or equal to);
- ! = (not equal to).

To test if a value is an element in a container, the operator **in** can be used as follows: *value* **in** *cont*, while the absence can be tested as: *value* **not in** *cont*.

Some examples are given next:

```
>>> x = 23.4
>>> y = 32.3
>>> y > x
True
>>> y <= x
False
# contained in list
>>> x = [1, 2, 3, 4, 7]
>>> 2 in x
True
>>> 5 in x
False
>>> 8 not in x
True
# contained in string
>>> "cd" in "abcdef"
True
>>> "g" in "abcdef"
False
```

2.2.6 Type Conversion

In some situations, it is necessary to convert variables from one type to another. The function **type** provides information on the data type of the variable passed as argument. Functions with the names of the corresponding data types provide the conversion of a variable to the required data types, namely: **int**, **float**, **bool**, **str**, **list**, **dict** and **set**. Let's check some examples:

```
# string to numeric
>>> int("123")
123
# integer to float
>>> float(123)
123.0
# provides integer part of float
>>> int(123.5)
123
# numeric to boolean. 0, null or empty objects to False.
# all other values correspond to True.
>>> bool(0)
False
# string representation of the variable
>>> str(123)
'123'
# string to list
>>> list("list")
['l', 'i', 's', 't']
# list to set
>>> set(["A","B","A"])
{'A', 'B'}
#list of tuples to dictionary
>>> dict([("one",1),("two",2)])
{'one': 1, 'two': 2}
```

A common operation on floating numbers is to round to a certain number of decimals. The function **round** can be used for that purpose:

```
# round to one decimal
>>> round(123.456,1)
123.5
```

Table 2.3 summarizes the functions to declare and convert variables to different data types.

2.3 Developing Python Code

2.3.1 Indentation

Before looking at some algorithmic structures and their Python implementation, it is important to check the set of indentation syntax rules of the language, which allow for a more

Table 2.3: Functions for data type conversion.

Function	Description
int(*x*)	converts string or float *x* to integer
float(*x*)	converts string or int *x* to float (real value)
str(*obj*)	string representation of an object or variable *obj*
tuple(*elems*)	returns tuple given its elements
list(*iter*)	empty list (if no argument passed) or list initialized with an iterable object *iter*
dict(*iter*)	empty dictionary (if no argument passed) or dictionary initialized with iterable object with name-value tuples
set(*iter*)	converts iterable object to set
type(*obj*)	returns the type of an object *obj*
repr(*obj*)	canonical string representation of an object

concise and clear coding. Being syntactically relevant, changes in indentation may affect the logic of the code. These rules can be summarized as follows:

- Code begins in the first column of the file.
- All lines in a block of code are indented in the same way, i.e. aligned by a fixed spacing. No brackets are required to delimit the beginning and the end of the block.
- A colon (`:`) opens a block of code.
- Blocks of code can be defined recursively within other blocks of code.

The following pseudo-code represents a cascade of three nested blocks of code, where *block_1* has N statements, *block_2* has M statements and *block_3* has K statements.

```
statement preceding block_1:
    statement_1 within block_1
    statement_2 within block_1
    .
    .
    statement_N within block_1 preceding block_2:
        statement_1 within block_2
        statement_2 within block_2
        .
        .
        statement_M within block_2 preceding block_3:
            statement_1 within block_3
            .
            .
            .
```

```
        statement_K within block_3
statement after block_1
```

2.3.2 User-Defined Functions

We have seen a number of pre-defined Python functions. Let us now proceed to defining our own functions. These are simply defined by the `def` keyword, the function name and a list of arguments, followed by a block of statements after the colon.

The `return` statement is used to provide a result for the function, and typically is the last statement, although with more complex code this might not be the case. In case there is nothing to return, `None` can be returned. In case multiple values need to be returned, a tuple with the results can be returned.

It is good practice to include at the beginning of the function one or more lines describing its purpose and usage. Documentation text is enclosed by triple quotes "' "'. These lines are called documentation string (*docstring*). Programs that generate automatic code documentation use this information to document the different functions.

```
def function_name([arguments,]):
    '''Function documentation'''
   —some_statements_here—
      (...)
    return result
```

As an example, a function that computes the square of an inputted value can be defined as follows:

```
def square(x):
    return x * x
```

Or, more generally, we can define a function that receives the base of the exponential expression and the power to raise the base.

```
def power(x, y):
    return x ** y
```

The function syntax allows defining default values for its arguments. In that case, if the argument is omitted when calling the function, the default value is assumed for that argument. Next, we show an example where the exponent is defined by default to be 2.

```
def power(x, y=2):
    '''returns x to the power of y assuming a default value of 2 for
  y'''
        return x ** y
```

Note that variables declared and used within the functions are local to the function definition and only exist during the function call, i.e. when the function code is being executed, not being available when the function terminates. This is termed the *scope* of a variable, i.e. where it can be used. In general, variables defined within function definition blocks are local to these blocks. If they share the name with other variables outside the function, they are strictly independent and do not affect each other; in this case, within the function definition block, the name will refer to the local variable.

A function is called by simply invoking the function name with the respective parameter values enclosed by parentheses, in the order they are provided in the function definition. The returned value can be captured by a variable for subsequent computation or directly used in further computation. When called directly in the Python console, the return value will be printed in the screen, as shown below.

```
>>> x = 3
>>> x_square = square(3)
>>> x_square
9
>>> power(3)
9
```

2.3.3 Conditional Statements

Whenever the execution of a block of code depends on the result of a certain logical condition, an `if` statement can be used. The simplest case occurs when two different alternatives arise depending on the value of a condition, being represented by the pseudo-code below:

```
if logical_condition:
    statement_if_true_1
    statement_if_true_2
    (...)
else:
    statement_if_false_1
    statement_if_false_2
    (...)
```

In this case, the statements of the first block (below the `if`) are executed when the condition is true, while the statements in the second block (below the `else`) are executed otherwise. Note that the `else` block may not exist if there are no statements to execute if the condition is false.

If there are more than two alternative blocks of code, several `elif` (with the meaning else if) branches may exist with additional logical conditions, while a single final `else` clause exists, for the case when all previous conditions fail. The pseudo-code below represents the case with multiple conditions:

```
if logical_condition1:
    statement_1_condition1
    (...)
elif logical_condition_2:
    statement_1_condition2
    (...)
(elif ...)
else:
    statement_1_else
    (...)
```

A condition can be either a Boolean variable, a function or operator returning a Boolean result or an expression including those. A common case is the use of comparison operators, presented above in Section 2.2.5. An example of Python code where the score of an exam is tested in a cascade of `if/elif/else` statements is provided next:

```
score = 45  # or some other value
exam_result = ""
if score < 50:
    exam_result = "failed"
elif score > 90:
    exam_result = "outstanding"
elif score > 70 and score <= 90:
    exam_result = "excellent"
else:
    exam_result = "good"

print ("Exam result for a score of " + str(score) + " was " +
   exam_result)
```

A more compact notation to test the logical value of a numerical variable can be used, by including only the name of the variable in the test condition: `if` *var*. The first test will hold true, if *var* is different from zero. Also, a variable with value `None` always holds false.

```
>>> x = 1
>>> if x == 1:
...     print "Yes"
... else:
...     print "No"
...
Yes

>>> if x:
...     print "Yes"
... else:
...     print "No"
...
Yes

>>> x = 0
>>> if x:
...     print "Yes"
... else:
...     print "No"
...
No
```

As an additional example, let us define and test a function to calculate the largest numerical value between two inputs.

```
def maximum_two(x,y):
    if x>y: return x
    else: return y

print(maximum_two(3,4))
print(maximum_two(5,4))
print(maximum_two(3,3))
```

2.3.4 Conditional Loops

If a statement needs to be executed multiple times (zero or more times) depending on a test condition holding true, the use of a `while` statement can be appropriate. The pseudo-code of the block of statements within a `while` cycle is the following:

```
while condition:
    statement1_inside_while
    statement2_inside_while
    (...)
next_statement_after_while
```

In this case, the statements in the block will be executed while the condition holds true. When the condition switches to false, the loop ends, and the program flow follows with the next statement. This means that, to avoid infinite cycles, the programmer needs to insure that the condition will be false at some point in the program execution.

In the following example, the value of the variable a is printed, while it is smaller than 100. At each iteration its value is incremented by 10, thus insuring the cycle terminates:

```
>>> a = 0
>>> while a < 100:
...     print(a)
...     a = a+10
...
```

As an illustration of the potential of `while` cycles, in the next example, we develop a function that searches if a given element is present in a list of numbers. The function returns the position of the first occurrence of that element, or −1 if the element does not occur in the list.

```
def first_occurrence(lst, elem):
    ind = 0
    found = False
    while ind < len(lst) and not found:
        if lst[ind] == elem:
            found = True
        else:
            ind += 1
    if found: return ind
    else: return -1
```

```
l = [1,3,5,7,9]
print(first_occurrence(l, 5))
print(first_occurrence(l, 2))
```

2.3.5 Iterative Loop Statements

If we know in advance that we need to execute a block of statements a fixed number of times, a `for` loop can be used. This control structure provides an iterative loop through all the elements of an iterator, which can be retrieved from a container variable or using a function that yields these values. An *iterable* object is any object that can be iterated over, i.e. that provides a mechanism to go through all its elements. Strings or lists are examples of such objects. These objects are particularly suitable to be iterated in `for` loops. Indeed, iteration through a range of values is one of the most common tasks in programming.

In the following example, the code iterates through all the characters in a string and increments the value of the variable *seq_len* for each of them, obtaining in the end the length of the string.

```
my_seq = "ATACTACT"
seq_len = 0
for c in my_seq:
    seq_len += 1

print ("Sequence length " + str(seq_len))
```

There are also functions that return iterators, which can be used directly in these loops. Python offers a function **range** to generate an immutable sequence of integers between a *start* and a *stop* value, with an increment *step* value. The general syntax *range*([*start*,], *stop*, [, *step*]) allows a more compact notation where only the stop value needs to be provided. In that case, the *start* value is assumed to be zero and *step* to be one. By considering a *step* with a negative value, sequences of decreasing values can be generated. Note that in the generated sequence of values, the *stop* value is not included:

```
>>> b = 10
>>> for a in range(10):
...     print (b * a)
0
10
...
90
```

The following example iterates through a string and prints pairs of values with the index and the respective character found in that position:

```
my_seq = "ATACTACT"
idx = 0
for idx in range(len(my_seq)):
    print (str(idx) + " " + my_seq[idx])
```

The **enumerate** function returns an iterable object that simultaneously provides access to the index and the respective element. Thus, the previous code can be re-written as follows:

```
for idx, val in enumerate(my_seq):
    print (str(idx) + " " + my_seq[idx])
# or in alternative
for idx, val in enumerate(my_seq):
    print (str(idx) + " " + val)
```

Although all previous examples here focus on iterating over strings, similar examples can be put forward considering lists. As an example, we will develop a function, similar to the one presented in the previous section, where in this case we take a list and an element, and return all positions where the element occurs in the list (as another list). If the element does not occur, the result will be the empty list.

```
def all_occurrences(lst, elem):
    res = []
    for ind in range(len(lst)):
        if lst[ind] == elem:
            res.append(ind)
    return res

l = [1,3,5,7,9,1,2,3]
print(all_occurrences(l, 1))
print(all_occurrences(l, 2))
```

These cycles can also be nested to work with multi-dimensional data structures. A simple example are matrices, that are normally processed by nesting two `for` cycles, one iterating over row indexes and the other over columns. The following example provides a script that creates a matrix and uses this strategy to calculate the sum of all its elements.

```
m = [[1,4,7],[2,5,8],[3,6,9]]
s = 0
for i in range(len(m)):
    for j in range(len(m[i])):
            s += m[i][j]
print(s)
```

In some situations, it may be necessary to alter the expected flow within the loop (including both **for** and **while** loops). Python provides two statements for loop control. The **break** statement forces an immediate exit of the loop. On the other hand, the **continue** statement forces the loop to jump to the next iteration.

2.3.6 List Comprehensions

The generation of new lists with elements that follow a mathematical or a logical concept is a frequent task in programming. Suppose that we want to generate a list with multiples of ten smaller than 200. This can be easily done creating a **for** loop:

```
>>> multiples_ten = []
>>> for x in range(1, 21):
...     multiples_ten.append(x*10)
...
```

Python offers a quite convenient way to create new lists from existing ones. This is called the list comprehension syntax and takes a general form of:

```
[expression for obj in iterable]
```

The example above can now be re-written as:

```
>>> multiples_ten = [10*x for x in range(1,21)]
>>> multiples_ten
[10, 20, 30, 40, 50, 60, 70, 80, 90, 100, 110, 120, 130, 140, 150,
   160, 170, 180, 190, 200]
```

In the next example, we will extract all sub-strings of length 3 from a given sequence. This is done using list comprehensions, where the iterable object corresponds to the indexes of the sequence (from 0 to the last possible, corresponding to the string length minus 2).

```
seq = "ATGCTAATGTACATGCA"
seq_substrings = [(seq[x:x+3]) for x in range(0, len(seq)-2)]
```

The list comprehension syntax can also include a conditional statement:

```
[expression for obj in iterable if condition]
```

Using this feature, in the following example, we will create a list with the square of all the odd numbers smaller than 20.

```
>>> [ x**2 for x in range(0, 20) if x % 2 != 0]
[1, 9, 25, 49, 81, 121, 169, 225, 289, 361]
```

In another example, we can select all the sub-strings from string *seq* that contain the letter "A":

```
>>> [s for s in seq_substrings if "A" in s]
['ATG', 'CTA', 'TAA', 'AAT', 'ATG', 'GTA', 'TAC', 'ACA', 'CAT', 'ATG'
 ]
```

These examples demonstrate that list comprehension syntax provide a very intuitive and concise way to generate lists.

2.3.7 Help

Documentation about a given function or object regarding the input arguments can be found by using in interactive mode the **help** function. When using **help** without arguments, an interactive help session utility is launched in the console. Besides the documentation on the built-in functions, it also provides information on the list of modules or keywords for the Python language.

```
>>> help()
help> keywords
```

Keywords refer to a set of words that have a special syntax meaning and are reserved for specific language instructions. Names of variables, functions, classes, or modules cannot be any of the keywords. The list of keywords is given in Table 2.4 that shows the result of the previous command. For the documentation on built-in container data types, the argument passed to the help function will be the constructor symbol. For instance, **help({})** shows the documentation for dictionaries or **help("")** for strings.

2.4 Developing Python Programs

Programs, which in the case of interpreted languages are typically called scripts, define a set of instructions including calls to built-in and to previously defined functions. These instructions define the flow of data required to achieve the proposed tasks.

Table 2.4: Python keywords.

False	def	if	raise
None	del	import	return
True	elif	in	try
and	else	is	while
as	except	lambda	with
assert	finally	nonlocal	yield
break	for	not	
class	from	or	
continue	global	pass	

A typical simple program will start by reading some data from the user in some way, process these data and make its results available to the user. More complex programs can execute several cycles of these steps, allowing for further user interaction.

2.4.1 Data Input and Output

In many situations, it is necessary to interact with the console where the Python code is interpreted. This is either to receive data from the user, for instance the value of an input variable, or to display the result of a computation or any other relevant message.

The **print** statement, which we have already use before in some examples, allows to display elements on the console with options to format the outputted string. It can handle strings, variables and expressions separated by commas, as well as additional arguments that define the separator (*sep*) to be used between arguments and the termination string (*end*).

```
>>> my_seq = 'ATACTACT'
>>> print ("Sequence", my_seq, "has length", len(my_seq))
Sequence ATACTACT has length 8
>>> print (1,2,3, sep=";", end = ".")
1;2;3.
```

In the previous example, we defined the string tokens one by one as independent arguments of the **print** function. Another possibility for string output is to pass the tokens in a tuple, and use the **%** operator to define the location of the tokens within the string defined on the left side of **%** operator. In the following example, the **%s** symbol within the quotes determines that a string will be included in that position. The value to include is given in the respective position within the tuple on the right side of the **%** symbol after the quoted string.

```
>>> print ("%s + %s = %s" % (1, 2, 3))
1 + 2 = 3
```

The operator **%** can also be used to format numbers into strings. The general format specification is given by *%width.precision datatype*. The *width* parameter is optional defining the number of columns to where the number is aligned. If we need zero-fill then the number should be preceded with 0. The *precision* parameter defines the number of precision digits used when printing floating point numbers. The *datatype* parameter is always required and defines the resulting data types: *d* (decimal integer), *f* (floating), *s* (string) and *e* (float in exponential notation). Some examples follow to illustrate the use of this syntax.

```
>>> ratio = 123/456
>>> print ("ratio:", ratio)
ratio: 0.26973684210526316
>>> print ("ratio:%3.2f" % ratio)
ratio:0.27
>>> x = 123
>>> print ("%d" % x)
123
>>> print ("%09d" % x)
000000123
>>> print("%e" % ratio)
2.697368e-01
```

Reading a string from the console can be done with the **input** function. The argument to be passed is an optional string to be printed in the console, typically indicating a message that provides the user with an indication that an action is required. The value that is read is returned by the function as a string. Thus, depending on the type of value to be read, further type conversion may be required. As an example, if the input is a number, then the input string needs to be converted to a numerical format.

```
>>> x = input("value of x: ")
value of x: 2
>>> x = int(x)
>>> print ("square of x:", square(x))
square of x: 4
```

2.4.2 Reading and Writing From Files

We have seen above that it is possible to pass user input data to Python programs. However, this strategy is not practical for large volumes of data. In that case, data can be saved in files

Table 2.5: Option for handling files.

Mode	Description
'r'	open for reading (default)
'w'	open for writing, truncating the file first
'x'	create a new file and open it for writing
'a'	open for writing, appending to the end of the file if it exists
'b'	binary mode
't'	text mode (default)
'+'	open a disk file for updating (reading and writing)

in the operating system and read from the program. Also, results of larger dimension can be written by the program to existing or new files.

Reading and writing to files is quite easy in Python. This is basically a three-step procedure:

1. Open a stream to the file given its name and path, and obtain a file handler to access the contents of the file.
2. Read or write text in blocks or by lines.
3. Close the connection to the file.

Files can be either in text or binary format. In the text format, we have a human readable representation of the data, while in the binary format we have a cryptic, but typically more efficient representation.

The **open** function creates a stream to the file, taking as arguments the *filename*, which corresponds to the name given to the file, and the *open_mode* that specifies the way in which the file is open. There are several possibilities for this parameter, as described in Table 2.5. The default mode is the reading mode in text format represented by the letter 'r' or equivalently 'rt'. The more commonly used modes are: read 'r', write 'w', and append 'a'.

When writing to a file in the write mode, if the file already contains any contents, these will be overwritten and the new contents will be written starting from the beginning of the file. If the append mode is used, the new contents will be added to the end of the file, keeping any previous contents intact.

Among the optional arguments of the **open** function, encoding is of particular relevance. If not specified, the encoding mode will be assumed to be the one defined in the platform where the program is being run. Alternatively, it can take values such as *utf*8, *ascii* or *latin*1 allowing to specify the proper character encoding in text files.

The general format for the **open** function is the following:

```
file_handler = open(file_name, open_mode [, arguments])
```

The **open** function looks for the file named *file_name* in the current directory. In case the file is present in another location of the file system, both relative and absolute paths can be specified for the file to be found. Once the file is opened and the file handler is created, there are several options to read and write to a file.

Starting with the reading mode, the following methods are available:

- **read**(n), to read a string corresponding to the next block of n characters (or if it is omitted, a string up to the end of the file);
- **readline**(n), returns the next n characters from the next line (or if the parameter is omitted the entire line);
- **readlines**(), returns a list of strings with all the lines in the file.

Note that every time we want to iterate the file, the file needs to be opened again, so the file handler is repositioned at the beginning of the file. Consider a text file called "test.txt" with the following lines:

```
line number 1
line number 2
line number 3
line number 4
line number 5
```

In the following examples, the second call to the **readlines** function or **read** functions returns an empty string, since in both cases the end of the file was reached in the first call.

```
>>> fh = open("test.txt", "r")
>>> print (fh.readlines())
>>> print (fh.readlines())

>>> fh = open("test.txt", "r")
>>> print (fh.read())
>>> print (fh.read())
```

An iterative **for** loop can be used to perform the computation on a line-by-line basis. The following code template is commonly used:

```
with open(file_name) as fh:
    for line in fh:
        statements
        (...)
```

For our previous example file, we can scan all its lines and print with indentation proportional to the respective line number:

```
my_file_name = "test.txt"
prefix = ""
with open (my_file_name) as fh:
    for line in fh:
        print (prefix + line)
        prefix += "..."
```

To write to a file, the function **write(*s*)** writes a string *s* to the file, while **writelines(*lst*)** writes all the elements in the list of strings *lst* as lines in the file. The final operation consists in closing the connection to the file. This allows the previous operations on the file to take full effect and free the file for future use. This is done with the **close** function.

The following example opens the file in append mode and writes a line to the end of the file.

```
my_file_name = "test.txt"
fh = open(my_file_name, "a")
fh.write("\nlast line in file")
fh.close()
```

As a final example, we show how to use the function **writelines** to append additional multiple lines to the end of the file:

```
last_lines = ["\njust to finish", "\ntwo more lines"]
fh = open(my_file_name, "a")
fh.writelines(last_lines)
fh.close()
```

A useful method for file management is **flush** that immediately stores in the file the contents from previous write operations.

2.4.3 Handling Exceptions

During the execution of a program, errors may occur for which the interpreter may not know how to handle and cause it to abort. If we expect that an error may occur we can try to handle it by capturing the statement that originates the error and propose an alternative to proceed with the execution of the program. This is done with **try-except** blocks that have the following structure:

```
try:
    statements
    (...)
except Exception_type [variable]:
    statements
    (...)
else:
    statements
    (...)
```

The `try` block contains the normal processing block where we expect the error to occur. The `except` block contains the statements that the program should execute in case the error occurs.

The *Exception_type* refers to the type of exception that is raised. There are nearly 50 exception types, including for instance: *Warning*, *KeyboardInterrupt*, *ZeroDivisionError*, *RuntimeError* , *OverflowError*, *IndexError*, *AssertionError*, *ArithmeticError*. The choice of the error type to handle may not always be straightforward. One possibility to identify the appropriate exception type is to generate the error in the interactive console and use the raised exception shown by the trace-back message. For instance, a division by zero raises the following message:

```
>>> 5/0
Traceback (most recent call last):
  File "<stdin>", line 1, in <module>
ZeroDivisionError: division by zero
```

The following code shows how to use the **try-except** to handle possible divisions by zero:

```
x = 5
y = 1
try:
    r = x/y
except ZeroDivisionError:
    print ("Division by zero detected")
else:
    print ("ratio:", r)
```

In the previous code, the `try` block should only contain the statements susceptible to generate the error. This allows isolating the origin of the exception. If other exceptions are expected to occur then they should be handled with different `try-except` branches. The `else` block is optional and will contain the statements to be run in case the try block runs successfully.

2.4.4 Modules

As the complexity of the programs increases, the number of programmed functions also grows. In these cases, it is important to keep a good code organization. This will allow more efficient program maintenance and code reusability, therefore saving time and developing programs less prone to errors.

It is often the case that a function developed for one program can be of use in another program. Modules are one of the Python mechanisms for code organization and reuse. If we have a set of functions, constants or classes that share a common aspect, these can be gathered in a Python source file, which basically consists in a file with a ".py" extension. The **import** statement can then be used to load to the current program specific functions or all the functions of the module. Once loaded, these functions can be called as if they were part of the program.

Besides the user-defined modules, it is also possible to install and import modules developed by other programmers. When these modules are organized in a bundle of software, possibly with dependencies to other modules, these are referred as *packages*. Within the Python standard library, several modules are available and can be immediately used. Among commonly used modules we can find *os* for operating system interface and file system access; *time* for time-related functions; *sys* for functions that interact with the interpreter including command line arguments, module search path or standard input/output/error file objects; *re* for regular expressions. Along the subsequent chapters of this book, we will make extensive use of different packages for distinct tasks.

The syntax for importing all functions in a module is as follows:

```
# imports all functions from module
>>> import module_name1
```

The functions of the imported module will be available with a name that is obtained by joining the *module_name*1 and the *function_name*, separated by a dot. As an example, let us see how to import and use functions from the package *os*:

```
>>> import os, sys
>>> print (os.name)
>>> print (os.getcwd())
```

If we are only interested in specific functions from the module these can be selected using the `from ...import` statement:

```
# imports specific functions in module
>>> from module_name import function_name1, ...
```

Table 2.6: Methods on module math.

Function	Description
e, pi, tau, inf, nan	mathematical constants and symbols
cos(*x*), **sin**(*x*), **tan**(*x*), **acos**(*x*), **acosh**(*x*), **asin**(*x*), **asinh**(*x*), **atan**(*x*), **atanh**(*x*)	different trigonometric functions
sqrt(*x*)	returns the square root of *x*
pow(*x,y*)	returns x to the power of *y*
exp(*x*)	returns e raised to the power of *x*
log2(*x*), **log10**(*x*), **log**(*x, [base]*)	returns the logarithm of *x* in different bases
ceil(*x*)	returns the smallest integer greater than or equal to *x*
factorial(*x*)	returns factorial of *x*
floor(x)	returns the integer part of *x*
hypot(*x,y*)	returns the Euclidean distance between *x* and *y*
degrees(*x*)	convert angle *x* from radians to degree
radians(*x*)	convert angle *x* from degrees to radians

In this last case, if the * symbol is placed after the import keyword, all functions of the module will be imported. It is important to note that in this case, the functions are called only by their names without the module name. While this is more readable, it can bring problems if two functions have the same name in different modules, so this option should only be used with care when this problem can be avoided.

As we have seen before, Python offers a set of basic mathematical and arithmetical operators. To have access to a more extended library of mathematical functions one can import the *math* module, already briefly mentioned above.

Table 2.6 lists functions and variables available from the module *math*. Other packages of interest for mathematical computation include *statistics*, *random*, *decimal*, *numpy* or *fractions*.

Other packages can be easily installed with package manager software. Examples of such software include: *pip*, which after Python 3.4 is present by default in the Python installation; *setuptools*, a library to facilitate packaging Python projects or *conda*, package manager for Anaconda Python installations. In order to get the list of all currently installed packages, the help interactive environment can be open with **help()** followed by a `modules` statement.

2.4.5 Putting It All Together

Once all the functionality necessary for the program to accomplish the required tasks is implemented, the code should be saved in a Python script file. This file should have a ".py" extension. Assuming that we have a script named *my_script.py*, it can be executed by calling the interpreter from the operating system command line:

```
> python my_script.py
```

Alternatively, under Unix based operating systems, including Mac OSX, the first line in our script can be used to indicate the path to the Python interpreter. This should appear after the symbols #!, as in #!*/path_in_my_os/bin/python*, where *path_in_my_os* is the path to the binaries folder that contains the Python interpreter. In this case, the program can be called directly without invoking the interpreter:

```
> my_script.py
```

Also, if using a proper IDE, there are options to write the script, save it and run it. In many cases, the program will run and show its results within a panel included in the IDE's interface.

Whenever a Python script is run, the code from the imported modules is interpreted and executed. In order to prevent immediate execution of the imported code within a module, a conditional statement can be used. When a file is run, the special variable *__name__* is set to "*__main__*". With the code below, when the module is run directly, the function **main()** is called and the respective code executed. When, on the other hand, the module is imported by some other program, the execution of the module's code is prevented. This feature is particularly useful for testing purposes.

```
if __name__ == "__main__":
    main()
```

With the elements presented in this section, you should be able to start tackling your programming challenges and write our own Python programs. To provide an example, let us check a simple program that reads a string, representing a DNA sequence, and computes the frequency of each nucleotide, also checking if there are non-valid characters.

```
def count_bases (seq):
    dic = {}
    seqC = seq.upper()
    errors = 0
    for b in seqC:
        if b in "ACGT":
            if b in dic: dic[b] += 1
            else: dic[b] = 1
        else: errors += 1
    return dic, errors

def print_perc_dic (dic):
```

```
    sum_values = sum(dic.values())
    for k in sorted(dic.keys()):
        print(" %s ->" % k, " %3.2f" % (dic[k]*100.0/sum_values), "%"
    )

## main program
seq = input("Input DNA sequence: ")
freqs, errors = count_bases(seq)
if errors > 0:
    print ("Sequence is invalid with ", errors , "invalid characters"
    )
else: print("Sequence is valid")
print("Frequencies of the valid characters:")
print_perc_dic (freqs)
```

Notice that the whole code can be put in a single file, or in alternative, the two functions can be part of a module (let's say called *sequences.py*) and in this case, the main program would start with the line:

```
from sequences import count_bases, printPercDic
```

2.5 Object-Oriented Programming

2.5.1 Defining Classes and Creating Objects

Object-oriented programming (OOP) is a popular paradigm that is based in the concepts of objects, classes and inheritance. This paradigm provides increased modularity, allowing the developed code to be encapsulated and more easily reused. Python supports OOP, allowing the definition of new classes by the programmer. Also, it provides a number of pre-defined classes, some of which were already presented in previous sections, including lists and dictionaries.

Classes are central concepts in OOP, representing an entity to be modeled. Classes allow to model objects that represent entities within the programs. They can be used to model entities such as strings, biological sequences, or more complex entities such as databases or networks.

Classes enclose two major components: one that specifies the data contents to be handled and a second component that specifies the behavior, i.e. the functionality that allows manipulating the respective contents. In the object-oriented terminology these functions are called *methods*, while information reflecting the state of the object is stored in variables that are called *attributes*.

As a convention for class naming, we will use *CamelCase* where each letter of a word in class name is represented in uppercase. Attributes are written in lowercase with underscores as for variable names.

Classes can be created with the instruction `class` according to the following syntax:

```
class ClassName:
    """ Optional documentation """
    —body_of_the_class—
```

In the body of the class, methods are defined, while attributes are used and defined implicitly. For each newly created class, we will need to define a constructor method called **__init__** (note the double underscores as prefix and suffix). This method is automatically invoked when new objects of this class are declared and defines the initial state of the objects that are instances of this class.

As an example, the following code implements a very simple class to represent and process biological sequences. This class contains two attributes *seq* and *seq_type*, which represent, respectively, a sequence and its biological type (protein, DNA, RNA). The constructor receives as input the sequence (a string) and assumes "DNA" as the default type, although in the constructor it can also be set to a different value ("RNA" or "protein"):

```
class MySeq:
    """Biological sequence class"""

    def __init__(self, seq, seq_type = "DNA"):
        self.seq = seq
        self.seq_type = seq_type

    def print_sequence(self):
        print ("Sequence: " + self.seq)

    def get_seq_biotype (self):
        return self.seq_type

    def show_info_seq (self):
```

```
        print ("Sequence: " + self.seq + " biotype: " + self.seq_type
 )

    def count_occurrences(self, seq_search):
        return self.seq.count(seq_search)
```

In the methods of the class, *self* always appears as the first argument. It refers to the newly created object instance in the constructor, or to the object over which the method is being called in the other methods. As you can see from Table 2.4, *self* is not a reserved keyword but a strong convention of the language.

While a class defines a template, an object is an instance of a class, i.e. a variable that follows the rules defined by the class. Objects of the same class have the same set of attributes, while the specific values for those may be different, and implement the same set of methods. Object instantiation, i.e. creating new objects/variables of a class, has the following syntax structure:

```
object_var = ClassName(arguments)
```

Also, methods defined within a class can be invoked with the following syntax:

```
object_var.method_name(arguments)
```

If a class is defined as above, we can easily create an object instance from that class, access the values of its attributes and invoke the defined methods using the following statements (either on the console or within a script in the same file of the class definition):

```
s1 = MySeq("ATAATGATAGATAGATGAT")
# access attribute values
print (s1.seq)
print (s1.seq_type)
# calling methods
s1.print_sequence()
print (s1.get_seq_biotype())
```

In Python, the attributes of a class can be directly modified. The programmer should evaluate if an attribute should be directly modified or if this modification should be validated before it occurs and, therefore, be made through a method.

In the next code snippet, an example is shown of a syntactically valid attribute modification, but that is semantically incorrect and would not be desirable. In alternative, a new method called *set_seq_biotype* is defined to update the attribute, where the validity of the modification is evaluated before it takes place.

```
# the type of the sequence is updated to an invalid biotype
# by direct alteration of the attribute
s1.seq_type = "time series"

# safer alternative: class method to validate update
def set_seq_biotype (self, bt):
    biotype = bt.upper()
    if biotype == "DNA" or biotype == "RNA" or biotype == "PROTEIN":
        self.seq_type = biotype
    else:
        print "Non biological sequence type!"

# testing the update of the attribute
s1.set_seq_biotype("time series")
s1.set_seq_biotype("dna")
```

Attributes can also be deleted through the **del** function. If an attribute is accessed after being deleted, an error message indicating that the class instance has no attribute is raised. Private attributes, which cannot be accessed except from inside the class, do not exist in Python. Instead, there is a naming convention that establishes that attributes with names starting with the prefix __ (double underscores) should be treated as non-public, i.e. not directly accessible from outside the class.

2.5.2 Special Methods

We have seen before that the method **__init__** has a special meaning and serves as a constructor for objects of the class. There are several methods that are shared across different classes. For instance, the method **len** is common to a number of classes such as strings, lists or dictionaries.

Some of the methods that implement this type of generic functionality can be redefined for new classes. This implies re-implementing a set of special methods that have a very specific behavior and follow a pre-defined naming, all of them starting and ending with double underscores __. The following code implements some of these methods, namely string representation of the object, length of the object, item access and slicing behavior.

The implementation of the method **__len__***(self)* determines the result of the application of the function **len** on an object of this class. Calling **len***(object)* is equivalent to *object.***__len__***()*. The method **__str__***(self)* determines the representation of the object as a string, where in the same way **str**(*object*) is equivalent to *object.*__**str**__(), being this also used by the **print** function.

On the other hand, the functions __**getitem**__ and __**getslice**__ allow to define how the operator [] will work, defining, respectively, what will be the result of indexing an object of the class with a single index and a slice (two indexes with the symbol :).

```
class MySeq:
    —some_code_here—

    def __len__(self):
        return len(self.seq)

    def __str__(self):
        return self.seq_type + ":" + self.seq

    def __getitem__(self, n):
        return self.seq[n]

    def __getslice__(self, i, j):
        return self.seq[i:j]

    —some_code_here—
```

Some examples for the use of these methods follow:

```
s1 = MySeq("MKKVSJEMSSVPYW", "PROTEIN")
print(s1)
print(len(s1))
print(s1[4])
print(s1[2:5])
```

2.5.3 Inheritance

If the class to be implemented has a very similar behavior and information than an existing class, but represents a more specialized version, then the class inheritance mechanism can be used. In the inheritance process, we name as child the class that inherits from another class, which is named the parent class.

Methods and attributes from the parent class are automatically available to the child class. However, the child class can introduce new methods and attributes or redefine some of these methods. In the class definition, to inherit from another class, we need to simply include the name of the parent class in the class statement.

The code below shows how we can extend the class **MySeq** for a class of numerical sequences, named **MyNumSeq**. The method **super** used in the constructor refers to the parent class explicitly. This code also shows how we override the method **set_seq_biotype** that is inherited from the parent class.

```
class MyNumSeq(MySeq):
    def __init__(self, num_seq, seq_type="numeric"):
        super().__init__(num_seq, seq_type)

    def set_seq_biotype (self, st):
        seq_type = st.upper()
        if seq_type == "DNA" or seq_type == "RNA" or seq_type == "
  PROTEIN":
            self.seq_type = seq_type
        elif seq_type == "NUMERIC" or seq_type == "NUM":
            self.seq_type = seq_type
        else:
            print ("Non-biological or Non-numeric sequence type")
```

We can now create an instance of a numerical sequence. By default, its sequence type will be "DNA" since this is the default type in the parent class. This can be correctly updated to the numeric type with the redefined **set_seq_biotype** method in **MyNumSeq**. The method **print_sequence** from the parent class can also be called by this object instance.

```
>>> a =  MyNumSeq("123456789")
>>> a.seq_type
'DNA'
>>> a.set_seq_biotype("numeric")
>>> a.seq_type
'NUMERIC'
>>> a.print_sequence()
Sequence: 123456789
```

Attributes of a class are not limited to built-in data types and can also represent objects. This will allow a more elaborated design for our classes. The instances of the created class can be stored in lists or dictionaries like any other object. The code below shows an example on how to create 100 instances of a numerical sequence and store them in a list and later print their sequence.

```
# create 100 objects of MyNumSeq and store them in a list
list_of_NumSeqs = []
for i in range(100):
    list_of_NumSeqs.append(MyNumSeq(str(int(random.random()*1000000))
   ))

for i in range(len(list_of_NumSeqs)):
    list_of_NumSeqs[i].print_sequence()
```

2.5.4 Modularity

One important comment goes to code maintenance. In order to keep the code organized, it is a good practice to move the code of a class to a module. As an example the code of **MySeq** and **MyNumSeq** and other possibly related classes, could be saved in a file called *myseq.py*. Later, individual classes or all the classes from this module can be imported with the following statements:

```
# import specific classes from a module
from myseq import MySeq, MyNumSeq
# import all classes from a module
import myseq
```

2.6 Pre-Defined Classes and Methods

2.6.1 Generic Methods for Containers

We have seen in previous sections that lists, strings, tuples, sets or dictionaries are container types. All these types are indeed classes pre-defined in the Python language. For the built-in container types, with the exception of the set object, when creating instances the *ClassName* is replaced by the container symbols: “” for strings, () for tuples, [] for lists and {} for dictionaries.

Container data types share many features, thus several methods and functions are commonly applied to all these data types. Table 2.7 presents a list of the most important.

Table 2.7: Methods/functions applicable to containers.

Function	Description
len(*c*)	number of elements in container *c*
max(*c*)	maximum value from elements in container *c*
min(*c*)	minimum value from elements in container *c*
sum(*nc*)	sum of the numerical values in container *nc*
sorted(*c*)	list of sorted values in container *c*
value **in** *c*	membership operator **in**. Returns a Boolean value

Examples of the usage of these functions, in the case of a numeric list are provided below:

```
>>> x = [1, 7, 4, 3, 5, 2]
>>> len(x)
6
>>> max(x)
7
>>> min(x)
1
>>> sum(x)
22
>>> sorted(x)
[1, 2, 3, 4, 5, 7]
```

These apply also to strings, as shown in the following examples:

```
>>> sorted("acaebf")
['a', 'a', 'b', 'c', 'e', 'f']
>>> "b" in ["a","b",""]
True
>>> "b" in "abcdef"
True
>>> "b" in {"a":1,"b":2,"c":3}
True
```

Table 2.8 presents some of the functions that can be used to generate iterable structures and examples of their usage in iterative loops are shown below.

```
>>> for e in enumerate(["a","b","c"]):
...     print (e)
```

Table 2.8: Functions for iterable information.

Function	Description
range(*x*)	iterable object with *x* integer values from 0 to *x-1*
enumerate(*c*)	iterable object with *(index, value)* tuples
zip(*c1, c2, ..., cn*)	creates an iterable object that joins elements from *c1, c2, ... cn* to create tuples
all(*c*)	returns `True` if all elements in *c* are evaluated as true, and `False` otherwise
any(*c*)	returns `True` if at least one element in *c* is evaluated as true, and `False` otherwise

```
...
(0, 'a')
(1, 'b')
(2, 'c')
>>> for i in range(2, 20, 2):
...     print (i)
2
4
...
18

>>> for z in zip([1,2,3],["a","b","c"], [7,8,9]):
...     print (z)
...
(1, 'a', 7)
(2, 'b', 8)
(3, 'c', 9)
```

The functions **all** and **any** provide a logical test for all the elements in a container, returning `True` if all elements are true, and if at least one element is true, respectively. Notice that empty lists or strings and the number zero are here interpreted as `False`.

```
>>> all(["a","b"])
True
>>> all(["a","b",""])
False
>>> any(["a","b",""])
True
>>> all([1, 1])
True
>>> all([1, 1, 0])
False
```

Table 2.9: Functions for ordered sequence containers.

Function	Description
c * *n*	replicates *n* times the container *c*
c1 + *c2*	concatenates containers *c1* and *c2*
c.**count**(*x*)	counts the number of occurrences of *x* in container *c*
c.**index**(*x*)	index of the first occurrence of x in container *c*
reversed(*c*)	an iterable object with elements in *c* in reverse order

```
>>> any([0, 0, 1, 0])
True
>>> any([0, 0, 0, 0])
False
```

Table 2.9 lists five operations commonly performed on sequence containers, as lists and strings. They can be easily applied as exemplified in the following code:

```
>>> a = [1, 2, 3]
>>> a * 3
[1, 2, 3, 1, 2, 3, 1, 2, 3]
>>> b = [4, 5, 6]
>>> ab = a + b
>>> ab
[1, 2, 3, 4, 5, 6]
>>> c = [1, 2, 3, 2, 1]
>>> c.count(1)
2
>>> c.index(3)
2
>>> for x in reversed(a):
...     print (x)
...
3
2
1
```

2.6.2 Methods for Lists

Lists are specific sequence containers that can hold indexed heterogeneous elements. The list type is mutable and, therefore, its content is typically changed by the application of differ-

Table 2.10: Functions/methods working over on lists.

Function	Description
lst.**append**(*obj*)	append *obj* to the end of *lst*
lst.**count**(*obj*)	count the number of occurrences of *obj* in the list *lst*
lst.**index**(*obj*)	returns the index of the first occurrence of *obj* in *lst*. Raises **ValueError** exception if the value is not present
lst.**insert**(*idx, obj*)	inserts object *obj* in the list in position *idx*
lst.**extend**(*ext*)	extend the list with sequence with all elements in *ext*
lst.**remove**(*obj*)	remove the first occurrence of *obj* in the list. Raises **ValueError** exception if the value is not present
lst.**pop**(*idx*)	removes and returns the element at index *idx*. If no argument is given, the function returns the element at the end of the list. Raises **IndexError** exception if list is empty or *idx* is out of range
lst.**reverse**()	reverses the list *lst*
lst.**sort**()	sorts the list *lst*

ent methods. A few of the most important methods for lists are provided in Table 2.10, being illustrated in the next code block.

```
>>> x = [1, 7, 4, 3, 5, 2]
>>> x.append(6)
>>> x
[1, 7, 4, 3, 5, 2, 6]
>>> x.index(5)
4
>>> x.extend([9,8])
>>> x.insert(1,10)
>>> x
[1, 10, 7, 4, 3, 5, 2, 6, 9, 8]
>>> x.pop()
8
>>> x.reverse()
>>> x
[9, 6, 2, 5, 3, 4, 7, 10, 1]
>>> x.sort()
>>> x
1, 2, 3, 4, 5, 6, 7, 9, 10]
```

Notice that lists can work as queues, for instance if elements are inserted using **append** (in the end) and removed with **pop(0)** (from the beginning of the list). Also, stacks can be implemented by adding elements with **append** and removing with **pop()**.

Table 2.11: Functions/methods working over strings.

Function	Description
s.**upper**(), *s*.**lower**()	creates a new string from *s* with all chars in upper or lower case
s.**isupper**(), *s*.**islower**()	returns `True` when in *s* all chars are in upper or lower case, and `False` otherwise
s.**isdigit**(), *s*.**isalpha**()	returns `True` when in *s* all chars are digits or alphanumeric, and `False` otherwise
s.**lstrip**(), *s*.**rstrip**(), *s*.**strip**()	returns a copy of the string *s* with leading/trailing/both whitespace(s) removed
s.**count**(*substr*)	counts and returns the number of occurrences of sub-string substr in *s*
s.**find**(*substr*)	returns the index of the first occurrence of sub-string *substr* in *s* or −1 if not found
s.**split**(*[sep]*)	returns a list of the words in *s* split using *sep* (optional) as delimiter string. If *sep* is not given, default is any white space character
s.**join**(*lst*)	concatenates all the string elements in the list *lst* in a string where *s* is the delimiter

2.6.3 Methods for Strings

Strings are immutable ordered sequence containers, holding characters. A number of methods that can be applied over strings are described in Table 2.11, being their usage illustrated by the following examples.

```
>>> seq = 'AATAGATCGA'
>>> len(seq)
10
>>> seq[5]
'A'
>>> seq[4:7]
'GAT'
>>> seq.count('A')
5
>>> seq2 = "ATAGATCTAT"
>>> seq + seq2
'AATAGATCGAATAGATCTAT'
>>> "1" + "1"
'11'
>>> seq.replace('T','U')
'AAUAGAUCGA'
>>> seq[::2]
```

```
'ATGTG'
>>> seq[::-2]
'ACAAA'
>>> seq[5:1:-2]
'AA'
>>> seq.lower()
'aatagatcga'
>>> seq.lower()[2:]
'tagatcga'
>>> seq.lower()[2:].count('c')
1
>>> c = seq.count("C")
>>> g = seq.count("G")
>>> float(c + g)/len(seq)*100
30.0
```

Some of these methods are particularly useful to identify matches within sequences, as shown below.

```
>>> "TAT" in "ATGATATATGA"
True
>>> "TATC" in "ATGATATATGA"
False
>>> seq = "ATGATATATGA"
>>> "TAT" in seq
True
>>> "TATC" in seq
False
>>> seq.find("TAT")
4
>>> seq.find("TATC")
-1
>>> seq.count("TA")
2
>>> text = "Restriction enzymes work by recognizing a particular
   sequence of bases on the DNA."
>>> text_tokens = text.split(" ")
>>> text_tokens
['Restriction', 'enzymes', 'work', 'by', 'recognizing', 'a', '
   particular', 'sequence', 'of', 'bases', 'on', 'the', 'DNA.']
```

Table 2.12: Methods/functions working over sets.

Function	Description
s.**update**(s2)	updates the set *s* with the union of itself and set *s2*
s.**add**(*obj*)	adds *obj* to set
s.**remove**(*obj*)	removes *obj* from the set. If *obj* does not belong to set raises an exception **KeyError**
s.**copy**()	returns a shallow copy of the set
s.**clear**()	removes all elements from the set
s.**pop**()	removes the first element from the set. Raises the exception **KeyError** if the set *s* is empty
s.**discard**(*obj*)	removes *obj* from the set *s*. If *obj* is not present in the set, no changes are performed

```
>>> text_tokens.count("the")
1
>>> text_tokens.index("sequence")
7
```

We revisit a previous example from Section 2.3.6 to generate a tuple with all the sub-strings of length 3 of a given sequence. We then use methods over tuples to count occurrences or obtaining the first position of different sub-strings. Notice that tuples are ordered containers that unlike lists are immutable.

```
seq = "ATGCTAATGTACATGCA"
seq_words = tuple([(seq[x:x+3]) for x in range(0, len(seq)-3)])
>>> seq_words.count("ATG")
3
>>> seq_words.count("CAT")
1
>>> seq_words.index("TAA")
4
```

2.6.4 Methods for Sets

We have seen before several operators between two sets. These also exist as methods over objects of the class representing sets: **intersection**, **intersection_update**, **isdisjoint**, **issubset**, **issuperset**, **symmetric_difference**, **symmetric_difference_update**, **union** and **update**. We refer the reader to `help(set)` in interactive mode for more details on these methods. Table 2.12 lists other methods available for sets.

Examples of the use of those methods are provided below.

Table 2.13: Methods/function working over dictionaries.

Function	Description
d.**clear**()	removes all elements from dictionary *d*
d.**keys**()	returns list of keys in dictionary *d*
d.**values**()	returns list of values in dictionary *d*
d.**items**()	returns list of key-value pairs in *d*
d.**has_key**(k)	returns `True` if *k* is present in the list of keys, and `False` otherwise
d.**get**(*k,[defval]*)	returns the value corresponding to key *k*, or default value if *k* does not exist as key
d.**pop**(*k,[defval]*)	removes entry corresponding to key *k* and returns respective value (or default value if key does not exist)

```
>>> A = set([2, 3, 5, 7, 11, 13])
>>> B = set([2, 4, 6, 8, 10])
>>> A | B
{2, 3, 4, 5, 6, 7, 8, 10, 11, 13}
>>> A & B
{2}
>>> A - B
{3, 5, 7, 11, 13}
>>> C = set([17, 19, 23, 31, 37])
>>> A.update(C)
>>> A
{2, 3, 5, 37, 7, 11, 13, 17, 19, 23, 31}
>>> A.add(35)
>>> A.pop()
2
>>> A.discard(35)
>>> A
{3, 5, 37, 7, 11, 13, 17, 19, 23, 31}
```

2.6.5 Methods for Dictionaries

As seen above, dictionaries are mapping data structures, also implemented as a container class. The main methods used to work with dictionaries are listed in Table 2.13, while their usage is exemplified by the examples provided below.

```
>>> dic = {"Dog":"Mammal", "Octopus":"Mollusk", "Snake":"Reptile"}
>>> dic['Dog']
```

```
'Mammal'
>>> dic['Cat']= 'Mammal'
>>> dic
{'Dog': 'Mammal', 'Octopus': 'Mollusk', 'Snake': 'Reptile', 'Cat': '
   Mammal'}
>>> len(dic)
4
>>> dic.keys()
dict_keys(['Dog', 'Octopus', 'Snake', 'Cat'])
>>> list(dic.keys())
['Dog', 'Octopus', 'Snake', 'Cat']
>>> "Human" in dic
False
>>> "Dog" in dic
True
>>> del dic["Snake"]
>>> dic
{'Dog': 'Mammal', 'Octopus': 'Mollusk', 'Cat': 'Mammal'}
>>> list(dic.values())
['Mammal', 'Mollusk', 'Mammal']
>>> for k in dic.keys():
...      print (k + " is a " + dic[k])
...
Dog is a Mammal
Octopus is a Mollusk
Cat is a Mammal
```

2.6.6 Assigning and Copying Variables

A distinction between an assignment and a copy of variables needs to be made. In an assignment, the new variable name will be pointing to the existing object or value. Changes in the original object will affect both variables.

A copy, on the other hand, only occurs if it is explicitly demanded. It can be further differentiated into shallow or deep copying. This difference will only be noticeable for objects containing other objects, such as lists or class instances. In both types of copy, a new object is created from the existing object and both become independent. In the case of shallow copy, if an element of the existing object being copied is of an immutable type then the element is copied integrally; if it is a reference to another object, then the reference is copied. In the case of deep copy, all elements are copied integrally, even the objects that are referred in the existing objects.

A shallow copy can be made, for instance, by slicing a list:

```
>>> x = [1, 2, 3, 4, 7]
>>> y = x[:]
```

Here, *x* and *y* are independent variables and any change in one variable will not affect the other. In case we just assign our variable to another name (*z*), any change made in one of the variables will affect the status of the other, as shown below.

```
>>> z = x
>>> z = x
>>> x.pop()
>>> z
[1, 2, 3, 4]
```

In the next example, notice that slicing can be used to alter multiple values in a list:

```
>>> x[1:-1] = [ -2, -3]
>>> x
[1, -2, -3, -4]
# remove values:
>>> del x[1:-1]
>>> x
[1, 4]
>>> y
[1, 2, 3, 4, 7]
>>> z
[1, 4]
```

The previous examples become more complex when the existing objects contain other objects, like for instance a list of lists. For those cases, we can take advantage of the package *copy* that contains two functions for shallow (**copy**) and deep copy (**deepcopy**) of container variables.

Bibliographical Notes and Further Reading

In this chapter, we have introduced the most important concepts of the Python language. We have discussed aspects that go from syntax indentation, primitive and container built-in datatypes to more advanced topics of object-oriented programming. Since this was not intended to be an in-depth introduction to the language, many specific aspects may have not been covered here.

There are currently many good textbooks and resources that provide a detailed overview to this programming language [2–5], which may be used to complement this chapter. The details of the full documentation of the latest distribution are available in `https://docs.python.org/3/`, including the Python and the standard library references. These are useful resources to clarify any doubts about the behavior of the different built-in instructions and pre-defined functions. In the site, you can also find a number of useful How To's and other relevant information.

There are also many important resources in algorithms and data structures that can be used to learn a lot more about programming principles, which include the seminal work by N. Wirth [155], and more recent books by Dasgupta et al. [42], and Sedgewick and Wayne [138]. The book by Phillips et al. is one of the many ways to learn a lot more about OOP in Python [128].

Exercises and Programming Projects

Exercises

1. Explore the Python shell by defining variables of different types (numerical, strings, lists, dictionaries, etc) and applying the functions and methods described along the chapter.
2. Install and explore the *Jupyter Notebooks* environment, running some of the examples from the previous exercise.
3. Write small programs, with the input-process-output structure, for the following tasks:
   a. Reads a value of temperature in Celsius degrees (ºC) and converts it into a temperature in Fahrenheit degrees (ºF).
   b. Reads the length of the two smallest sides of a right triangle and calculates the length of the hypotenuse (the largest side, opposite to the right angle).
   c. Reads a string and converts it to capital letters, printing the result.
   d. Adapt the previous program to read the string from a file, whose name is entered by the user.
   e. Reads a string and checks if it is a palindrome, i.e. if it reads the same when it is reversed. Implement different versions using functions over strings, and cycles (for/while).
   f. Reads three numerical values from the standard input, and calculates the largest and the smallest value.

   g. Reads two numerical intervals (defined by lower and upper range), and outputs their union and their intersection.
   h. Reads a numerical interval (defined by lower and upper range), and calculates the sum of all integer values includes in the interval.
   i. Reads a sequence of integer (positive) values, terminated by value 0, and outputs their sum, mean and largest value.
   j. Reads a sequence of integer (positive) values, terminated by value 0, and outputs the same sequence in decreasing order.
4. Repeat the previous exercise, now creating functions for the different tasks and calling those functions within your programs.
5. Define a class to represent a rectangle, where the attributes should be *height* and *length*.
   a. Implement the following methods: constructor; calculate area; calculate perimeter; calculate length of the diagonal.
   b. Test your class defining instances of different sizes.
   c. Implement a sub-class (child) that extends this class, to represent squares.
6. Extend the class for handling sequences developed in this chapter, defining a sub-class to handle DNA sequences. Implement a method to validate if the sequence is valid (i.e. if it only contains the symbols "A", "C", "G", or "T"). Add other methods that you think may be useful in this context.

Programming Projects

1. Write a module in Python including a set of functions working over a list with numerical values, passed as an argument to the function, with the following aim (avoid using pre-defined methods over lists), validating with a script that tests the functionality of these functions:
   a. calculate the sum of the values in the list;
   b. indicate the largest value in the list;
   c. calculate the mean of the values in the list;
   d. calculate the number of elements in the list larger than a threshold passed as argument;
   e. check if a given element (passed as an argument) is present in the list, returning the index of its first occurrence, or -1 if the element does not exist;
   f. return a list of all positions of an element in the list (empty list if it does not occur);
   g. return the list resulting from adding a new element in the end of the list;
   h. return the list resulting from summing the elements of the list with the ones of another list with the same size passed as argument;
   i. return the list resulting from ordering the original list by increasing order.

2. Write a module in Python including a set of functions working over matrices. The matrix (represented as a list of lists) will be passed as the first argument. Some functions to include may be the following:
   a. calculate the sum of the values in the matrix;
   b. indicate the largest (smallest) value in the matrix;
   c. calculate the mean of the values in the matrix;
   d. calculate the mean (or sum) of the values in each row (or column); the result should be a list;
   e. calculate the multiplication of the values in the diagonal;
   f. check if the matrix is square (same number of rows and columns);
   g. multiply all elements by a numerical value, returning another matrix;
   h. add two matrices (assuming they have the same dimension);
   i. multiply two matrices.
3. Develop a class to keep numerical matrices. The attributes of the class should be the number of rows, number of columns, and a list of lists keeping the elements of the matrix. Implement a constructor to create an empty matrix given the number of rows and columns. Implement methods with a functionality similar to the ones listed in the previous question.

CHAPTER 3

Cellular and Molecular Biology Fundamentals

In this chapter, we review the major concepts in cellular and molecular Biology relevant for the Bioinformatics algorithms covered in this book. These fields look inside the cell and try to understand its mechanisms by studying how its molecular components co-exist and interact. We will start by providing an overview on the composition and organization of cells and their different types. Then, we will discuss characteristics of the genetic material and how the genetic information flows along different cellular processes. Next, we present the notion of gene, a discrete unit of genetic information and discuss details of its codification in the genetic material. We provide an outline of the major milestones in the history of the human genome and give examples on how its study is providing us insights in the understanding of human diversity and disease. Finally, we address some important resources on biological data in particular for biological sequences.

3.1 The Cell: The Basic Unit of Life

Cells are the basic units of life. All living things are made of cells. In organisms like plants or animals, the first cell of an organism, called the primordial germ cell, is obtained by the fusion of the sperm and the egg that are contributed by the parents. This first cell undergoes multiple rounds of division and differentiation, resulting in cells of different nature. This way, each cell will derive from another cell and contain all the necessary information to replicate itself.

Some features are common to all cells, but, in multicellular organisms as humans, through differentiation mechanisms, many different cell types arise, eventually showing very distinct characteristics. Cells of a given type have similar functions and aggregate to form tissues. Tissues reflect their cell type composition and can also be very distinct between them. For instance, kidney or heart are very distinct tissues from skin or stomach. Organs are made of tissues and shape the complexity and aspect of a complex organism. Multicellular organisms, which include animals, plants or most fungi, are the combined result of many cells. But not all organisms have many cells. In fact, unicellular organisms, such as *Saccharomyces cerevisiae*, a fungi also known as the budding or baker's yeast, or bacteria only contain one cell.

A cell is mostly composed by water. For instance, a bacterial cell has a composition by weight of approximately 70% of water and 30% of chemical origin. From these, 7% are small-

Bioinformatics Algorithms. DOI: 10.1016/B978-0-12-812520-5.00003-1

molecules including amino acids and nucleotides and 23% are macro-molecules including proteins, lipids or polysaccharides [120].

According to their internal structure, cells can be divided in two major categories: *prokaryotic cells*, that have no nucleus or internal membranes, and *eukaryotic cells* that have a defined nucleus, internal membranes and functional elements called organelles that have varying shapes and size and play specific functions.

The cell type is the factor that drives organism categorization as prokaryotes or eukaryotes. Most of the unicellular organisms are prokaryotes, including bacteria and archea species. But, eukaryotic unicellular organisms also exist. Several fungi species, such as *Saccharomyces cerevisiae* or *Schizosaccharomyces pombe* are good examples. The more complex multicellular species are all eukaryotes.

At the structural level, all cells are surrounded by a structure called cell membrane or plasma membrane. This membrane is permeable to molecules that cells need to absorb from or excrete to the outside medium. Within the cell, we find the cytoplasm, a gel-like substance, largely composed of water, which serves as the medium for the cell.

Among the molecules with a biological role, we can find nucleic acids. These encode and express the genetic code that is kept within the cell. There are two major types of nucleic acids: deoxyribonucleic acid (DNA) and ribonucleic acid (RNA). As we will see, although chemically very similar, these molecules have different purposes. DNA contains the information necessary to build a cell and keep it functioning. In eukaryotic cells, DNA will be found in the nucleus, while in prokaryotic cells it will be in the cytoplasm. By reflecting the information encoded in the DNA, the RNA will be used as an intermediate in the definition of the composition and, therefore, the nature of the cell.

Amino acids are another important type of bio-molecule. They are the building blocks of *proteins*, which are macromolecules that perform most of the functions in the cell. Proteins have a broad range of functions, spanning from catalytic to structural functions. Enzymes, for instance, are a type of abundant proteins that promote chemical reactions, and convert certain molecules into other types of molecules required for the functioning of the cell. Other important organic macromolecules include: carbohydrates that serve as energy storage, both for immediate or long-term energy demands; and, lipids that are part of the plasma membrane and their function involves signaling and energy storage.

The cell also contains other components of varying complexity. Of notice, the mitochondria and the chloroplasts, which are organelles involved in the production of energy and the ribosomes that are large and complex molecules composed by a mixture of genetic material. Their function is to assemble proteins and, as we will see later, they play a central role in the flow of genetic information.

3.2 Genetic Information: Nucleic Acids

The DNA is a polymer composed of four nucleic acid units, called *nucleotides* or *base pairs*. Nucleotides have a similar chemical composition, but can be distinguished by the respective nitrogenous bases: *Adenine*, *Guanine*, *Thymine* or *Cytosine*. Adenine and Guanine are part of the purines group, while Cytosine and Thymine are pyrimidines. Nucleotides are usually referred by their first letter: A, G, T or C.

The DNA is a molecule composed of two complementary strands that form and stick together due to the connections established between the nucleotides in both strands. This complementarity is possible due to a chemical phenomenon where Adenine bonds only to Thymine nucleotides as a result of two hydrogen connections ($A = T$), and Guanine bonds only to Cytosine by three hydrogen bonds ($C \equiv G$). This results in two complementary and anti-parallel strands (connected in opposite directions). Knowing the sequence of the nucleotides in one of the strands it is possible to obtain the sequence in the opposite strand by taking the complement of its nucleotides. The two opposite strands are also read in reverse directions, and therefore we say that one strand is the reverse complement of the other. The existence of these two strands is essential for passing the genetic information to new cells and to produce proteins. Due to this complementarity and redundancy, it is therefore standard to describe the DNA through only one of its strand sequences using the four-letter alphabet: $\{A, G, T, C\}$.

DNA molecules can form sequences of hundreds of thousands or millions of nucleotides. Each of these individual and long DNA molecules are called *chromosomes*. Within chromosomes, we find *genes* that are functional regions of the DNA and encode the instructions to make proteins. The complete set of DNA forms the genetic material of the organism and is called *genome*. The size of the genome and the corresponding number of chromosomes in a cell is variable and depends on the species. For instance, certain bacterial species have a genome that contains around a few hundred or a few million base pairs in a single chromosome, while mice or humans have a genome with approximately 3 billion (3×10^9) base pairs, but there are other species with even larger genomes.

An organism or a cell is *haploid* if it only contains a single set of unpaired chromosomes. If it contains two complete sets of chromosomes it is called *diploid*. Humans, for instance, are diploids and contain 23 chromosome pairs. The twenty-third chromosome pair (*sexual chromosome*) determines the gender of the individual. If it contains two copies of the X chromosome, it results in a female, while if it contains a chromosome X and chromosome Y, it results in a male individual. So, in total, the human genome has 24 distinct chromosomes. The remaining 22 chromosome pairs are called *autosomal chromosomes*. Each chromosome contains many genes, and in the human genome they can sum up to nearly 21,000 coding for proteins. As we will see later, different types of genes can be found in the human genome.

Depending on the type of the cell, the genome is organized differently. In a prokaryotic cell, the genome exists in the form of a circular chromosome and it is located in the cytoplasm. On the other hand, in an eukaryotic cell, the genome is found in the nucleus and tightly packed into linear chromosomes. The chromosome organization is highly hierarchical and consists of a DNA-protein complex called *chromatin*, that is organized in an array of sub-units called *nucleosomes*. In these intermediate sub-structures, the DNA structure wraps around proteins called *histones*. This organization allows accommodating a long DNA molecule in a small space. It also provides a structure to regulate the expression of genes.

The genetic composition of an organism comprised on its genome is called the *genotype*. The *phenotype* is the set of physical and observable traits of an individual. It results from the combination of the genotype and the environment. The height of an individual or the color of the eyes are phenotypic traits that are totally or partially encoded on the genotype.

In most multicellular organisms, each cell contains the same genetic information. At a certain point, the cell divides into two daughter cells, passing a copy of its DNA to each of its child cells. The process of copying the DNA is called *DNA replication*. It is important that this process is accurate to ensure that the new cells contain the same genetic material of the mother and result in healthy cells. Since DNA has a double-helix structure, for DNA replication to take place, it is necessary that the two strands separate. Then, proteins (enzymes) called *DNA polymerases* will synthesize a new DNA strand by adding nucleotides that complement each of the nucleotides in the original strand, also called the template strand.

So far, we have seen that the DNA contains the genetic information required for the cells to function and replicate and that genes, which are regions found along the DNA, encode the information necessary to synthesize proteins and other molecular products. The process of using the information encoded in a gene to produce a functional gene product is called *gene expression*. But, what we have not discussed yet, is how the information flows from DNA to proteins. DNA, RNA and proteins are the central elements on this flow of genetic information that occurs in two steps: *transcription* and *translation*, in what is also called the *central dogma* of molecular and cell Biology.

RNA is a single strand molecule that in contrast to DNA does not form an helix structure. It is also composed of four nucleotides, but it contains a fifth type of nucleic acid called Uracil (U) that is not present in the DNA. So, while thymine only exists in DNA, uracil is only found in RNA. The other three bases are common to DNA and RNA. The sequence alphabet of the RNA is: {A, G, U, C}.

3.2.1 Transcription: RNA Synthesis

Transcription is the first step required to produce a protein. In this step, the nucleotide sequence of a gene from one of the DNA strands is transcribed, i.e. copied into a complementary molecule of RNA. The complementarity of the genetic code allows recovering the information encoded in the original DNA sequence, a process performed by an enzyme called *RNA polymerase*. Additional steps of RNA processing, including stabilizing the elements at the end of the molecule, are performed by different protein complexes.

After these steps, which occur within the nucleus of the cell, an RNA molecule called *mature messenger RNA* or *mRNA* is obtained. The mature mRNA is then transported to the cytoplasm, where it will be used by the cellular machine to guide the production of a protein. Fig. 3.1 depicts the different steps involved in the flow of genetic information, including transcription and translation, covered in the next section, as well as replication, aforementioned.

3.2.2 Translation: Protein Synthesis

Proteins are cellular entities that have either a structural function, participating in the physical definition of the cell, or a chemical function being involved in chemical reactions occurring in the cell. In order to function as expected, a protein needs to acquire the appropriate structure. This structure is often decomposed at different complexity levels. The primary structure is defined by the chain of amino acids and is called a *polypeptide*. These polypeptides will coil and fold, forming regular and local sub-structures, called secondary structures of the protein. One or more combined polypeptides with the appropriate structure will form a fully functional protein. It is the sequence of amino acids and the current cellular conditions that will determine how the polypeptide will fold and the protein acquires its structure.

Translation is the process in which the nucleotide sequence of an mRNA molecule is transcribed into a chain of amino acids forming a polypeptide, that will consist in part or the totality of a protein. This process is performed by the ribosomes that attach and scan the mRNA from one end to the other, in groups of *nucleotide triplets* or *codons*. In each position of the triplet, we have one of four nucleotides, so there are $4 \times 4 \times 4 = 64$ possible triplets. To each codon in the mRNA sequence corresponds an amino acid in the polypeptide chain. Some of these codons represent specific signals that indicate the initiation or the termination of the translation process. Once the ribosome detects an initiation codon, it starts the formation of the amino acid chain, and when it scans the stop codon it stops the translation and detaches from the mRNA molecule.

There are 20 types of amino acids used to form polypeptides, much less than the 64 possible codons. Therefore, more than one codon corresponds to a type of amino acid. This mapping between codons and amino acids is provided in Table 3.1 and is commonly called *genetic code*.

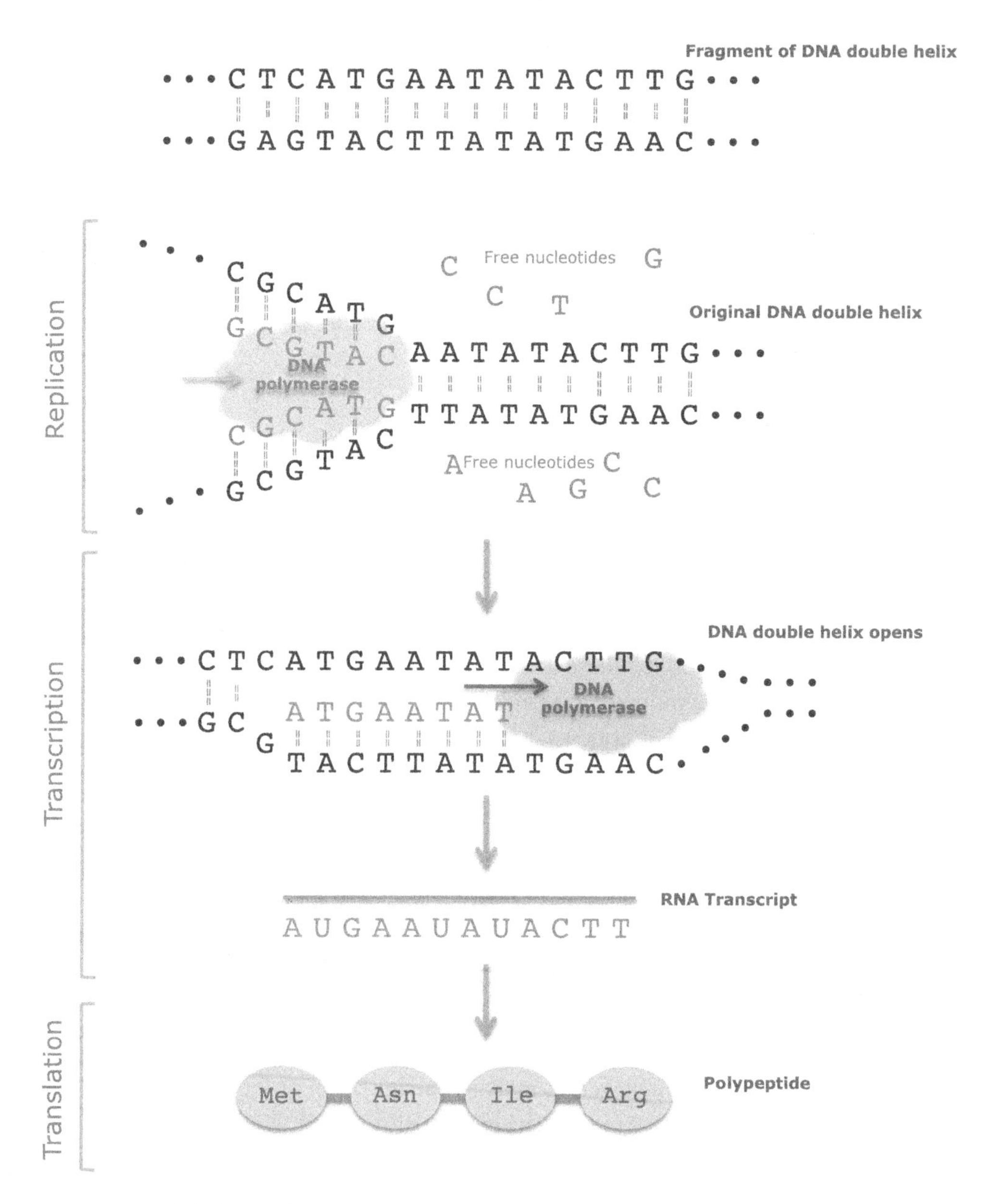

Figure 3.1: Genetic information flow. Example of a fragment of a DNA double helix is shown at the top. Replication of DNA forms two identical DNA double helices. Transcription allows the synthesis of an mRNA transcript. Translation synthetizes a polypeptide from the mRNA transcript.

Table 3.1: Genetic code: Mapping between codons and amino acids. Different notation of amino acids is presented.

Nucleotides	Amino acids
UUU, UUC	Phenylalanine/ Phe / P
UUA, UUG, UCU, UCA, UCC, UCG	Leucine / Leu / L
AUU, AUC, AUA	Leucine / Leu / L
AUG	Methionine / Met / M (start)
GUU, GUC, GUA, GUG	Valine / Val / V
UCU, UCC, UCA, UCG, AGA, AGG	Serine / Ser / S
CCU, CCC, CCA, CCG	Proline / Pro / P
ACU, ACC, ACA, ACG	Threonine / Thr / T
GCU, GCC, GCA, GCG	Alanine / Ala / A
UAA, UAC	Tyrosine / Tyr / Y
UAA, UAG, UGA	Stop codons
CAU, CAG	Histidine / His / H
CAA, CAG	Glutamine / Gln / Q
AAU, AAC	Asparagine / Asn / N
AAA, AAG	Lysine / Lys / K
GAU, GAC	Aspartic Acid / Asp / D
GAA, GAG	Glutamic Acid / Glu / G
UGU, UGC	Cysteine / Cys / C
UGG	Arginine / Arg / R
CGU, CGC, CGA, CGG, AGA, AGG	Glycine / Gly / G

During the translation process, a type of small RNA molecule, called *transfer RNA* or *tRNAs*, will bring to the ribosome the amino acids of the corresponding type, which will be complementary to the mRNA codon that is currently being scanned. Each mRNA molecule can be scanned multiple times by different ribosomes giving rise to multiple copies of the polypeptide.

The genetic code is an example where nature has developed a clever and robust cellular mechanism. While there are exceptions, the genetic code constitutes a standard language common to most cells from the simpler to the more complex organisms. With its redundancy, where more than one codon encodes an amino acid, the genetic code encloses a very efficient code-correction mechanism that minimizes the impact of errors in the nucleotide sequence occurring in DNA replication.

During translation, the parsing of the mRNA sequence by the ribosome may start at different nucleotides. Given that a codon is composed of three nucleotides, the mRNA sequence may have three possible interpretations. Let's consider the example in Table 3.2 and the possibilities in which the sequence can be translated.

These three ways of parsing the sequence are called *reading frames*. Note that in the above reading frames, Stop is just an indicative signal and that when a stop codon is found the trans-

Table 3.2: An example of an mRNA sequence and the three reading frames.

AAUGCUCGUAAUUUAG
AAU-GCU-CGU-AAU-UUA → Asn-Ala-Arg-Asn-Leu
AUG-CUC-GUA-AUU-UAG → Met-Leu-Val-Ile-Stop
UGC-UCG-UAA-UUU → Cys-Ser-Stop-Stop

lation stops and no other amino acid is added to the polypeptide chain. A reading frame with a sufficient length and with no stop codons is called *open reading frame (ORF)*. In genetics, when the sequence of the genome is known, but the location of the genes has not been annotated yet, ORFs are of particular importance as they indicate a part of the genome where genes are potentially encoded.

So far, we have considered a simplistic view where the genetic information is found in one of the DNA strands. Given the double-helix structure, genes can be found in both strands. Thus, the information in DNA should be read from both strands. To identify the directionality of the DNA strands, the ends of each strand have been named as 3'-end and 5'-end. Since the information of both strands is complementary, typically only the sequence of a single strand of DNA is given. By convention, this represents the sequence read from the 5'-end to the 3'-end, also represented as 5' → 3'.

The DNA strand from where the mRNA coding for a protein is read, is called *DNA sense*, *positive* or + *strand*. Its complementary strand will be called *DNA antisense*, *negative* or − *strand*. Genes can be found in both strands of DNA. Thus, when translating an mRNA sequence, the three previous reading frames actually become six possible reading frames.

In this section, we have seen that, DNA and RNA molecules can be represented as sequences written in four nucleotide symbols, while protein molecules can be represented as sequences of amino acids written in a twenty letter alphabet. Transcription is the process that creates a complementary copy of one the DNA strands of a gene. The result is a one-dimensional sequence of nucleotides called messenger RNA (mRNA). The mRNA serves as an intermediary between the nuclear RNA and the ribosomes in the cytoplasm. Translation is the process that uses the mRNA information to encode the polypeptide or protein chain of amino acids. This way, proteins are not directly synthesized by DNA. In the translation process, the AUG codon has a double function of encoding a start codon indicating the initiation of the translation or the amino acid methionine. The translation stop signal is encoded by three codons and never included in the translated sequence. In the next section, we will discuss how the genetic information to encode proteins is organized within a gene.

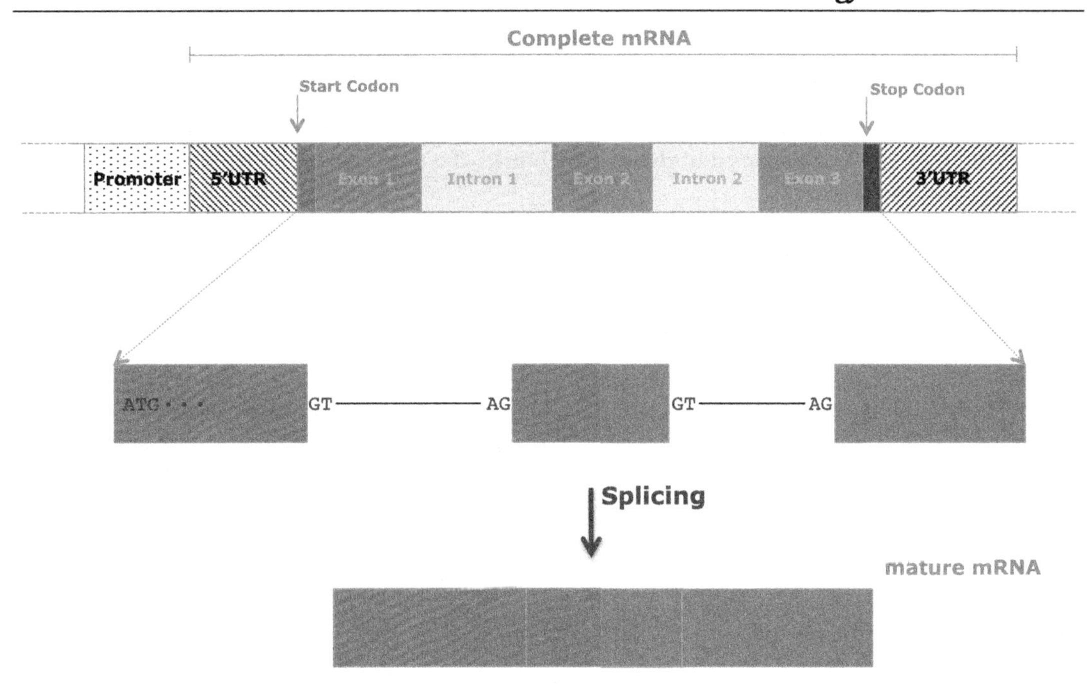

Figure 3.2: General gene structure. Different regions that are present in genes of eukaryotic species. The intermediate scheme shows some of the signals that allow to define exon/intron structure. After splicing, a mature mRNA is formed only with exonic regions.

3.3 Genes: Discrete Units of Genetic Information

3.3.1 Gene Structure

We have seen that genes form discrete units of genetic information located on the chromosomes and consist of DNA that encodes the instructions to produce the polypeptides that will form proteins. Given the differences in the complexity of prokaryotic and eukaryotic cells, it is expected that genes from these two types of cells also show different complexity.

However, regardless of the different cell complexity, genes share a common general structure. At one end of the gene there is a region of transcription initiation. In this region, different signals where regulatory proteins bind to regulate gene transcription can be found. These signals work as switches where transcription is regulated both in terms of amount of produced mRNA and periodicity. At the other end of the gene there is a region that contains signals that determine the end of transcription.

In between these two regions there is the transcribed region as shown in Fig. 3.2. It is in this region where differences between prokaryotic and eukaryotic genes are more noticeable. In

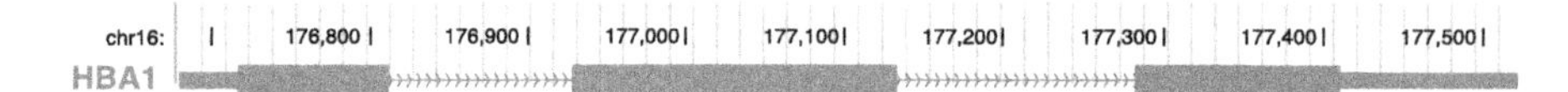

Figure 3.3: Exon structure for gene HBA1. Darker solid regions represent exons. Larger blocks indicate the coding part of exons and the thinner blocks the untranslated regions. Dashed regions represent exons.

prokaryotic species, like for instance bacteria, genes are organized as a continuous stretch of DNA, and the totality of the transcribed mRNA sequence will be used to build the polypeptide.

In eukaryotes, on the other hand, the transcribed region of a gene consists of a combination of coding segments called *exons* interspersed with non-coding segments called *introns*. While, in transcription, the pre-mRNA will contain the sequence of both exons and introns, a subsequent step called *splicing*, will remove the introns and concatenate the exons to form the mature mRNA. Splicing occurs in the nucleus and, therefore, before translation. This means that effectively, only the nucleotide sequence corresponding to the exons will be used in the translation process. Fig. 3.2 depicts the process.

The gene structure in eukaryotes represents an increased complexity. The DNA contains signals that easily allow the delineation of this exon/intron structure. On average, exons are significantly shorter than introns and marked by short and conserved nucleotide sequences. The exon/intron borders include signal sequences that mark the end of the exon and the start of the intron, the so-called *5' splice site* or *donor site*, and the end of the intron and the start of an exon, called *3' splice site* or *acceptor site*. These will be the regions where, during the splicing process, the intronic sequence is cut out and removed from the pre-mRNA. Additional signals, for instance the polypyrimidine tract, can be found in introns or in exons to help recognizing the exon/intron borders and splice the introns, as shown in Fig. 3.2.

We have seen, in the previous section, that translation only starts when the ribosome detects the start codon AUG and stops when one of the stop codons UAA, UAG or UGA is found. The region within the start and the stop codon of the pre-mRNA, after properly spliced, is called *coding region*. This name comes from the fact that only this nucleotide sequence will be used to code for the protein sequence. The remaining regions before the start codon and the stop codon, although part of the pre-mRNA, will not be translated into amino acids. Thus, they are called respectively *5'* and *3'-untranslated regions* or for short *5'-UTR* or *3'-UTR*.

In Fig. 3.3, we can find a representation of the structure of the human gene hemoglobin alpha 1 (HBA). This gene is part of a cluster of alpha globin genes spanning about 30 KB and located in the chromosome 16 of the human genome. This gene is involved in oxygen transport from the lung to the different peripheral tissues. The pre-mRNA contains around 900 nucleotides and consists of three exons. The resulting protein is 144 amino acids long.

We have assumed that a gene corresponds to a genomic region that harbors a sequence that contains the information necessary to produce a protein. However, this is not always the case. In fact, some genes do not code for proteins, i.e. the respective RNA sequence is not translated into an amino acid chain. While the first type of genes are called *protein-coding genes*, the second ones are called *non-coding RNAs (ncRNAs)*. ncRNAs are characterized as being a class of relatively short RNAs, many of them are known to be fully functional and playing regulatory roles within the cell, even though the function of the majority of them is currently unknown. There are several types of ncRNAs including microRNAs, snoRNAs and long non-coding RNAs. Some of these ncRNAS share several properties of protein-coding genes [46].

For simpler eukaryotic species, protein-coding genes tend to cover a large fraction of the genome [37]. For instance, budding yeast (*S. cerevisiae*) contains a genome of 12 million base pairs, with roughly 6600 genes corresponding to approximately 70% of the genome being used as protein coding sequence. For genomes of higher animals and plants, this percentage becomes much lower.

This is the case of the human genome, with its 3.2×10^9 bases pairs and with $\sim$21,000 protein coding genes, where only 3% of the genome corresponds to protein-coding sequences. So, it is often the case that a large proportion of the genome sequence of different species appear to not contain any gene. This is called the non-coding genome and the regions free of genes are intergenic regions. For some time, it was considered that these regions had no particular functional interest. With the advance of high-throughput sequencing technologies, several landmark studies like Encode [27,49,51] and Phantom [13,33] projects are revealing that these regions harbor many different functional signals that are used in cellular regulatory process mostly controlling gene expression in different ways.

Surprisingly, across diverse species, the number of protein coding genes does not totally correlate with the genome size or the overall cellular complexity. Consider for instance the species in Table 3.3. The fission yeast has a longer genome than the budding yeast but a smaller number of genes. The fruit fly and the tobacco plant (model organism for plants) have a genome of similar length, but the plant has almost twice more genes than the fly. Mouse and human although very different species have a similar number of genes. This implies that there are several mechanisms, other than the total number of genes, that through evolution of the species have contributed to this differential complexity.

Alternative splicing is one of the mechanisms that contribute to differential complexity. Given its combinatorial nature its represents an interesting and challenging topic to be approached from the computational point of view. Alternative splicing is the process in which the exons of the pre-mRNAs are spliced forming different combinations. This allows that from one

Table 3.3: Genomes by the numbers. Genome size, number of chromosomes and number of protein coding genes for different model species. Numbers were retrieved from [129].

Organism	Genome size (Base Pairs)	Protein coding genes	Number of chromosomes
Budding yeast - *S. cerevisiae*	12 Mbp	6600	16
Fission yeast - *S. pombe*	13 Mbp	4800	3
Fruit fly - *D. melanogaster*	140 Mbp	14,000	8
Tobacco plant - *A. thaliana*	140 Mbp	27,000	10 (2n)
Mouse - *M. musculus*	2.8 Gbp	20,000	40 (2n)
Human - *H. sapiens*	3.2 Gbp	21,000	46 (2n)

pre-mRNA sequence several distinct mature mRNAs are produced, leading to structurally and functionally protein variants. This process, most extensively used in higher eukaryotes allows for an increased macromolecular and cellular complexity.

Recent studies, with high-throughput RNA sequencing, discovered that, in human cells, alternative splicing is a very common event with 90% to 95% of human multi-exon genes producing mRNA transcripts that are alternatively spliced [122,151]. Alternative splicing has a combinatorial nature where exons and introns are fully or partially combined in a multitude of ways [89]. Another important aspect of alternative splicing is cell and tissue specificity, allowing that in higher eukaryotes different types of cells generate from the same cell a differentiated repertoire of protein isoforms [24,111].

3.3.2 Regulation of Gene Expression

We have seen that genes encode for proteins, that proteins play a multitude of functions and that the concerted actions of these proteins determine the function of the cell. Gene expression refers to the process of expression of thousands of genes within the cell at a given moment. Given the dynamic and adaptive nature of cells, the protein amounts required by the cell changes through time. The available amount of a certain protein results from the balance of its synthesis and its degradation.

Regulation of gene expression refers to the combination of processes that determine how much of a particular protein should be produced. As expected, these regulatory processes are much more complex in eukaryotic than in prokaryotic cells.

Since prokaryotic cells do not have a nucleus, the DNA flows in the cytoplasm. Transcription and translation occur almost simultaneously [36]. Whenever the production of a protein is no longer necessary, transcription stops. Given the immediate effect on the amount of produced protein, control of gene expression is essentially determined at the level of transcription. Another interesting aspect of some prokaryotic species like bacteria is the organization of their genes in clusters of co-regulated genes, called *operons*. Genes from an operon are close in the genome and under a common control mechanism that allows them to be transcribed in a coordinated way. This gives the species the capacity to rapidly react to different external stimuli [131].

In eukaryotic cells, gene regulation is a multi-step process controlled by multiple mechanisms. Given that transcription and translation are decoupled, regulation may occur in the nucleus during transcription and RNA processing or in the cytoplasm during translation. It may even occur after the protein being produced, where biochemical modifications to the protein result in *post-translational regulation* [36,131].

3.4 Human Genome

We will now focus a little bit on our species. The human body contains an estimated number of 75 to 100 trillion (10^{12}) cells [120,129]. Each cell contains a genome and apart from some abnormalities like for instance cancer, the genetic information carried throughout all the somatic cells is the same. Humans have diploid cells with two copies of each chromosome.

As seen before, human somatic cells contain 46 chromosomes, consisting of two sets of 23 chromosomes. Male and Female individuals are distinguished by the thirty-third chromosome set. While females have two copies of chromosome X, males have a copy of chromosome X and a copy of chromosome Y. Each chromosome contains an array of genes. Current estimates point to approximately 21,000 protein-coding genes. Nevertheless, other types of genes have been discovered and current gene annotation efforts annotate more than 50,000 genes in the human genome [72].

The history of modern genetics started more than 150 years ago with the pioneering work of the botanist and scientist Gregor Mendel [121]. He has been considered the father of the modern genetics due to its discoveries of the basic principles of heredity made through experiments in plant hybridization. In the middle of the 20th century, the work of Rosalind Franklin, Maurice Wilkins, James Watson and Francis Crick revealed the double helix structure of the DNA. This was a major breakthrough that boosted the research on the human genome. In 1961, Marshall Nirenberg deciphered the genetic code for protein synthesis that we have seen in Table 3.1.

In the late 1970's and early 1980's, two techniques have revolutionized modern molecular biology. These were the rapid DNA sequencing technique developed by Frederick Sanger and polymerase chain reaction (PCR) for DNA amplification by Kary Mullis. Also, during this period, the first genetic diseases started to be mapped.

The Human Genome Project started in 1990. The goal was to characterize both physically and functionally the human genome. Its impact on so many different aspects of human life makes it one of the greatest achievements of human history. In 1998, the Celera Genomics Corporation joins the race of sequencing the human genome in an effort lead by Craig Venter. In 1999, chromosome 22 was the first human chromosome to be sequenced.

During this time, other efforts were also being made to sequence other species. In 1995, *Haemophilus influenzae* was the first bacterium to have its genome sequenced. Early years of the decade of 2000 were exciting years for molecular biology. In 2000, the genome sequence of the model organism fruit fly was released, followed by the first draft of the human genome in 2001. In 2002, the mouse model organism was the first mammalian to have its genome sequence revealed and in 2003 a more complete version of the human genome was announced with the end of this project.

Since then, improvements have been made to the genome sequence. Other large-scale international efforts have helped us to better understand the human genome. Examples of such projects include the ENCODE Project [27,49,51] that had the goal of providing a detailed view of the functional elements in the human genome and the 1000 Genomes Project [15] that mapped the genetic variation in humans.

Despite certain phenotypic differences that humans may have, like height, color of the skin, hair or color of the eyes, we are all very similar at the genetic level. In fact, on average, any two humans share more than 99% of their DNA sequence [98] with the amount of genetic variation due to differences in populations being modest [156]. The remaining fraction of the genome that varies is important, since in combination with the environment is what makes us unique. There are different ways in which this variation can occur.

Point mutations that consist in the change of one nucleotide by another are the most frequent type of genomic alteration. When they occur with a certain frequency, for instance in more than 1% of the population, they are called *polymorphisms*. So *Single Nucleotide Polymorphisms*, *SNP* for short, are variations of a single nucleotide in the DNA sequence among a group of individuals.

Other forms of variation include the insertion or deletion of a small number, typically less than a dozen, of nucleotides generally called *indels*. Loss or gain of large portions of the genome or other genomic rearrangements may also occur.

Fortunately, most of these variations have little or no impact on our health, as they may for instance occur in non-coding genomic regions. When these alterations hit the regions that code for genes they may have an important functional impact. By changing the DNA and the respective RNA sequence, such alterations may result in variations to the amino acid sequence leading to damage proteins or to an abnormal increase or decrease of its production [95,113].

With the development of technologies to characterize genetic variability along the entire genome, many studies have been made to assess the association of the genotype with a certain phenotype. In order to obtain statistically robust results, these studies are made in large groups of individuals that present the studied phenotype, which could be for instance an anatomical phenotype or a disease phenotype [1,26]. These are called *Genome-Wide Association Studies* or *GWAS*. From these studies, we have learned that some genetic variants influence certain human behaviors like the risk of disease or the response to drugs.

3.5 Biological Resources and Databases

In Section 3.4, we have seen that the molecular biology field went through great and quick advances in the last quarter of the previous century. It certainly benefited from the development and dissemination of the Internet that was also occurring during this period. In those early days, the amount of sequence information was very limited. It was shared and communicated by printed pages or fitting in text files. The development of new sequencing techniques created an exponential growth of sequence data and this prompted the need to develop efficient, scalable and consistent ways of transferring and sharing the data.

One of the first efforts to collect sequence data was made by Margaret O. Dayhoff that compiled the first comprehensive collection of protein sequences published from 1965 to 1978. This gave raise to the Protein Information Resource (PIR, `pir.georgetown.edu/`) created in 1984.

The European Molecular Biology Laboratory (EMBL)-bank was created in 1982 as the first international database for nucleotide sequences. Also, in that year, there was the public release of Genbank (`www.ncbi.nlm.nih.gov/genbank/`), now maintained by the National Center for Biotechnology Information (NCBI) in the United States, which contained an annotated collection of public available nucleotide sequences and respective sequence translations. In 1986, the Swiss-Prot (now part of the UniProtKB) database was presented, containing non-redundant and curated protein sequence data complemented with other high level information and interconnected with other sequence resources.

In 1988, the International Nucleotide Sequence Database Collaboration (INSDC) was launched, a joint effort of EMBL-EBI in Europe, NCBI in the United States and DDBJ (`www.ddbj.nig.ac.jp/`) in Japan to collect and disseminate nucleotide sequences. It currently involves the databases of DNA Data Bank of Japan, GenBank and the European Nucleotide Archive (ENA, `www.ebi.ac.uk/ena`).

During the years of the Human Genome Project, independent on-line browsers were created to share and provide a graphical display of the sequence assembly of the human genome.

The Ensembl genome database (`www.ensembl.org/`), a joint initiative of the European Bioinformatics Institute and the Wellcome Trust Sanger Institute, and the genome browser at University of California Santa Cruz (UCSC, `genome.ucsc.edu/`) are the most well known browsers for genomic information. These browsers have evolved to very complete resources, integrating nowadays many different types of genomic information from several different species.

In 2002, the UniProt (`www.uniprot.org/`) consortium was created joining the protein sequence databases of EBI, Swiss-Prot and PIR in a collection of curated and non-curated sequences with high-level of annotation. With the massive growth of sequence data from many different formats, including those from high-throughput sequencing, lead to the replacement of EMBL-Bank with the European Nucleotide Archive integrating nucleotide sequence, associated data and the respective experiment annotations.

Sequence data is being generated in a multitude of laboratories worldwide. Most of the current sequence databases allow for scientists to submit their own generated data. While this allows a quicker and generalized dissemination of the data, it also creates some problems in terms of consistency, completeness and redundancy of the submitted data.

The reader should be aware that biological sequence databases can then be divided into primary and secondary databases. Primary databases contain sequence data submitted by the researchers that may not be fully processed and it is not curated. Redundancy often exists in the form of multiple entries for the sequence of a gene or transcript submitted by different people working on related topics. Another critical issue is that, since data is shared by different databases, an error or an inaccuracy in the data of the primary databases may easily propagate to other databases that use data from these sources. Examples of primary databases include the ENA database, Genbank or DDBJ.

Secondary databases contain data, which may be obtained from primary databases and that have been curated by specialists in terms of consistency and completeness. The data entries are complemented and annotated with metadata and additional information. Examples include the NCBI RefSeq (`www.ncbi.nlm.nih.gov/refseq/`) database that has been curated from Genbank or the UniProtKB/Swiss-Prot (`www.uniprot.org/uniprot/`).

We now describe some of the databases that are of relevance in the context of this book and where the user can access to obtain data for its own experiments.

ENA (includes EMBL-bank) – www.ebi.ac.uk/ena
GenBank – www.ncbi.nlm.nih.gov/GenBank
DDBJ – www.ddbj.nig.ac.jp

These are the three main primary databases of publicly available nucleotide sequences. They integrate the INSDC consortium and share data among them, being periodically updated. Each database has its own data format.

NCBI Gene – http://www.ncbi.nlm.nih.gov/gene

This is a gene centric database containing data from multiple species. Beyond the sequences, it integrates aspects like genotypic variation of the gene, associated phenotypes or molecular pathways in which the gene is involved.

NCBI RefSeq – http://www.ncbi.nlm.nih.gov/refseq

This is a secondary database that processes data from the primary database GenBank to deliver a comprehensive database of curated information integrating data from the genome, transcriptome and proteome. The RefSeq dataset comprises an important reference to be used in genome annotation and characterization studies being widely used in species comparison or in gene expression analysis.

Gencode – www.gencodegenes.org

The Gencode annotation started as an effort within the ENCODE project [27,49,51] to provide a fully integrated annotation of the human genome. It provides a comprehensive annotation of the human gene set that has been used by ENCODE and many other different projects. It contains information not only on protein-coding genes, but also on many other different RNA types. Currently, it is expanding to include the annotation of the mouse genome.

UCSC Genome Browser – https://genome.ucsc.edu/
Ensembl – http://www.ensembl.org/

These are two genome browsers that allow online viewing and download of genomic data from multiple species. Here, we can visualize, by zooming in and out on the genome, the genomic sequence, the gene annotation, the conservation of the sequence across species, regulatory and disease data. These browsers also allow researchers to upload their own data for visualizations within the selected genomic context. The browsers allow the download of full datasets of sequence and annotation data that have been previously processed and are easy of use.

NCBI Protein – http://www.ncbi.nlm.nih.gov/protein

This database aggregates protein data from multiple sources including sequence translations from genes in Genbank or RefSeq to curated entries retrieved from Swiss-Prot, PIR or PDB.

UniProt – http://www.uniprot.org/

UniProt is an integrated and comprehensive repository on sequence and functional protein information. It gathers data from multiple other databases including Swiss Prot, TrEMBL and PIR-PSD. Within UniProt we can find three databases: UniParc, UniProtKB and UniRef. UniParc is a non-redundant and comprehensive database of publicly available protein sequences. UniRef provides clustered sets of protein sequences from UniProtKB and UniParc. UniProtKB, the most important resource, provides functional information on the proteins, containing a curated database (SwissProt) and a non-curated component (TrEMBL).

Protein Data Bank (PDB) – http://www.rcsb.org/

PDB is a database that contains structural data of proteins, nucleic acids and other complex assemblies. It provides functionalities for data deposit and download and tools for data visualizations in multiple data formats. It contains information organized by protein, containing from the sequence, annotations of secondary structure, tri-dimensional coordinates and views, similarity search at sequence and structure level and the details of the experiment

dbSNP – www.ncbi.nlm.nih.gov/snp
dbVar – www.ncbi.nlm.nih.gov/dbvar

These are two databases from NCBI that contain the annotation of short genetic and large structural variations within the human genome and other species. dbSNP is mostly focused on point mutations, microsatellites, and small insertions and deletions. It contains information on the mutated alleles, their sequence context, visualization of their occurrence within the gene sequence, frequency in populations and also connects with other databases to show information on clinical significance. dbVar entries contain a view and details of the genomic region where the variation occurs, complemented experimental evidence and validation, publication where its was first reported and clinical associations.

ClinVar – www.ncbi.nlm.nih.gov/clinvar/

ClinVar is a database that provides information and supporting evidence on the association of human genetic variation and phenotypes. It is particularly useful in the clinical and health context since it reports variants found in patient samples along with assertions made by the researchers or the clinicians that submitted data about the clinical relevance of these variants.

Gene Expression Ominbus (GEO) – www.ncbi.nlm.nih.gov/geo/

GEO is a database from NCBI that collects gene expression datasets obtained either with micro-array or sequencing technologies. The database is organized into datasets that may contain multiple samples. In each entry a reference to the platform in which the data was generated, the raw and the processed data and the article in which the data was published are reported. It is possible to search the datasets by keyword.

PubMed – www.ncbi.nlm.nih.gov/pubmed/

PubMed is a database from NCBI that indexes information of scientific articles related to biomedical and life sciences research. It contains currently pointers to more than 27 million articles and books. Article entries contain links to publisher website. Advanced article search can be made based on keywords, title or author names.

The exponential growth of sequence data urged the development of multiple efforts to catalogue, share and disseminate this data through online databases. Many different databases have been proposed and are currently available containing comprehensive repositories of nucleotide and protein sequences often complemented and integrated with additional information from other databases. Many other databases have been developed to integrate and complement with the genetic information.

Bibliographic References and Further Reading

This chapter contains a very basic approach to some major concepts in cellular and molecular Biology. Many different textbooks cover these concepts, and many other, in a deeper way [10, 37,120].

Exercises

1. Sort, by increasing order of organizational structures, the following cellular elements: cell, nucleotide, chromosome, gene, DNA.
2. Consider a word of length 5. Indicate the number of possible word combinations:
   a. based on the DNA alphabet?
   b. based on the RNA alphabet?
   c. based on the protein alphabet?
3. Consider a DNA sequence with 12 nucleotides:
   a. Indicate the number of codons that can be derived from direct reading of this sequencc?
   b. Consider that, in the beginning of the sequence, we find the start codon and in the end one of the stop codons. What is the total number of codons that can be derived from this reading?
   c. Indicate the total number of reading possibilities, i.e. all the reading frames?
4. The following sequence represents part of a DNA sequence where in upper case are represented exons and lower case the introns.

```
> Exon-Intron-Exon sequence
ACTCTTCTGGTCCCCACAGACTCAGAGAGAACCCACCATGGTGCTGTCTC
CTGCCGACAAGACCAACGTCAAGGCCGCCTGGGGTAAGGTCGGCGCGCAC
GCTGGCGAGTATGGTGCGGAGGCCCTGGAGAGgtgaggctccctcccctg
ctccgacccgtgctcctcgcccgcccggacccacaggccaccctcaaccg
tcctggccccggacccaaaccccacccctcactctgcttctccccgcagG
ATGTTCCTGTCCTTCCCCACCACCAAGACCTACTTCCCGCACTTCGACCT
GAGCCACGGCTCTGCCCAGGTTAAGGGCCACGGCAAGAAGGTGGCCGACG
CGCTGACCAACGCCGTGGCGCACGTGGACGACATGCCCAACGCGCTGTCC
GCCCTGAGCGACCTGCACGCGCACAAGCTTCGGGTGGACCCGGTCAACTT
CAAG
```

   - Based on the previous sequence indicate the RNA sequence that will be obtained after mRNA splicing.
   - Discuss what would happen if an SNP occurs in the border of the first exon to the intro, g $\rightarrow$ t. Propose a possible alternative for the mRNA sequence after splicing?
5. Suppose that you are studying a very short protein sequence. We would like to know the sequence of nucleotides that gave rise to this protein sequence: Met-Ala-His-Trp.

a. How many mRNA sequences can code this protein?
b. What is the length of the mRNA?
c. How many nucleotides from the mRNA are critical to code for this protein, i.e. how many of the nucleotides will have a direct impact in the amino acid sequence?

Hint: i) Use the genetic code table provided in Table 3.1. ii) Remember that mRNAs coding for proteins contain a stop codon. iii) One amino acid can be coded by multiple codons.

6. From the NCBI Gene, retrieve the genomic sequence of the TP53 gene in human. Obtain additional information on this gene by looking for the entry in the Genbank database. Retrieve the list of protein sequences derived from each of the isoforms of the gene.
   Hint: FASTA is a format to organize biological sequences. Search for the entry that contains this file.
7. From the NCBI RefSeq, find the coding sequence and the protein sequence for the TP53 gene in human, chimp and mouse.
   Hint: Search for the gene name, select the gene entry corresponding to each of the species and use the "Send to" to obtain the nucleotide or the protein file for the coding sequence.
8. From PDB, retrieve the files with the primary structure and the tri-dimensional structure of one of the versions of the Transthyretin protein in human.
9. In the UCSC Genome Browser, visualize the available information for the MDM2 gene in human.
   a. For one of the Gencode isoforms, obtain the genomic, mRNA and protein sequence. (Hint: click on the transcript/isoform and go to the "Sequence and Links to Tools Database" section.)
   b. Visualize the RNA-seq expression data from GTEx and find in which tissues this gene has higher expression.
   c. Find all the alternate gene symbols for this gene.
10. In the Ensembl Genome Browser, visualize the available information for the MDM2 gene in human.
    a. For the longest transcript, obtain the coding sequence (click on the gene structure of the longest transcript).
    b. How many splice variants do you find for this gene? How many are protein coding? (See transcript table.)
    c. Compare the evolutionary tree of MDM2. (Use the Comparative Genomics/Gene Tree tool.)
    d. Find a germline and a somatic SNP occurring within the coding sequence of the gene. Somatic variants are identified by *COSM* prefix and germline variants by an *rs* prefix. (Use the Genetic Variation/Variant table tool.)

CHAPTER 4

Basic Processing of Biological Sequences

In this chapter, we address the computational representation of biological sequences and cover basic algorithms for their processing. We will address the implementation of the processes related to gene expression, covering transcription, translation and the identification of open reading frames. We also cover the implementation of a class for biological sequences (including DNA, RNA and protein sequences). Finally, we will review a set of classes from the *BioPython* package to store and process sequences, together with their annotations, allowing their loading from databases and reading/writing from files in different formats.

4.1 Biological Sequences: Representations and Basic Algorithms

As previously discussed, in Chapter 3, in biological systems, the genetic information is encoded into DNA molecules. For many practical purposes, in Bioinformatics algorithms and tools, these molecules can be represented as one-dimensional sequences of nucleotides.

Since in DNA (or RNA) molecules, there are four distinct types of nucleotides, the computational representation of these sequences consists of strings (i.e. sequences of characters) defined over an alphabet of four distinct symbols. For DNA, these symbols are the letters A, C, G, and T, which represent, respectively, adenine, cytosine, guanine and thymine, while, in the case of RNA, the T gives place to a U, representing uracil.

Although the basic alphabet for DNA only contains the symbols for the four nucleotides, the *International Union of Pure and Applied Chemistry* (IUPAC) has defined an extended set of symbols that allow ambiguity in the identification of a nucleotide, useful for instance in the results of a sequencing process in positions where there is uncertainty in the identification of the nucleotide, or in the design of Polymerase Chain Reaction (PCR) primers. Table 4.1 shows the symbols in this IUPAC alphabet and their meaning.

The other most important biological sequences are proteins, which are typically represented by sequences of aminoacids. In the standard genetic code, there are 20 different encoded aminoacids that can occur in protein sequences, and thus the strings are defined in an alphabet of 20 characters (see Table 4.2). In some cases, it is important to add to this alphabet a symbol representing the decoding of a stop codon, to be used for instance when running automatic tools for DNA translation, which we will cover in more detail in the next sections. The most used symbols for the stop codon are the underscore (_) or the asterisk (*).

Bioinformatics Algorithms. DOI: 10.1016/B978-0-12-812520-5.00004-3

Table 4.1: IUPAC symbols for nucleotides.

Symbol	Name	Nucleotides represented
A	Adenine	A
C	Cytosine	C
G	Guanine	G
T	Thymine	T
U	Uracil	U
K	Keto	G, T
M	Amino	A, C
R	Purine	A, G
S	Strong	C, G
W	Weak	A, T
Y	Pyrimidine	C, T
B	Not A	C, G, T
D	Not C	A, G, T
H	Not G	A, C, T
V	Not T	A, C, G
N	Any base	A, C, G, T

Table 4.2: IUPAC symbols for aminoacids.

Symbol	Name
A	Alanine
C	Cysteine
D	Aspartic acid
E	Glutamic acid
F	Phenylalanine
G	Glycine
H	Histidine
I	Isoleucine
L	Lysine
M	Methionine
N	Asparagine
P	Proline
Q	Glutamine
R	Arginine
S	Serine
T	Threonine
V	Valine
T	Tryptophan
Y	Tyrosine

Since DNA, RNA, and protein sequences can be represented as strings, we can use many of the features and functions that were shown on Chapter 2 to process these sequences and extract meaningful biological results. As an example, let us define a function that checks if an inputted DNA sequence is valid, using some of the functions over strings discussed in Section 2.6.3.

Note that the first line provides a documentation for the function, providing a comment that states the function's purpose, detailing the expected inputs and results. This will be used in most functions developed throughout the book to improve its readability.

In the function code, valid characters are counted and their sum is compared to the length of the string to check if all characters are valid. Similar functions could be easily built for RNA or protein sequences.

```
def validate_dna (dna_seq):
    """ Checks if DNA sequence is valid. Returns True is sequence is
   valid, or False otherwise. """
    seqm = dna_seq.upper()
    valid = seqm.count("A") + seqm.count("C") + seqm.count("G") +
   seqm.count("T")
    if valid == len(seqm): return True
    else: return False

>>> validate_dna("atagagagatctcg")
True
>>> validate_dna("ATAGAXTAGAT")
False
```

Another useful example is the ability to calculate the frequency of the different symbols in the sequences, which will be the aim of the function shown in the next code block. The result of the function will be a dictionary, where the keys are the symbols and the corresponding values are frequencies. Note that this can be applied to any of the sequence types defined above (DNA, RNA and proteins).

```
def frequency (seq):
    """ Calculates the frequency of each symbol in the sequence.
   Returns a dictionary. """
    dic = {}
    for s in seq.upper():
        if s in dic: dic[s] += 1
        else: dic[s] = 1
    return dic

>>> frequency("atagataactcgcatag")
```

```
{'A': 7, 'C': 3, 'G': 3, 'T': 4}
>>> frequency("MVVMKKSHHVLHSQSLIK")
{'H': 3, 'I': 1, 'K': 3, 'L': 2, 'M': 2, 'Q': 1, 'S': 3, 'V': 3}
```

This function will be used in the next program to calculate the frequency of the aminoacids in a sequence read from the user's input. The frequency of the aminoacids will be printed from the most frequent to the least frequent. Note, in this example, the use of the `lambda` notation that allows to define functions implicitly within blocks of code. In this case, this notation is used to select the second element of a tuple, applied to each element of the list of (key, value) tuples coming from the dictionary when applying the method **items**.

```
seq_aa = input("Protein sequence:")
freq_aa = frequency(seq_aa)
list_f = sorted(freq_aa.items(), key=lambda x: x[1], reverse = True)
for (k,v) in list_f:
    print("Aminoacid:", k, ":", v)
```

To end this section, let us define a new function for a task that is in many cases quite useful. In this case, we will compute the GC content of a DNA sequence, i.e. the percentage of 'G' and 'C' nucleotides in the sequence.

```
def gc_content (dna_seq):
    """ Returns percentage of G and C nucleotides in a DNA sequence.
    """
    gc_count = 0
    for s in dna_seq:
        if s in "GCgc": gc_count += 1
    return gc_count / len(dna_seq)
```

In many cases, for instance when looking for genes or exons, we need to compute the GC content of the different parts of a sequence. The next function provides a solution for this task, computing the GC content of the non-overlapping sub-sequences of size *k* of the inputted sequence.

```
def gc_content_subseq (dna_seq, k=100):
    """ Returns GC content of non-overlapping sub-sequences of size k
    . The result is a list. """
    res = []
    for i in range(0, len(dna_seq)-k+1, k):
        subseq = dna_seq[i:i+k]
```

```
        gc = gc_content(subseq)
        res.append(gc)
    return res
```

As a practical note, the functions developed in this section (and the next ones) can be put together in a file (e.g. names "sequences.py"), implementing a Python module. This allows to define the main program (e.g. for testing) either in this file, typically in the end, or in other files in the same folder, by using the `import` command (whose behavior was explained in the previous chapter).

4.2 Transcription and Reverse Complement

The transcription of DNA to RNA molecules is a fundamental step within the whole process of protein synthesis from the genetic information contained within DNA molecules. This process has been explained in Section 3.2.1 regarding its molecular details. Here, we will cover the aspects related to information processing, providing some examples of code that allow to implement this process with Python functions and programs.

To understand the functions that compute the transcription, we need to take in mind that the DNA molecules have two complementary strands that are read in reverse directions. When the transcription occurs, these two strands are split, and the new RNA molecule will be created as a complement to one of the strands. Thus, its sequence will be similar to the one of the other DNA strand (which we call the *template* strand), but with 'U' instead of 'T' nucleotides. In this context, if we consider the template DNA strand as input, the resulting RNA sequence will be obtained by simply replacing 'T's by 'U's.

This is exactly what is done by the function provided next. Note that the `assert` clause is used to test if the function's input is a valid DNA sequence, raising an exception otherwise. The `return` statement will provide the processed sequence by using the **replace** method to change all occurrences of 'T's by 'U's, returning a new string, the result of the function. In this function, as well as in others within this chapter, we will always convert the sequences to upper case letters and work with those in the DNA processing tasks.

```
def transcription (dna_seq):
    """ Function that computes the RNA corresponding to the
   transcription of the DNA sequence provided. """
    assert validate_dna(dna_seq), "Invalid DNA sequence"
    return dna_seq.upper().replace("T","U")
```

Given what was previously explained, it is clear that there is the need to compute the content of one of the strands, given the other (complementary). Indeed, in Bioinformatics databases, it is common that only one of the strands is provided, and therefore we need to compute the other for a number of DNA processing tasks.

The function provided next achieves this task, taking as input a DNA sequence, and providing as result a new DNA sequence that corresponds to the reverse complement of the original one. Note that the 'A' nucleotide pairs with the 'T', while the 'C' pairs with a 'G'. Also, the reverse of the sequence is obtained by building the result with a `for` cycle, always appending the new symbol in the head of the string.

```
def reverse_complement (dna_seq):
    """ Computes the reverse complement of the DNA sequence. """
    assert validate_dna(dna_seq), "Invalid DNA sequence"
    comp = ""
    for c in dna_seq.upper():
        if c == 'A':
            comp = "T" + comp
        elif c == "T":
            comp = "A" + comp
        elif c == "G":
            comp = "C" + comp
        elif c== "C":
            comp = "G" + comp
    return comp
```

4.3 Translation

In cells, proteins are synthesized by creating chains of aminoacids, according to the information contained within messenger RNA molecules, in a process named translation. This complex biological process was detailed in Section 3.2.2 and, as before, we will cover here mostly the information processing details and implement those as Python functions.

Since, as seen in the previous section, the conversion from DNA to messenger RNA (transcription) is trivial, we will here consider the direct translation from DNA to proteins. The computational process works as follows: (i) the DNA sequence to be translated is split into non-overlapping sub-sequences of size three (named codons); (ii) for each codon, the corresponding aminoacid is determined by a conversion table (provided as Table 3.1); (iii) the aminoacids are gathered, keeping their order, in the resulting protein sequence.

The core data structure is the conversion table from codons to aminoacids, which will be here implemented in Python as a dictionary, where keys will be the codons (64 in total), and the corresponding values will be the aminoacids. The following function implements this conversion, using an internal dictionary to keep the conversion table. In the case an invalid codon is passed as an argument, the function returns None. Also note that we will be using the '_' symbol to denote the decoding of a stop codon.

```
def translate_codon (cod):
    """Translates a codon into an aminoacid using an internal
   dictionary with the standard genetic code."""
    tc = {"GCT":"A", "GCC":"A", "GCA":"A", "GCG":"A",
      "TGT":"C", "TGC":"C",
      "GAT":"D", "GAC":"D",
      "GAA":"E", "GAG":"E",
      "TTT":"F", "TTC":"F",
      "GGT":"G", "GGC":"G", "GGA":"G", "GGG":"G",
      "CAT":"H", "CAC":"H",
      "ATA":"I", "ATT":"I", "ATC":"I",
      "AAA":"K", "AAG":"K",
      "TTA":"L", "TTG":"L", "CTT":"L", "CTC":"L", "CTA":"L", "CTG":"L
   ",
      "ATG":"M", "AAT":"N", "AAC":"N",
      "CCT":"P", "CCC":"P", "CCA":"P", "CCG":"P",
      "CAA":"Q", "CAG":"Q",
      "CGT":"R", "CGC":"R", "CGA":"R", "CGG":"R", "AGA":"R", "AGG":"R
   ",
      "TCT":"S", "TCC":"S", "TCA":"S", "TCG":"S", "AGT":"S", "AGC":"S
   ",
      "ACT":"T", "ACC":"T", "ACA":"T", "ACG":"T",
      "GTT":"V", "GTC":"V", "GTA":"V", "GTG":"V",
      "TGG":"W",
      "TAT":"Y", "TAC":"Y",
      "TAA":"_", "TAG":"_", "TGA":"_"}
    if cod in tc: return tc[cod]
    else: return None
```

One aspect to take into account is the non-universal nature of the standard genetic code that is provided by the dictionary used in the previous function. Indeed, in some organisms and organelles (for instance in eukaryotic mitochondria), other alternative codon tables are used in the translation process. Regarding the previous implementation, a solution to consider other tables is to alter the previous function to receive the translation table (dictionary) as an argument (in this case, the standard table could be considered if the argument was `None`). The implementation of this alternative solution is left as an exercise for the reader.

Using the previous function, it is now possible to write a function that translates a whole DNA sequence. This function will receive as inputs the DNA sequence and the initial translation position, and will return the corresponding aminoacid sequence. Note that the split of the sequence into codons is done by generating the appropriate indexes using the **range** function, where the last argument allows to define a step value. The translation of the individual codons is, as expected, achieved by calling the previously developed function.

```
def translate_seq (dna_seq, ini_pos = 0):
    """ Translates a DNA sequence into an aminoacid sequence. """
    assert validate_dna(dna_seq), "Invalid DNA sequence"
    seqm = dna_seq.upper()
    seq_aa = ""
    for pos in range(ini_pos,len(seqm)-2,3):
        cod = seqm[pos:pos+3]
        seq_aa += translate_codon(cod)
    return seq_aa
```

Note that since the genetic code is redundant, there will be repeated values, i.e. a single aminoacid will be encoded by different codons. This is normally called the *codon usage*, and provides interesting statistics when applied to the genes of different species.

Let us write a function that computes the codon usage, providing an aminoacid and a DNA coding sequence. The result will be a dictionary, where keys will be codons and values will represent the percentage of usage for each of those codons.

```
def codon_usage(dna_seq, aa):
    """Provides the frequency of each codon encoding a given
   aminoacid, in a DNA sequence ."""
    assert validate_dna(dna_seq), "Invalid DNA sequence"
    seqm = dna_seq.upper()
    dic = {}
    total = 0
    for i in range(0, len(seqm)-2, 3):
        cod = seqm[i:i+3]
        if translate_codon(cod) == aa:
            if cod in dic:
```

```
                dic[cod] += 1
            else: dic[cod] = 1
            total += 1
    if total >0:
        for k in dic:
            dic[k] /= total
    return dic
```

4.4 Seeking Putative Genes: Open Reading Frames

In the previous section, we looked at translation without any concern about some of the rules that can be observed in real life. Indeed, the translation of a protein always begins with a specific codon (the start codon – the "ATG" in the standard table), which codes for the aminoacid Methionine ('M'). This aminoacid is always the first in a protein, but can also occur in other positions. Also, the translation process terminates when a stop codon is found.

Thus, the translation function developed in the previous section can only be used if the initial position coincides with a start codon, and the sequence terminates in a stop codon, i.e. the DNA sequence passed is a coding DNA sequence.

However, in many cases, we are given a DNA sequence (e.g. from a genome sequencing project) and we do not know in advance where the coding regions are. In these cases, we may be interested in scanning the DNA sequence to search for putative genes or coding regions.

The first step to search for these regions of interest is to take a DNA (or RNA) sequence and compute the *reading frames*. A reading frame is a way of dividing the DNA sequence into a set of consecutive non-overlapping triplets (possible codons) (see Section 3.2.2 and Table 3.2). A given sequence has three possible reading frames, starting in the first, second, and third positions. Adding to these three, considering that there is another complementary strand, we should also compute the other three frames corresponding to the reverse complement.

In this context, the following function computes the translation of the six different reading frames, given a DNA sequence. This allows to scan for all possibilities for protein coding regions given a region of the genome.

```
def reading_frames (dna_seq):
    """Computes the six reading frames of a DNA sequence (including
   the reverse complement."""
    assert validate_dna(dna_seq), "Invalid DNA sequence"
    res = []
```

```
    res.append(translate_seq(dna_seq,0))
    res.append(translate_seq(dna_seq,1))
    res.append(translate_seq(dna_seq,2))
    rc = reverse_complement(dna_seq)
    res.append(translate_seq(rc,0))
    res.append(translate_seq(rc,1))
    res.append(translate_seq(rc,2))
    return res
```

Having the reading frames computed, the next obvious step is to find possible proteins which can be encoded within these frames. This task is to find the so-called *open reading frames* (ORF), which are reading frames that have the potential to be translated into protein.

The next function starts to deal with this problem, by extracting all possible proteins from an aminoacid sequence. Note that this function goes through the sequence, when an 'M' is found starts a putative protein (kept in the *current_prot* list), and when a stop symbol is found all proteins in this list are added to the result.

```
def all_proteins_rf (aa_seq):
    """Computes all possible proteins in an aminoacid sequence.
   Returns list of possible proteins. """
    aa_seq = aa_seq.upper()
    current_prot = []
    proteins = []
    for aa in aa_seq:
        if aa == "_":
            if current_prot:
                for p in current_prot:
                    proteins.append(p)
                current_prot = []
        else:
            if aa == "M":
                current_prot.append("")
            for i in range(len(current_prot)):
                current_prot[i] += aa
    return proteins
```

This function, together with the previous one, allows to compute all putative proteins in all reading frames. It starts by computing the translation of the reading frames and then goes through the six aminoacid sequences and searches for all possible proteins using the previous function, gathering those into a resulting list.

```
def all_orfs (dna_seq):
    """Computes all possible proteins for all open reading frames."""
    assert validate_dna(dna_seq), "Invalid DNA sequence"
    rfs = reading_frames (dna_seq)
    res = []
    for rf in rfs:
        prots = all_proteins_rf(rf)
        for p in prots: res.append(p)
    return res
```

Since the lists returned by this last function may contain a large number of proteins in real world scenarios, it is useful to order the list by the size of the putative proteins and to filter this list considering a minimum size. This makes sense, since small putative proteins may occur frequently by chance, while a protein pattern with over a couple of tens of aminoacids is not likely to occur by chance. Indeed, notice that the probability of a stop codon is 3/64, around 5%, and therefore a stop codon is expected roughly every 20 aminoacids.

The next function is, therefore, an improved version of the previous that considers the ordered insertion of the proteins in the list, considering their size. This is achieved by the auxiliary function provided below, which inserts each protein in the list in the right position considering its size, by keeping the resulting lists ordered by decreasing size.

```
def all_orfs_ord (dna_seq, minsize = 0):
    """Computes all possible proteins for all open reading frames.
   Returns ordered list of proteins with minimum size."""
    assert validate_dna(dna_seq), "Invalid DNA sequence"
    rfs = reading_frames (dna_seq)
    res = []
    for rf in rfs:
        prots = all_proteins_rf(rf)
        for p in prots:
            if len(p) > minsize: insert_prot_ord(p, res)
    return res

def insert_prot_ord (prot, list_prots):
    i = 0
    while i < len(list_prots) and len(prot) < len(list_prots[i]):
```

```
        i += 1
    list_prots.insert(i, prot)
```

4.5 Putting It All Together

Given all the resources we have built in this chapter, and assuming those are all defined in a Python module (a file with extension .py), we can now use the defined functions to create a script/program. In the next example, we define a program that reads a DNA sequence from the keyboard, checks its validity and then, if the sequence is valid, applies a few of the functions defined earlier, including those that allow to check all putative proteins in its six reading frames. Note that this program can either be defined in the same file, after the function definitions, or in another file, in this case using the **import** statement (see Section 2.4.4). In the next code chunk, it is assumed that the previous functions are kept in a file *sequences.py*.

```
from sequences import *

seq = input("Insert DNA sequence:")
if validate_dna (seq):
    print ("Valid sequence")
    print ("Transcription: ", transcription (seq))
    print("Reverse complement:", reverse_complement(seq))
    print("GC content (global):", gc_content(seq))
    print("Direct translation:" , translate_seq(seq))
    print("All proteins in ORFs (decreasing size): ", all_orfs_ord(
   seq))
else: print("DNA sequence is not valid")
```

The usability of the previous program can be quite difficult in scenarios where the sequence to read is large and not very practical to introduce by hand. This is the case with common real scenarios, where these sequences result from DNA sequencing projects. Thus, it is desirable to have the possibility of reading sequences from files, as well as writing the results of our programs to files.

The following functions allow to read a sequence from a text file, where the sequence may be given in multiple lines, and to write a sequence to a text file.

```
def read_seq_from_file(filename):
    """ Reads a sequence from a multi-line text file. """
    fh = open(filename, "r")
```

```
    lines = fh.readlines()
    seq = ""
    for l in lines:
        seq += l.replace("\n","")
    fh.close()
    return seq

def write_seq_to_file(seq, filename):
    """ Writes a sequence to file. """
    fh = open(filename, "w")
    fh.write(seq)
    fh.close()
    return None
```

Using these functions, we can redefine the previous program to read the input sequence from a given file and write the resulting proteins in another file. This is shown in the next code block, where each of the computed putative proteins is written in a separate file, where the file name starts with "orf" and has a sequential number appended, being the extension ".txt". In this case, the file "orf-1.txt" will have the largest protein, and so on.

```
fname = input("Insert input filename:")
seq = read_seq_from_file(fname)
if validate_dna (seq):
    print ("Valid sequence")
    print ("Transcription: ", transcription (seq))
    print("Reverse complement:", reverse_complement(seq))
    print("GC content (global):", gc_content(seq))
    print("Direct translation:" , translate_seq(seq))
    orfs = all_orfs_ord(seq)
    i = 1
    for orf in orfs:
        write_seq_to_file(orf, "orf-"+str(i)+".txt")
        i += 1
else: print("DNA sequence is not valid")
```

4.6 A Class for Biological Sequences

In Section 2.5.1, in the context of the illustration of object-oriented programming concepts, we defined a first prototype of a class which allows to represent biological sequences, the class **MySeq**. We will now work over this class to show how object-oriented programming can help in turning the code developed so far into a more organized, modular e reusable software.

Let us recover the definition of the class **MySeq** in the next code chunk, which includes the class definition and a set of special methods, whose meaning is further explained in Section 2.5.1, including the constructor. Also, we include methods to check the sequence type and the full information on a sequence.

```
class MySeq:
    """ Class for biological sequences. """

    def __init__ (self, seq, seq_type = "DNA"):
        self.seq = seq.upper()
        self.seq_type = seq_type

    def __len__(self):
        return len(self.seq)

    def __getitem__(self, n):
        return self.seq[n]

    def __getslice__(self, i, j):
        return self.seq[i:j]

    def __str__(self):
        return self.seq

    def get_seq_biotype (self):
        return self.seq_type

    def show_info_seq (self):
        print ("Sequence: " + self.seq + " biotype: " + self.seq_type
    )
```

Let us now add to this class other useful methods. First, let us define a method to validate sequences, taking into account the sequence type. For that purpose, a function to return the allowed alphabet, given the sequence type, will also be defined.

```
    def alphabet (self):
        if (self.seq_type=="DNA"): return "ACGT"
        elif (self.seq_type=="RNA"): return "ACGU"
        elif (self.seq_type=="PROTEIN"): return "ACDEFGHIKLMNPQRSTVWY
  "
        else: return None

    def validate (self):
        alp = self.alphabet()
        res = True
        i = 0
        while i < len(self.seq) and res:
            if self.seq[i] not in alp: res = False
            else: i += 1
        return res
```

Notice that the set of functions defined in previous sections along this chapter can now be redefined within the context of this class. This allows a more coherent and reusable code, using the *seq_type* to provide easier validation of inputs and outputs. Let us see three examples considering methods to allow for the transcription, reverse complement and translation of an inputted DNA sequence. Note that the methods all return objects of the **MySeq** class, but with different types, according to the function's biological meaning.

```
    def transcription (self):
        if (self.seq_type == "DNA"):
            return MySeq(self.seq.replace("T","U"), "RNA")
        else:
            return None

    def reverse_comp (self):
        if (self.seq_type != "DNA"): return None
        comp = ""
        for c in self.seq:
            if (c == 'A'): comp = "T" + comp
            elif (c == "T"): comp = "A" + comp
            elif (c == "G"): comp = "C" + comp
            elif (c== "C"): comp = "G" + comp
        return MySeq(comp, "DNA")
```

```
    def translate (self, iniPos= 0):
        if (self.seq_type != "DNA"): return None
        seq_aa = ""
        for pos in range(iniPos,len(self.seq)-2,3):
            cod = self.seq[pos:pos+3]
            seq_aa += translate_codon(cod)
        return MySeq(seq_aa, "PROTEIN")
```

Although we could redefine here other methods provided as functions in previous sections, adding those to this class, we believe that the previous examples are sufficient to allow the reader to address this task as an exercise.

To close this section, let us see a simple example of how to use this class in programs/scripts.

```
s1 = MySeq("ATGTGATAAGAATAGAATGCTGAATAAATAGAATGACAT")
s2 = MySeq("MKVVLSVQERSVVSLL", "PROTEIN")
print(s1.validate(), s2.validate())
print(s1)
s3 = s1.transcription()
s3.show_info_seq()
s4 = s1.reverse_comp().translate()
s4.show_info_seq()
```

4.7 Processing Sequences With BioPython

The *BioPython* package [7] (`http://www.biopython.org/`) gathers a set of open-source software resources written in Python to address a large number of Bioinformatics tasks, including also an abundant documentation that facilitates its use, including a very complete tutorial and cookbook [8]. It is one of the projects which are members of the *Open Bioinformatics Foundation* (OBF), that also includes a number of other Bioinformatics free software written in other languages, such as Perl, Ruby or Java.

The package can be easily installed using one of the available tools for package management, which were covered in Section 2.4.4, being instructions provided in the package web page. We will cover many of the functions available in this package in different chapters within this book, addressing in each the topics related to the presented algorithms.

In this section, we will start by addressing the way *BioPython* handles biological sequences. Since it is written using the object-oriented features of the Python language, we will show here its main classes related to sequence processing. Here, the **Seq** class is the core, allowing to work over biological sequences of different types. One first observation to notice is that the object instances of this class are immutable, i.e. their content cannot be changed after creation.

The following example shows how to create a simple sequence only passing a string as an argument to the constructor. All objects of the class **Seq** have an associated alphabet defining allowed symbols in their content, which is also an object of one of the classes defining possible alphabets in *BioPython*. In this case, the alphabet will be an instance of class **Alphabet**, that defines the most generic case.

Note that the examples will be given in interactive mode, to be run in a Python console, although they could be inserted within runnable scripts with minor changes. The first line allows to load the corresponding module, serving also to test if your installation was done correctly.

```
>>> from Bio.Seq import Seq
>>> my_seq = my_seq = Seq("ATAGAGAAATCGCTGC")
>>> my_seq
Seq('ATAGAGAAATCGCTGC', Alphabet())
>>> print(my_seq)
ATAGAGAAATCGCTGC
>>> my_seq.alphabet
Alphabet()
```

The following examples show how to define sequences specifying the desired alphabets according to the intended use. In this case, we define two sequences, one of DNA and the other of a protein, indicating compatible alphabets, that in this case are similar to the ones defined in the context of the previous functions in this chapter, only including non-ambiguous symbols.

```
>>> from Bio.Alphabet import IUPAC
>>> my_seq = Seq("ATAGAGAAATCGCTGC", IUPAC.unambiguous_dna)
>>> my_seq
Seq('ATAGAGAAATCGCTGC', IUPACUnambiguousDNA())
>>> my_seq.alphabet
IUPACUnambiguousDNA()
>>> my_prot = Seq("MJKLKVERSVVMSVLP", IUPAC.protein)
>>> my_prot
Seq('MJKLKVERSVVMSVLP', IUPACProtein())
```

Other alphabets are illustrated by the following example, where we can also learn how to check the content of each alphabet type. Note that in these examples, we assume the previous modules were imported in the above code chunks.

```
>>> IUPAC.unambiguous_dna.letters
'GATC'
>>> IUPAC.ambiguous_dna.letters
'GATCRYWSMKHBVDN'
>>> IUPAC.IUPACProtein.letters
'ACDEFGHIKLMNPQRSTVWY'
>>> IUPAC.ExtendedIUPACProtein.letters
'ACDEFGHIKLMNPQRSTVWYBXZJUO'
```

The objects of the **Seq** class can be treated, for many effects, as strings, since they support many of the functions and operators of lists and strings. Among these are indexing, splicing, concatenating (with + operator) , `for` cycles and other iterators, **`in`** operator, as well as functions as **len**, **upper**, **lower**, **find**, **count**, among others. Let us see some examples:

```
>>> for i in my_seq: print(i)
A
T ...
>>> len(my_seq)
16
>>> my_seq[2:4]
'AG'
>>> my_seq.count("G")
4
>>> "GAGA" in my_seq
True
>>> my_seq.find("ATC")
8
```

It is important to note that, regarding concatenation, only sequences with compatible alphabets will be merged. Let us see two examples, where firstly we have two sequences from the same alphabet, and then two sequences from compatible alphabets to be merged. Notice that in the second case, the resulting alphabet is a third one, in this case, one that is more generic than the two original ones.

```
>>> seq1 = Seq("MEVRNAKSLV", IUPAC.protein)
>>> seq2 = Seq("GHERWKY", IUPAC.protein)
```

```
>>> seq1+seq2
Seq('MEVRNAKSLVGHERWKY', IUPACProtein())
>>> from Bio.Alphabet import generic_nucleotide
>>> nuc_seq = Seq("ATAGAGAAATCGCTGC", generic_nucleotide)
>>> dna_seq = Seq("TGATAGAACGT", IUPAC.unambiguous_dna)
>>> nuc_seq + dna_seq
Seq('ATAGAGAAATCGCTGCTGATAGAACGT', NucleotideAlphabet())
```

The **Seq** class also implements a number of biologically relevant functions, many of those similar to the ones covered in previous sections. In the next example, we can observe how to calculate the transcription and the reverse complement of a DNA sequence.

```
>>> coding_dna = Seq("ATGAAGGCCATTGTAATGGGCCGC", IUPAC.
    unambiguous_dna)
>>> template_dna = coding_dna.reverse_complement()
>>> template_dna
Seq('GCGGCCCATTACAATGGCCTTCAT', IUPACUnambiguousDNA())
>>> messenger_rna = coding_dna.transcribe()
>>> messenger_rna
Seq('AUGAAGGCCAUUGUAAUGGGCCGC', IUPACUnambiguousRNA())
```

Also, there are several functions that allow to do translation over both DNA and RNA sequences, as shown below.

```
>>> rna_seq = Seq('AUGCGUUUAACU', IUPAC.unambiguous_rna)
>>> rna_seq.translate()
Seq('MRLT', IUPACProtein())
>>> coding_dna = Seq("ATGGCCATTGTAATGGGCCGCTGAAAGGGTGCCCGATAG", IUPAC
    .unambiguous_dna)
>>> coding_dna.translate()
Seq('MAIVMGR*KGAR*', HasStopCodon(IUPACProtein(), '*'))
>>> coding_dna.translate(table="Vertebrate Mitochondrial")
Seq('MAIVMGRWKGAR*', HasStopCodon(IUPACProtein(), '*'))
```

Note that when doing translation, stop codons may occur in the sequence, which will be represented as '*' symbols. This leads to alphabets where this symbol is included, as it is shown in the example.

In the last example above, the translation was done using a codon translation table different from the standard one. In this case, the table used was the one that occurs in the translation of proteins synthesized in the mitochondria. There are several different codon tables in *BioPython*, including the "Standard", the "Vertebrate Mitochondrial", and the "Bacterial", which are used in different scenarios.

The following examples show how to check some of the contents of each table, highlighting, in this case, some differences between the standard and the mitochondrial tables.

```
>>> from Bio.Data import CodonTable
>>> standard_table = CodonTable.unambiguous_dna_by_name["Standard"]
>>> mito_table = CodonTable.unambiguous_dna_by_name["Vertebrate
   Mitochondrial"]
>>> print (standard_table)
( ... )
>>> mito_table.stop_codons
['TAA', 'TAG', 'AGA', 'AGG']
>>> mito_table.start_codons
['ATT', 'ATC', 'ATA', 'ATG', 'GTG']
>>> mito_table.forward_table["ATA"]
'M'
>>> standard_table.forward_table["ATA"]
'I'
```

4.8 Sequence Annotation Objects in BioPython

As already discussed, the usability of functions working over biological sequences is greatly enhanced if we can work with sequences stored in files, given the typical dimensions of these sequences. The **SeqIO** class provides a number of functions that allow to read and write sequences from/to files, in a wide range of different formats, not only containing the sequences themselves, but also meta-information, in the form of annotations. These annotations allow to add information to the whole sequence, or parts of the sequence, that can be related to its biological function or other relevant knowledge.

The **SeqRecord** class is the basic container of sequences and their annotation in *BioPython*, therefore being used to keep the results of the reading functions. A **SeqRecord** object has the following fields:

- *seq*: the sequence itself, an object of the **Seq** class;
- *id*: the sequence identifier;
- *name*: the sequence name;
- *description*: a description of the sequence;

- *annotations*: global annotations for the whole sequence (a dictionary where the keys are annotation types – unstructured properties – and the values are the specific values of those properties for the sequence);
- *features*: structured features, a list of **SeqFeature** objects, which can apply to the whole sequence or parts of it;
- *letter_annotations*: possible annotations for each letter (position) in the sequence;
- *dexrefs*: references to databases.

To understand the organization of the features, we need to take a look at the **SeqFeature** class, which allows to keep structured information about the annotations of a sequence. The structure of this object is deeply oriented towards the content of the records in the Genbank and EMBL databases.

The main attributes of a **SeqFeature** object are:

- *location* – indicates the region of the sequence affected by this annotation as a **Feature-Location** object;
- *type* – string stating the feature type;
- *qualifiers* – additional information about the feature, kept as a property-value dictionary.

The **SeqLocation** class allows to represent regions of the sequences in a flexible manner, allowing to define fixed or fuzzy ranges of positions. Positions can be either exact or fuzzy, and in this last case, there are many options, like the **AfterPosition**, **BeforePosition** or the **BetweenPosition**, which specify a value larger, lesser or in between defined values. The following example can help to understand some of the possibilities, which we will not explore fully here.

```
>>> from Bio import SeqFeature
>>> start = SeqFeature.AfterPosition(10)
>>> end = SeqFeature.BetweenPosition(40, left=35, right=40)
>>> my_location = SeqFeature.FeatureLocation(start, end)
>>> print (my_location)
[>10:(35^40)]
>>> int(my_location.start)
10
>>> int(my_location.end)
40
```

The last part of the example creates a **FeatureLocation** object from the defined positions. The **int** function forces the fuzzy positions to be considered as a single value.

The definition of feature locations is quite useful in many scenarios, for instance to extract, from the whole sequence, the part to which a given feature applies. The next example shows that scenario, where a *gene* feature is defined in part of a sequence, more concretely referring to the reverse complement strand. The method **extract** is used here to access the nucleotides between positions 5 and 18 of the reverse complement strand.

```
>>> from Bio.SeqFeature import SeqFeature, FeatureLocation
>>> example_seq = Seq("ACCGAGACGGCAAAGGCTAGCATAGGTATGAGACTT")
>>> example_feat = SeqFeature(FeatureLocation(5, 18), type="gene",
    strand=-1)
>>> feature_seq = example_feat.extract(example_parent)
>>> print (feature_seq)
AGCCTTTGCCGTC
```

It is possible to create a **SeqRecord** object and fill it field by field, including features and annotation, as shown in the next example.

```
>>> from Bio.Seq import Seq
>>> seq = Seq("ATGAATGATAGCTGAT")
>>> from Bio.SeqRecord import SeqRecord
>>> seq_rec = SeqRecord(seq)
>>> seq_rec.id = "ABC12345"
>>> seq_rec.description = "My own sequence."
>>> seq_rec.annotations["role"] = "unknown"
>>> seq_rec.annotations
{'role': 'unknown'}
```

However, it is more common and convenient to have this as the result of a reading function. The **SeqIO** class provides two main functions to read information from files in different formats: the **read** function reads a single **SeqRecord** object, with one sequence and possibly its annotations, while the function **parse** is able to read multiple records, returning an iterator over a **SeqRecord** container.

To exemplify the use of the methods from the **SeqIO** class, we will use the complete sequence, and its annotation, from a plasmid (pPCP1) of the *Yersinia pestis* bacterium, more specifically from the strain *Yersinia pestis biovar Microtus str. 91001*, with the accession number *NC_005816* in Genbank. The corresponding record, from the RefSeq database, can be found in the following URL: `https://www.ncbi.nlm.nih.gov/nuccore/NC_005816`. We will save the contents of this record in two different formats: FASTA and Genbank.

The first example shows how to read a single sequence from a file in the FASTA format. This format allows a minimal representation of the sequences, where for each sequence there is a header row (initiated by the > symbol) with some meta-information (typically identifiers and a brief description) and followed by a number of rows with the sequence itself. In the example, the file was named "NC_005816.fna". We can verify that the sequence has 9609 nucleotides.

```
>>> from Bio import SeqIO
>>> record = SeqIO.read("NC_005816.fna", "fasta")
>>> record
SeqRecord(seq=Seq("TGTAACGAACGGTGCAAT..."))
>>> len(record.seq)
9609
>>> record.id
gi|45478711|ref|NC_005816.1|
>>> record.description
gi|45478711|ref|NC_005816.1| Yersinia pestis biovar Microtus str.
   91001 plasmid pPCP1, complete sequence
>>> record.annotations
{}
>>> record.features
[]
```

When a minimal format as FASTA is used, most of the information within the resulting records will be empty, as can be easily checked from the results in previous examples (for instance, there are no annotations or features), as this information is not available in this format. Also, the *id* and *description* fields are filled a bit by guessing, using some rules that are common in certain tools when saving FASTA files, but are not accepted universal standards.

A different scenario occurs if a richer format is provided. Let us illustrate this by saving the previous record in the Genbank format, that is able to store all annotations and features from the provided sequence, i.e. the information that is displayed when accessing the URL defined above.

In this case, the file name is similar but with the extension ".gb". Notice that the format needs to be supplied as the second argument of the **read** function. In this case, while the *seq* field, i.e. the sequence itself, is the same, the other fields show different content. To start, *id*, *name*, and *description* fields can now be filled correctly. Also, the *annotations* and *features* fields have now gained a richer content loaded from the RefSeq record.

```
>>> from Bio import SeqIO
>>> record = SeqIO.read("NC_005816.gb", "genbank")
>>> record.seq
Seq('TGTAACGAACGGTGCAATC...CTG', IUPACAmbiguousDNA())
>>> print(len(record.seq))
9609
>>> record.id
'NC_005816.1'
>>> record.name
'NC_005816'
>>> record.description
'Yersinia pestis biovar Microtus str. 91001 plasmid pPCP1, complete
   sequence.'
>>> len(record.annotations)
11
>>> len(record.features)
29
```

The *annotations* fields, as stated above, is a dictionary that provides a number of properties for the sequence as a whole. Let us check, in the next example, some of the information contained in this case.

```
>>> record.annotations["source"]
'Yersinia pestis biovar Microtus str. 91001'
>>> record.annotations["taxonomy"]
['Bacteria', 'Proteobacteria', 'Gammaproteobacteria', '
   Enterobacteriales', 'Enterobacteriaceae', 'Yersinia']
>>> record.annotations["date"]
23-MAY-2013
>>> record.annotations["gi"]
45478711
```

To explore further the 29 features in this record, firstly, we will count how many of the features correspond to annotated genes:

```
>>> feat_genes = []
>>> for i in range(len(record.features)):
......if record.features[i].type == "gene": feat_genes.append(record.
   features[i])
>>> len(feat_genes)
10
```

Let us now explore a bit more these genes, getting their locus tag, strand, and location. Note that the strand 1 is the one represented by the sequence itself, while -1 stands for the reverse complement.

```
>>> for f in feat_genes: print(f.qualifiers['locus_tag'], f.strand, f
    .location)
['YP_pPCP01'] 1 [86:1109](+)
['YP_pPCP02'] 1 [1105:1888](+)
['YP_pPCP03'] 1 [2924:3119](+)
...
```

Using the **extract** function we can try to find which are the proteins encoded by each of these genes.

```
>>> for f in feat_genes: print(f.extract(record.seq).translate(table=
    "Bacterial", cds=True))
MVTFETVMEIKILHKQGMSSRAIARELGISRNTVKRYLQAKSEPP ...
```

It is left as an exercise for the reader to check that there are equally 10 features of the type "CDS" (coding sequence), which have a qualifier key named *translation* that keeps the encoded protein, and therefore should be equal to the result obtained above.

There are some cases where we need to read several sequences from a single file, which can be done by using the **parse** function. The file mentioned in the example below can be obtained from the book's web site or the BioPython's tutorial [8], and contains different sequences of ribosomal rRNA genes from different species of orchids.

In the example, we go through all the records, printing their description and collecting the organisms in a list.

```
>>> all_species = []
>>> for seq_record in SeqIO.parse("ls_orchid.gbk", "genbank"):
...:        print (seq_record.description)
...:        all_species.append(seq_record.annotations["organism"])
C.irapeanum 5.8S rRNA gene and ITS1 and ITS2 DNA.
C.californicum 5.8S rRNA gene and ITS1 and ITS2 DNA.
C.fasciculatum 5.8S rRNA gene and ITS1 and ITS2 DNA.
...
>>> print (all_species)
['Cypripedium irapeanum', 'Cypripedium californicum', ... , '
    Paphiopedilum barbatum']
```

It is also important to mention that *BioPython* has a number of functions which allow to retrieve sequences directly from databases and process them afterwards. In the next example, we show how to retrieve a set of sequences from NCBI, providing the GI identifiers in a list.

```
>>> from Bio import Entrez
>>> from Bio import SeqIO
>>> Entrez.email = "example@gmail.com"
>>> handle = Entrez.efetch(db="nucleotide", rettype="gb", retmode="
    text", id="6273291, 6273290, 6273289")
>>> for seq_record in SeqIO.parse(handle, "gb"):
...:        print (seq_record.id, seq_record.description[:100], "...")
...:    print ("Sequence length: ", len(seq_record))
>>> handle.close()
AF191665.1 Opuntia marenae rpl16 gene; chloroplast gene for
    chloroplast product, partial intron sequence. ...
Sequence length:  902
...
```

To conclude this section, and this chapter, we will check how to write records to file. The **write** method from the **SeqIO** class allows to write the contents of one (or several) **SeqRecord** objects to file in different formats, receiving as arguments a list of **SeqRecord** objects, the file name and a string defining the format. On the other hand, the **convert** function can be used to directly convert records from one format to another, a very useful task in a bioinformatician's daily work. So, the two code blocks below have an equivalent behavior.

```
records = SeqIO.parse("ls_orchid.gbk", "genbank")
count = SeqIO.write(records, "my_example.fasta", "fasta")
```

```
count = SeqIO.convert("ls_orchid.gbk", "genbank", "my_example.fasta",
    "fasta")
```

Exercises and Programming Projects

Exercises

1. Write a program that reads a DNA sequence, converts it to capital letters, and counts how many nucleotides are purines and pyrimidines.
2. Write a program that reads a DNA sequence and checks if it is equal to its reverse complement.

3. Write and test a function that, given a DNA sequence, returns the total number of "CG" duplets contained in it.
4. Write and test a Python function that, given a DNA sequence, returns the size of the first protein that can be encoded by that sequence (in any of the three reading frames). The function should return −1 if no protein is found.
5. Write and test a function that given an aminoacid sequence returns a logic value indicating if the sequence can be a protein or not.
6. Write and test a Python function that, given a DNA sequence, creates a map (dictionary) with the frequencies of the aminoacids it encodes (assuming the translation is initiated in the first position of the DNA sequence). Stop codons should be ignored.
7. Write a program that reads an aminoacid sequence and a DNA sequence and prints the list of all sub-sequences of the DNA sequence that encode the given protein sequence.
8. Write a function that, given a sequence as an argument, allows to detect if there are repeated sub-sequences of size k (the second argument of the function). The result should be a dictionary where keys are sub-sequences and values are the number of times they occur (at least 2). Use the function in a program that reads the sequence and k and prints the result by decreasing frequency.

Programming Projects

1. Taking as your basis the class **MySeq** developed above, implement sub-classes for the three distinct types of biological sequences: DNA, RNA, and proteins. In each define an appropriate constructor. Redefine the methods from the parent class, in the cases where you feel this is necessary or useful. Adapt the types of the outputs in each method accordingly.
2. The package *random* includes a number of function that allow to generate random numbers. Using some of those functions, build a module that implements the generation of random DNA sequences and the analysis of mutations over these sequences. You can include functions to generate random sequences of a given size, to simulate the occurrence of a given number of mutations in a DNA sequence in random positions (including insertions, deletions, and substitutions), and functions to study the impact of mutations in the encoded proteins of those sequences.

CHAPTER 5

Finding Patterns in Sequences

In this chapter, we discuss how to find patterns in sequences and the importance of this task in Bioinformatics. We put forward the basic algorithms for pattern finding and discuss their complexity. Heuristic algorithms are presented to lower the average computational complexity of this task, by suitable pre-processing of the patterns to search. Also, we present regular expressions as a way to find more flexible patterns in sequences, showing how these can be implemented in Python, and discussing their biological relevance with some examples.

5.1 Introduction: Importance of Pattern Finding in Bioinformatics

Finding specific patterns within biological sequences is one of the most common tasks bioinformaticians need to deal with, in their daily lives. Since relevant biological information is, as we have seen in the previous chapters (most notably in Chapter 3), kept in sequences, the analysis of local patterns in those sequences is quite relevant.

Indeed, these patterns are in many cases closely connected to specific functions of the molecules represented by those sequences. Some examples include specific patterns corresponding to protein domains with a given function (e.g. binding sites for ligands in enzymes or for DNA molecules in regulatory proteins), or nucleotide patterns corresponding to parts of DNA sequences, which may be associated to binding sites for regulation of gene expression (e.g. promoters, enhancers, transcription factors).

Thus, the identification of specified patterns in different types of sequences (DNA, RNA, proteins) is deeply connected to the association of sequences and sequence locations to diverse biological functions. Also, the existence of numerous repeats, of different sizes, in the genomes of living beings, and most pronouncedly in complex organisms (as humans), brings an added importance to pattern finding.

Given the nature of the patterns we typically need to find, there are diverse tasks in Bioinformatics related to pattern finding, which diverge in the type of pattern to find, as well as in the size and number of patterns and target sequence(s). Therefore, in this book, we will devote a few chapters to cover problems related to biological patterns in sequences, and the related concept of sequence motifs.

Here, we will start by discussing algorithms for finding fixed patterns in larger sequences, addressing the complexity of these algorithms and trying to make them more efficient. These

Bioinformatics Algorithms. DOI: 10.1016/B978-0-12-812520-5.00005-5

will be the aims of the next three sections. We will put forward ways to pre-process the pattern to make the search more efficient, with added advantages in the cases where the same pattern needs to be found in a large number of sequences. Lastly, in this chapter, we will discuss regular expressions, as a way to search for more flexible patterns in sequences.

5.2 Naive Algorithm for Fixed Pattern Finding

A naive approach to search for the occurrence of a pattern p (of length k) in a sequence s (of length $N > k$) is to consider all possible sub-sequences of s, of size k, and compare each of those, position by position, to p. Once a mismatch is found, we can move to test the next sub-sequence. If the sub-sequence matches the pattern in all positions, an occurrence of p is found. If the purpose is to find all occurrences of p in s, we must continue even in case of success, while if it is only required to find the first occurrence (or simply if the pattern occurs), we can stop when the first sub-sequence matches p.

Both variants are implemented in the Python functions provided in the following code block. The **search_first_occ** function uses an outer `while` cycle that finishes when the pattern is found, or the last sub-sequence is tested. The function returns the position of the first occurrence of the pattern, or if the pattern does not occur returns -1.

The **search_all_occurrences** function uses a `for` cycle that tests for all possible sub-sequences (note that there are $N - k + 1$ possible positions where the pattern may occur), and returns a list with all initial positions of the pattern's occurrences (the list is empty if the pattern does not occur).

```
def search_first_occ(seq, pattern):
    found = False
    i = 0
    while i <= len(seq)-len(pattern) and not found:
        j = 0
        while j < len(pattern) and pattern[j]==seq[i+j]:
            j = j + 1
        if j== len(pattern): found = True
        else: i += 1
    if found: return i
    else: return -1

def search_all_occurrences(seq, pattern):
    res = []
    for i in range(len(seq)-len(pattern)+1):
```

```
        j = 0
        while j < len(pattern) and pattern[j]==seq[i+j]:
            j = j + 1
        if j == len(pattern):
            res.append(i)
    return res

seqDNA = "ATAGAATAGATAATAGTC"
print( search_first_occ(seqDNA, "GAAT") )
print( search_first_occ(seqDNA, "TATA") )
print( search_all_occurrences(seqDNA, "AAT") )
```

These functions might also be used in a more general way for user inputted patterns and sequences. An example is provided below.

```
def test_pat_search():
    seq = input("Input sequence: ")
    pat = input("Input pattern: ")
    print(pat, " occurs in the following positions:", )
    print( search_all_occurrences(seq, pat) )

test_pat_search()
```

In the previous functions, the inner `while` cycle may be replaced by simply putting, in the condition of the following `if`, the test $pattern == seq[i : i + k]$. The reason to explicitly put the cycles is to show the whole number of comparisons done and better understand the complexity of these algorithms.

Thus, the previous naive algorithm, in the worst case scenario, implies testing the character equality a total number of $(N - k + 1) \times k$ times. Although the worst case scenario does not occur frequently in practice, this algorithm is still slow when working over large sequences, since it needs to go back to the position $i + 1$, when a failure occurs in matching the subsequence starting at i. We will see in the next section an algorithm that seeks to improve this.

Before finishing this section, it is important to remind the reader that there are pre-defined functions over strings, seen in Section 2.6.3, that can perform similar tasks to the ones defined above, namely *s*.**find**(*p*), which finds the first occurrence of a pattern *p* in *s* (the function **rfind** is similar but returns the last occurrence), and *s*.**count**(*p*), which counts the number of occurrences of *p* in *s*. Also, the operator **in** allows to check if a pattern exists in a string. Note that there is no direct way of replacing the function **search_all_occurrences** for one pre-defined alternative defined over strings.

5.3 Heuristic Algorithm: Boyer-Moore

As we saw in the previous section, the complexity of the naive algorithm can be penalizing in its computational efficiency. There are alternative algorithms which seek to improve the average computational efficiency of pattern searching, trying to use the structure of the pattern to speed up the search saving a significant part of the pairwise symbol comparisons.

Although there are a number of alternatives for these algorithms, we will cover here only the Boyer-Moore algorithm, that although having in the worst case scenario a complexity similar to the naive algorithm, in most cases allows significant gains in performance. The algorithm is based on two rules that allow to move forward more than one position in the target sequence in some situations.

As in the naive algorithm, the target sequence is scanned from the beginning to the end (left to right), but in this case the comparison of the pattern, with the sub-sequence in the text, is done from right to left. When there is a mismatch between the target sequence and the pattern, two rules can be applied to check if the process can be more efficient, by moving forward more than one position in the sequence.

The first rule that can be applied is the *bad-character rule*, which states that we can advance the pattern to the next occurrence (in the pattern) of the symbol in the sequence at the position of the mismatch. If no occurrences of that symbol exist in the pattern, we can move forward the maximum number of symbols (until the end of the pattern).

Fig. 5.1A shows examples of three cases of a possible application of this rule. In the first example, the symbol in the sequence where the mismatch occurs (T) does not occur in the pattern, and therefore we can move forward the number of positions corresponding to placing the pattern's first symbol in the position following the mismatch. In the second example, the symbol occurs in the pattern, and, thus, we can move the pattern so that the rightmost occurrence of the symbol in the pattern matches the mismatched symbol in the sequence. The same happens in the third example, but in this case we only move forward one position.

The other rule that may be applied is the *good suffix rule*, which states that, in case of a mismatch, we can move forward to the next instance in the pattern of the part (suffix) that matched before (in the right) the mismatch.

Fig. 5.1B shows examples of the application of this rule. In the first case, the suffix "AC" matched and, therefore, the pattern moves to the next occurrence of this string in the pattern. The next case shows what happens when the suffix does not occur in the remaining of the pattern, allowing to move forward the pattern a number of positions equal to its length. The third example shows a special case, where the full matching suffix ("CAC") does not occur in the pattern, but a suffix does ("AC").

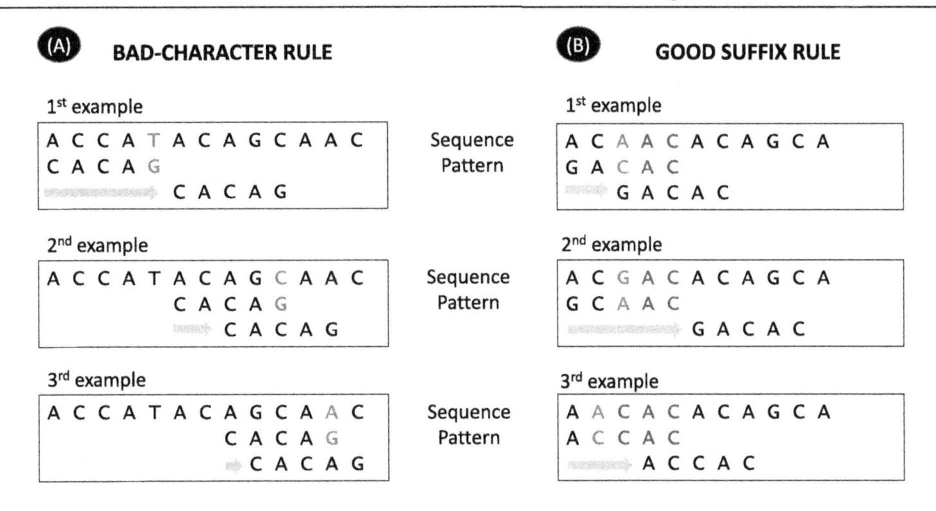

Figure 5.1: Examples of the application of the two rules used by the Boyer-Moore algorithm: (A) bad-character rule; (B) good suffix rule.

To make this algorithm efficient, and allow to rapidly verify which rule may be applied in each case, a pre-processing of the pattern needs to be conducted before the search itself, keeping relevant information in efficient data structures. Note that the number of positions to move forward depends only on the pattern and, therefore, this pre-processing applies only over the pattern to make the required information readily available, being independent of the target sequence. Thus, the pre-processing does not need to be repeated to search the same pattern in other target sequences.

Although this pre-processing has computational costs, these are normally worth since the pattern is typically much smaller than the sequence, and the costs are paid by the larger gains in the search process.

In the case of the bad-character rule, we create a dictionary with all possible symbols in the alphabet as keys, and with values defining the rightmost position where the symbol occurs in the pattern (-1 if the symbol does not occur). This allows to rapidly calculate the number of positions to move forward according to this rule by calculating the offset: *position of the mismatch in the pattern – value for the symbol in the dictionary*. Notice that this value might be negative and, in this case, this means the rule does not help and it will be ignored in that iteration. This process is done in the **process_bcr** function in the code given below.

The pre-processing for the good suffix rule is more complex and we will not explain here all the details (the code is given below and we leave a detailed analysis to the interested reader). The result of this process is to create a list that keeps the number of positions that may be moved forward, depending on the position of the mismatch on the pattern (list index). Notice that in this process, both the situations illustrated above need to be taken into account. This process is done in the **process_gsr** function in the code given below.

The implementation of the algorithm is given next as a Python class. The class allows to define an alphabet and a pattern in the constructor and does the pre-processing for both rules, according to the pattern, in the function **preprocess** called by the constructor.

The function **search_pattern** allows to use an initialized object of this class to search over target sequences for the given pattern. It is an adaptation of the naive algorithm given in the previous section, but which makes use of the data structures from the pre-processing, using the two rules to move forward the maximum number of allowed positions. In the worst case, it advances a single position (as in the naive algorithm), but in other cases it can use one of the rules to move forward more positions (the maximum of the values provided by each of the rules).

```
class BoyerMoore:

    def __init__(self, alphabet, pattern):
        self.alphabet = alphabet
        self.pattern = pattern
        self.preprocess()

    def preprocess(self):
        self.process_bcr()
        self.process_gsr()

    def process_bcr(self):
        self.occ = {}
        for symb in self.alphabet:
            self.occ[symb] = -1
        for j in range(len(self.pattern)):
            c = self.pattern[j]
            self.occ[c] = j

    def process_gsr(self):
        self.f = [0] * (len(self.pattern)+1)
        self.s = [0] * (len(self.pattern)+1)
        i = len(self.pattern)
        j = len(self.pattern)+1
        self.f[i] = j
```

```
        while i>0:
            while j<= len(self.pattern) and self.pattern[i-1] != self
    .pattern[j-1]:
                if self.s[j] == 0: self.s[j] = j-i;
                j = self.f[j]
            i -= 1
            j -= 1
            self.f[i] = j
        j = self.f[0]
        for i in range(len(self.pattern)):
            if self.s[i] == 0: self.s[i] = j
            if i == j: j = self.f[j]

    def search_pattern(self, text):
        res = []
        i = 0
        while i <= len(text) - len(self.pattern):
            j= len(self.pattern)- 1
            while j>=0 and self.pattern[j]==text[j+i]: j -= 1
            if (j<0):
                res.append(i)
                i += self.s[0]
            else:
                c = text[j+i]
                i += max(self.s[j+1], j- self.occ[c])
        return res

def test():
    bm = BoyerMoore("ACTG", "ACCA")
    print (bm.search_pattern("
   ATAGAACCAATGAACCATGATGAACCATGGATACCCAACCACC"))

test()
```

5.4 Deterministic Finite Automata

A *Deterministic Finite Automaton* (DFA) can be defined as a machine that processes a sequence of symbols from right to left, changing its internal state as it processes the symbols.

The new state will depend on the previous state and the symbol read. A DFA may be used to perform pattern search, by defining the appropriate alphabet, states and transition function for a given pattern. Indeed, a DFA can determine all occurrences of a pattern in a sequence, performing a single run over the sequence.

Formally, a DFA may be defined as a tuple $M = (Q, A, q_0, \delta, F)$, where Q is the set of states, A is the alphabet of allowed symbols, $q_0 \in Q$ is the initial state, $\delta : Q, A \mapsto Q$ is the transition function and F is the set of final states.

In the case of pattern matching, we define the set of states $Q = \{0, 1, \ldots, m\}$, where m is the size of the pattern. When the DFA is in state k, this means that the sequence matched the k first symbols of the pattern in the previous positions. So, it is clear that, in this case, we set $q_0 = 0$ and $F = \{m\}$, thus there will be a single final state corresponding to an occurrence of the pattern being detected.

The most important step in building a DFA for pattern matching is to define the transition function. The basic idea is the following: if we are in state $k-1$, then if the next symbol seen in the sequence is equal to the symbol in position k in the pattern, then we should move to state k. When this is not the case, a mismatch occurs. In this situation, we could be tempted to automatically return to state $q_0 = 0$. However, this is not always the case, as the previous symbols seen before the mismatch may overlap with the pattern in some way.

So, we need to test if the $k-1$ first symbols of the pattern, followed by the symbol in the sequence, overlap with the pattern. The number of characters in the maximum overlap of those sequences will be the next state of the DFA for that symbol. The maximum overlap of two sequences s and t is defined as the maximum value of x, such as the last x characters of s match the first x characters of t.

Therefore, the general rule for the DFA transition table is given by:

$$\delta(k, a) = max\_overlap(p_0 \ldots p_{k-1}a, p) \tag{5.1}$$

where p is the pattern and p_i is the i-th symbol of the pattern. The function **overlap**(*s,t*) provides the maximum overlap between sequences *s* and *t*, as defined above, being implemented in Python in the following way (note the implementation is a simple, not an efficient one):

```
def overlap(s1, s2):
    maxov = min(len(s1), len(s2))
    for i in range(maxov,0,-1):
        if s1[-i:] == s2[:i]: return i
    return 0
```

State	Symbol	Next state
0	A	1
0	C	0
1	A	1
1	C	2
2	A	3
2	C	0
3	A	1
3	C	2

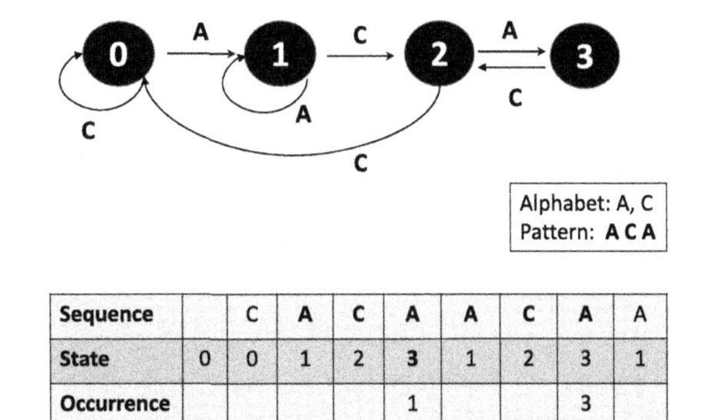

Sequence		C	A	C	A	A	C	A	A
State	0	0	1	2	3	1	2	3	1
Occurrence					1			3	

Figure 5.2: Example of an automata for pattern search, where the alphabet has two symbols (A and C), and the pattern is "ACA".

An example of a DFA is shown in Fig. 5.2 for an alphabet with two symbols and the pattern $p =$ "ACA". In the example, we can see the transition table and its graphical representation, and the result of applying the DFA to a sequence, showing the internal states of the DFA and the pattern occurrences found.

The DFAs for pattern searching are implemented in a Python class provided below. Notice the transition table is implemented as a Python dictionary, where keys are tuples with (previous state, symbol), and values represent the next state (last column in the table in the figure). The function **build_transition_table** creates the transition table according to the definition above, being called by the constructor of the class.

Notice that a DFA is created for a given alphabet and pattern, and can be used to search for that pattern in different sequences. The function **apply_seq** computes the list of states the DFA goes through when processing a sequence, while the **occurrences_pattern** function provides a list of the positions of the pattern in the sequence (note that when the DFA reaches the final state an occurrence is found, but the initial position needs to be calculated subtracting the size of the pattern).

```
class Automata:

    def __init__(self, alphabet, pattern):
        self.numstates = len(pattern) + 1
        self.alphabet = alphabet
        self.transition_table = {}
        self.build_transition_table(pattern)
```

```
    def build_transition_table(self, pattern):
        for q in range(self.numstates):
            for a in self.alphabet:
                prefix = pattern[0:q] + a
                self.transition_table[(q,a)] = overlap(prefix,
pattern)

    def print_automata(self):
        print ("States: " , self.numstates)
        print ("Alphabet: " , self.alphabet)
        print ("Transition table:")
        for k in self.transition_table.keys():
            print (k[0], ",", k[1], " -> ", self.transition_table[k])

    def next_state(self, current, symbol):
        return self.transition_table.get((current, symbol))

    def apply_seq(self, seq):
        q = 0
        res = [q]
        for c in seq:
            q = self.next_state(q, c)
            res.append(q)
        return res

    def occurrences_pattern(self, text):
        q = 0
        res = []
        for i in range(len(text)):
            q = self.next_state(q, text[i])
            if q == self.numstates-1:
                res.append(i - self.numstates + 2)
        return res

def test():
    auto = Automata("ACGT", "ACA")
    auto.print_automata()
    print (auto.apply_seq("CACATGACATG"))
    print (auto.occurrences_pattern("CACATGACATG"))
```

```
test()
```

5.5 Finding Flexible Patterns: Regular Expressions

5.5.1 Definitions and Regular Expressions in Python

Regular expressions (REs) are a programming concept, that exists in all modern programming languages, that allow to define patterns to search in strings in a flexible manner. REs are defined as strings, where some of the characters are not used as regular characters, but rather as *meta-characters* to represent patterns. While regular characters in an RE only match themselves in a search process, meta-characters may offer many alternatives for matching depending on their meaning.

One of the most used meta-characters is the dot (.), that matches any character in a string. Thus, the RE "..." will match any string of length 3. Some of the meta-characters are used to modify the predecessor character (or set of characters) by allowing repetitions of patterns, as follows:

- * – zero or more repetitions of the pattern to which it applies;
- + – one or more repetitions of the pattern;
- ? – zero or one repetitions (pattern may occur or not);
- {n} – exactly *n* repetitions, where *n* is an integer;
- {m,n} – between *m* and *n* repetitions, where *m* and *n* are integers and $n >= m$.

The previous "modifiers" may be applied to a single character or to a group of characters, that can be defined using regular brackets. On the other hand, square brackets are used to define possible lists of characters that match. This syntax is quite useful as the following examples show:

- [A-Z] – matches any upper-case letter;
- [a-z] – matches any lower-case letter;
- [A-Za-z] – matches any letter;
- [0-9] – matches any digit;
- [ACTGactg] – matches a DNA nucleotide symbol (upper or lower case);
- [ACDEFGHIKLMNPQRSTVWY] – matches an aminoacid symbol (upper case).

If the ^ symbol is put before the list, it is negated, i.e. it will match all characters not included in the list. Thus, "[^0-9]" matches with non-digits.

Combining these definitions with the previous ones, it is easy to define an RE for a natural number as the string "[0-9]*", or to define a DNA sequence as "[ACTGactg]*",

Table 5.1: Examples of regular expressions and matching strings.

RE	Matching strings
ACTG	ACTG
AC.TC	ACCTC, ACCTC, ACXTC, ...
A[AC]A	AAA, ACA
A*CCC	CCC, ACCC, AACCC, ...
ACC |G.C	ACC, GAC, GCC, ...
AC(AC){1,2}A	ACACA, ACACACA
[AC]3	CAC, AAA, ACC, ...
[actg]*	a, ac, tg, gcgctgc, ...

Table 5.2: Functions/methods working over regular expressions.

Function	Description
re.search(*regexp, str*)	checks if *regexp* matches *str*; returns results on the first match
re.match(*regexp, str*)	checks if *regexp* matches *str* in the beginning of the string
re.findall(*regexp, str*)	checks if *regexp* matches vstr; returns results on all matches as a list
re.finditer(*regexp, str*)	same as previous, but returns results as an iterator

or a protein sequence with between 100 and 200 aminoacids as "[ACDEFGHIKLMN PQRSTVWY]{100,200}". And, of course, the hypotheses are endless.

There are other ways to select groups of characters, using the \ symbol followed by a letter. Some examples of this syntax are given below:

- \s – includes all white space (spaces, newlines, tabs, etc);
- \S – is the negation of the previous, thus matches with all non-white-space characters;
- \d – matches with digits;
- \D – matches with non-digits.

Other important meta-characters include the | that works as a logical or (disjunction), stating that the pattern can match with either the expression on the left or the expression on the right, $ matches with the end of a line and ^ with the beginning of a line.

Some examples of strings representing regular expressions and possible matching strings are given in Table 5.1.

Python includes, within the package *re*, a number of tools to work with REs, allowing to test their match over strings. The main functions and their description are provided in Table 5.2.

In these functions, the result of a match is kept in a Python object that holds relevant information about the match. The methods *m*.**group**() and *m*.**span**(), applied over an object *m* returned from a match, allow to retrieve the matching pattern in the string and the initial and final positions of the match. Some examples run in the Python shell illustrate the behavior of these functions.

```
>>> import re
>>> str = "TGAAGTATGAGA"
>>> mo = re.search("TAT", str)
>>> mo.group()
'TAT'
>>> mo.span()
(5, 8)
>>> mo2 = re.search("TG.",str)
>>> mo2.group()
'TGA'
>>> mo2.span()
(0, 3)
>>> re.findall("TA.",str)
['TAT']
>>> re.findall("TG.",str)
['TGA', 'TGA']
>>> mos = re.finditer("TG.",str)
>>> for x in mos:
...     print x.group()
...     print x.span()
...
TGA
(0, 3)
TGA
(7, 10)
```

Using those functions and the methods to retrieve information, it is possible to define two new functions to get the first occurrence or gather all occurrences of a pattern in a sequence, now providing the pattern as a regular expression, which allows to define more flexible patterns. This is done in the next code chunk, where these functions are defined and a program is built to allow users to input desired sequences and patterns (through the **test** function).

```
def find_pattern_re (seq, pat):
    from re import search
    mo = search(pat, seq)
    if (mo != None):
        return mo.span()[0]
    else:
        return -1
```

```
def find_all_occurrences_re (seq, pat):
    from re import finditer
    mos = finditer(pat, seq)
    res = []
    for x in mos:
        res.append(x.span()[0])
    return res

def test():
    seq = input("Input sequence:")
    pat = input("Input pattern (as a regular expression):")

    res = find_pattern_re(seq, pat)
    if res >= 0:
        print("Pattern found in position: ", res)
    else:  print("Pattern not found")

    all_res = find_all_occurrences_re(seq, pat)
    if len(all_res) > 0:
        print("Pattern found in positions: ", all_res)
    else:  print("Pattern not found")

test()
```

This program may be used to test different REs and their occurrence in biological sequences (DNA, RNA, proteins).

One important limitation of the previous function to identify all occurrences of a pattern (**find _all_occurrences_re**) is the fact that it does not consider instances of the patterns that overlap. To illustrate this consider the following example of the previous program:

```
Input sequence:ATATGAAGAG
Input pattern (as a regular expression):AT.

Pattern found in position:  0
Pattern found in positions:  [0]
```

Note that the pattern occurs both in positions 0 ("ATA") and 2 ("ATG"), but only the first is identified by the function. One possible solution for this problem is to define the pattern using the lookahead assertion, i.e. we will match the pattern, but will not "consume" the characters allowing for further matches. This is done by using the syntax "(?=*p*)", where *p* is the pattern to match, a solution shown in the code example shown below. There is also another alternative, to use the more recent *regex* package, which already supports overlapping matches by simply defining a parameter in the matching functions.

```
def find_all_overlap(seq, pat):
    return find_all_occurrences_re(seq, "(?="+pat+")")

def test():
    seq = input("Input sequence:")
    pat = input("Input pattern (as a regular expression):")

    (..)

    ll_ov = find_all_overlap(seq, pat)
    if len(all_ov) > 0:
        print("Pattern found in positions: ", all_ov)
    else:
        print("Pattern not found")

test()
```

The behavior of the previous program is now what is expected:

```
Input sequence:ATATGAAGAG
Input pattern (as a regular expression):AT.

Pattern found in position:  0
Pattern found in positions:  [0]
Pattern found in positions (overlap):  [0, 2]
```

If the application of REs demands the search of the same pattern over many strings, there are ways to optimize this process by pre-processing the RE to make its search more efficient, in a process that is normally named as *compilation*. The computational process executed in this case involves transformation of the pattern into data structures similar to the ones discussed in the previous section.

The compilation process can be done using the **compile** function in the *re* package. Over the resulting object, the functions **match**, **search**, **findall** and **finditer** can be applied passing the target string as an argument, returning the same results as the homonymous ones defined above. The following code chunk shows an illustrative example.

```
>>> import re
>>> seq = "AAATAGAGATGAAGAGAGATAGCGC"
>>> rgx = re.compile("GA.A")
>>> rgx.search(seq).group()
'GAGA'
>>> rgx.findall(seq)
['GAGA', 'GAGA', 'GATA']
>>> mo = rgx.finditer(seq)
>>> for x in mo: print(x.span())
(5, 9)
(13, 17)
(17, 21)
```

Another important feature of REs is the possibility to define *groups* within the pattern to find, allowing to identify the match not only of the full RE, but also check which parts of the target string match specific parts of the RE. Groups in REs are defined by enclosing parts of the RE with parentheses.

The following code shows an example of the use of groups in REs.

```
>>> rgx = re.compile("(TATA..)((GC){3})")
>>> seq = "ATATAAGGCGCGCGCTTATGCGC"
>>> result = rgx.search(seq)
>>> result.group(0)
'TATAAGGCGCGC'
>>> result.group(1)
'TATAAG'
>>> result.group(2)
'GCGCGC'
```

5.5.2 Examples in Biological Sequence Analysis

REs can be useful in a huge number of Bioinformatics tasks, including some of the ones we have addressed in the previous chapter. One interesting example is its use to validate the content of specific sequences depending on its type.

The example below shows how to define such a function for a DNA sequence. Similar functions may be written for other sequence types, which is left as an exercise for the reader (these may be integrated in a general-purpose function that receives the sequence type as an input).

```
def validate_dna_re (seq):
    from re import search
    if search("[^ACTGactg]", seq) != None:
        return False
    else:
        return True

>>> validate_dna_re("ATAGAGACTATCCGCTAGCT")
True
>>> validate_dna_re("ATAGAGACTAXTCCGCTAGCT")
False
```

One other task that can be achieved with some advantages using REs is the translation of codons to aminoacids. Indeed, the similarity in the different codons that encode the same aminoacid can be used to define an RE for each aminoacid simplifying the conditions. This is shown in the function below, which may replace the **translate_codon** function presented in the previous chapter.

```
def translate_codon_re (cod):
    import re
    if re.search("GC.", cod): aa = "A"
    elif re.search("TG[TC]", cod): aa = "C"
    elif re.search("GA[TC]", cod): aa = "D"
    elif re.search("GA[AG]", cod): aa = "E"
    elif re.search("TT[TC]", cod): aa = "F"
    elif re.search("GG.", cod): aa = "G"
    elif re.search("CA[TC]", cod): aa = "H"
    elif re.search("AT[TCA]", cod): aa = "I"
    elif re.search("AA[AG]", cod): aa = "K"
    elif re.search("TT[AG]|CT.", cod): aa = "L"
    elif re.search("ATG", cod): aa = "M"
    elif re.search("AA[TC]", cod): aa = "N"
    elif re.search("CC.", cod): aa = "P"
    elif re.search("CA[AG]", cod): aa = "Q"
    elif re.search("CG.|AG[AG]", cod): aa = "R"
    elif re.search("TC.|AG[TC]", cod): aa = "S"
```

```
    elif re.search("AC.", cod): aa = "T"
    elif re.search("GT.", cod): aa = "V"
    elif re.search("TGG", cod): aa = "W"
    elif re.search("TA[TC]", cod): aa = "Y"
    elif re.search("TA[AG]|TGA", cod): aa = "_";
    else: aa = ""
    return aa
```

To finish this set of simple examples, let us recall the problem of finding a putative protein in a sequence of aminoacids. A protein may be defined as a pattern in the sequence, that starts with symbol "M" and ends with a "_" (the symbol representing the translation of the stop codon). Note that in the other intermediate symbols "M" can occur, but "_" cannot. So, the regular expression to identify a putative protein can be defined as: "M[^_]*_". This is used in the next code block to define a function which identifies the largest possible protein contained in an inputted aminoacid sequence.

```
def largest_protein_re (seq_prot):
    import re
    mos = re.finditer("M[^_]*_", seq_prot)
    sizem = 0
    lprot = ""
    for x in mos:
        ini = x.span()[0]
        fin = x.span()[1]
        s = fin - ini + 1
        if s > sizem:
            lprot = x.group()
            sizem = s
    return lprot
```

Note that the **finditer** function is used to get all matches of the RE defining a putative protein, but fails to identify overlapping proteins. This case may occur when a protein includes an "M" symbol, i.e. the protein is of the form "M ... M ... M ... _". In such situations, the only protein matching will be the one starting with the first "M" (the outer one), which in this case corresponds to the largest. So, since the purpose of the function is to identify the largest protein, there are no problems. If the aim is to find all putative proteins, including overlapping ones, the solutions for this issue presented in the previous section need to be used.

5.5.3 Finding Protein Motifs

As we mentioned in the introduction, several types of sequence patterns play a relevant role in biological functions. This is the case with DNA/RNA, but also with protein sequences, being normally called as motifs. These motifs are typically associated with conserved protein domains, that determine a certain tri-dimensional configuration, which leads to a given specific biological function.

We will cover protein (and DNA) motifs in different chapters of this book, addressing different types of patterns and tasks. Here, as an example of the usefulness of regular expression, we will discuss a specific type of patterns, which may be represented by regular expressions.

The Prosite database (`http://prosite.expasy.org/`) contains many protein motifs, represented with different formats. One of the most popular (they name as patterns) represents the possible content of each position, specifying either an aminoacid or a set of possible aminoacids. Also, there is the possibility of specifying segments of aminoacids of variable length. This is achieved using a specific representation language, using the 20 aminoacids symbols, but also a set of specific meta-characters.

Some of the syntax rules of this representation are the following:

- each aminoacid is represented by one letter symbol (see Table 4.2 in Chapter 4);
- a list of aminoacids within square brackets represents a list of possible aminoacids in a given position;
- the symbol "x" represents any aminoacid in a given position;
- a number within parenthesis after an aminoacid (or aminoacid list) represents the number of occurrences of those aminoacids;
- a pair of numbers separated by a comma symbol within parentheses, indicates a number of occurrences between the first and the second number (i.e. indicates a range for the number of occurrences);
- the "-" symbol is used to separate the several positions.

An example is the "Zinc finger RING-type signature" (PS00518) motif, which is represented by "C-x-H-x-[LIVMFY]-C-x(2)-C-[LIVMYA]". This means a pattern starting with aminoacid "C", followed by any aminoacid, aminoacid "H", any aminoacid, an aminoacid in the group "LIVMFY", aminoacid "C", two aminoacids, aminoacid "C" and an aminoacid in the group [LIVMYA].

We will provide here some examples of how to represent Prosite patterns using REs, and the way to define functions to search for these REs in given protein sequences. An example would be a function to search for the previous motif (PS00518) in a given sequence. This implies transforming the pattern representation into an RE and then finding the matches of the RE in the sequence. This function is given in the next code block.

```
def find_zync_finger(seq):
    from re import search
    regexp = "C.H.[LIVMFY]C.{2}C[LIVMYA]"
    mo = search(regexp, seq)
    if (mo != None):
        return mo.span()[0]
    else:
        return -1

def test():
    seq = "HKMMLASCKHLLCLKCIVKLG"
    print(find_zync_finger(seq))

test()
```

Note that, in this case, we transformed the given Prosite pattern into an RE (given by the variable *regexp* in the code above). A more interesting approach would be to create a general-purpose function where the Prosite pattern would also be given as an argument, thus allowing to search for multiple different patterns using the same function. In this case, we would need to have a way of transforming any Prosite pattern into the corresponding RE. This is done in the next code chunk, where the function is tested with the same example as above.

```
def find_prosite(seq, profile):
    from re import search
    regexp = profile.replace("-","")
    regexp = regexp.replace("x",".")
    regexp = regexp.replace("(","{")
    regexp = regexp.replace(")","}")
    mo = search(regexp, seq)
    if (mo != None):
        return mo.span()[0]
    else:
        return -1

def test():
    seq = "HKMMLASCKHLLCLKCIVKLG"
    print(find_prosite(seq,"C-x-H-x-[LIVMFY]-C-x(2)-C-[LIVMYA]"))

test()
```

Other examples of Prosite patterns may be found in the database website provided above. Note that we have not covered here all syntax rules of Prosite patterns. The full list of rules can be found in `http://prosite.expasy.org/scanprosite/scanprosite_doc.html`, and thus there might be cases that do not work with this function. The service provided in the page `http://prosite.expasy.org/scanprosite/` allows to search for motif instances within provided sequences, searching over all patterns in the database.

5.5.4 An Application to Restriction Enzymes

Restriction enzymes are proteins that cut the DNA in areas that contain specific sub-sequences (patterns or motifs). For instance, the *EcoRI* restriction enzyme cuts DNA sequences that contain the pattern "GAATTC", specifically between the "G" and the first "A". Note that the pattern is a *biological palindrome*, i.e. a sequence that is the same as its reverse complement. This means that a restriction enzyme cuts the sequence in both DNA chains, while leaving an overhang, since it does not cut exactly in the same position, that is useful in molecular biology for cloning and sequencing. Thus, restriction maps (the positions where a restriction enzyme cuts the sequence) are useful tools in molecular biology.

Databases of restriction enzymes, as it is the case with REBASE (`http://rebase.neb.com/`), keep restriction enzymes represented as strings in an alphabet of symbols that includes not only the nucleotide sequences, but also symbols that allow ambiguity, given that some enzymes allow variability in the target regions. The IUPAC extended alphabet, also known as IUB ambiguity codes, already given in Table 4.1, is normally chosen for this task.

Given strings in this flexible alphabet, and being the purpose to find their occurrences in DNA sequences, the first task is to convert strings written in this alphabet to regular expressions that can be used to search over sequences. The next function addresses this task.

```
def iub_to_RE (iub):
    dic = {"A":"A", "C":"C", "G":"G", "T":"T", "R":"[GA]", "Y":"[CT]"
    , "M":"[AC]", "K":"[GT]", "S":"[GC]", "W": "[AT]", "B":"[CGT]", "
   D":"[AGT]", "H":"[ACT]", "V":"[ACG]", "N":"[ACGT]"}

    site = iub.replace("^","")
    regexp = ""

    for c in site:
        regexp += dic[c]

    return regexp
```

```
def test():
    print(iub_to_RE("G^AMTV"))

test()
```

Note that, in this function, it is assumed that the symbol ^ is used to denote the position of the cut. To convert to an RE, this symbol is ignored, but will be necessary to determine the restriction map.

Given this function, we can now proceed to write functions to detect where a given enzyme will cut a given DNA sequence, and also to calculate the resulting sub-sequences after the cut (restriction map). These tasks are achieved by the functions **cut_positions** and **cut_subsequences**, respectively, provided in the code below.

```
def cut_positions (enzyme, sequence):
    from re import finditer

    cutpos = enzyme.find("^")
    regexp = iub_to_RE(enzyme)

    matches = finditer(regexp, sequence)
    locs = [ ]
    for m in matches:
        locs.append(m.start() + cutpos)

    return locs

def cut_subsequences (locs, sequence):
    res = []
    positions = locs
    positions.insert(0,0)
    positions.append(len(sequence))
    for i in range(len(positions)-1):
        res.append(sequence[positions[i]:positions[i+1]])
    return res

def test():
    pos = cut_positions("G^ATTC", "GTAGAAGATTCTGAGATCGATTC")
    print(pos)
```

```
    print(cut_subsequences(pos, "GTAGAAGATTCTGAGATCGATTC"))

test()
```

The former function defined will return a set of positions where the RE matches, i.e. the enzyme cuts the sequence, while the latter will use these positions to gather the respective sub-sequences resulting from the cut. Since a sequence cuts both chains of the DNA molecule, it is left as an exercise for the reader to write a function that can calculate the sub-sequences of the reverse complement sequence.

Bibliographic Notes and Further Reading

A more formal description and analysis of the complexity of the naive, Boyer-Moore, DFAs and other string matching algorithms can be found in [28]. The Boyer-Moore algorithm was firstly presented by its authors in [29]. The usage of DFAs for the pattern searching problems was introduced by Aho and colleagues in [9]. Other algorithms for this purpose were not covered in this textbook, as it is the case of the Knuth-Morris-Pratt algorithm, are described in detail in [38].

Regular expressions are covered in many other books and other resources, such as the book by Friedl et al. [66]. A more theoretical perspective on REs and DFAs can be found in the book by Hopcroft and colleagues [78].

As mentioned in the text, the Python package *regex* provides a more recent set of tools for REs in Python. The documentation of this package may be found in `https://pypi.python.org/pypi/regex/`.

As we mentioned in this chapter, the notion of *motif* is deeply connected to the definition of a pattern that is potentially related to a biological function. In subsequent chapters we will explore both the tasks of identifying known motifs from sequences (representing motifs in ways that extend the ones presented in this chapter by considering probabilities) and of discovering motifs as over-represented patterns in biological sequences, in Chapters 10 and 11. Also, in Chapter 16, we will address other algorithms for pattern searching, which are more appropriate to find many patterns over a single large sequence (e.g. a full chromosome or genome).

Exercises and Programming Projects

Exercises

1. Write a Python function that, given a DNA sequence, allows to detect if there are repeated sequences of size k (where k should be passed as an argument to the function).

The result should be a dictionary with sub-sequences as keys, and their frequency as values.

2. Most introns can be recognized by their consensus sequence which is defined as: GT ... TACTAAC ... AC, where ... mean an unknown number of nucleotides (between 1 and 10). Write a Python function that, given a DNA sequence, checks if it contains an intron, according to this definition. The result should be a list with all initial positions of the introns (empty list if there are none).
3. In many proteins present in the membrane, there is a conserved motif that allows them to be identified in the transport process of these protein by the endosomes to be degraded in the lysosomes. This motif occurs in the last 10 positions of the protein, being characterized by the aminoacid tyrosine (Y), followed by any two aminoacids and terminating in a hydrophobic aminoacid of the following set – phenylalanine (F), tyrosine (Y) or threonine (T).
   a. Write a function that, given a protein (sequence of aminoacids), returns an integer value indicating the position where the motif occurs in the sequence or -1 if it does not occur.
   b. Write a function that, given a list of protein sequences, returns a list of tuples, containing the sequences that contain the previous motif (in the first position of the tuple), and the position where it occurs (in the second position). Use the previous function.
4. Write a function that given two sequences of the same length, determines if they have at most two d mismatches (d is an argument of the function). The function returns `True` if the number of mismatches is less or equal to d, and `False` otherwise. Using the previous function, write another function to find all approximate matches of a pattern in a sequence. An approximate match of the pattern can have at most d characters that do not match (d is an argument of the function).
5. Write a function that reads a file in the FASTA format and returns a list with all sequences.
6. Files from UniProt saved in the FASTA format have a specific header structure given by:

   $$db|Id|Entry\ Protein\ \mathrm{OS} = Organism\ [\mathrm{GN} = Gene]\ \mathrm{PE} = Existence\ \mathrm{SV} = Version$$

   Write a function that using regular expressions parses a string in this format and returns a dictionary with the different fields (the key should be the field name). Note the part in right brackets is optional, the parts in italics are the values of the fields, while the parts in upper case are constant placeholders.

Programming Projects

1. Integrate the functions provided in this chapter with the classes for representing biological sequences developed in the previous chapter.

2. Consider the Genbank file format, used by NCBI to keep sequences and their annotations. Write functions to read a Genbank record and parse it, using regular expressions. From the record, you should extract the identifiers (ACCESSION number, GI, etc), the size of the sequence, its definition, the organism, the full sequence. Also, you should extract the information from the features, including for each feature, the source, the location, and a dictionary with the qualifiers. Check Section 4.8 to see how *BioPython* handles these files.

CHAPTER 6

Pairwise Sequence Alignment

In this chapter, we show that to assess pairwise sequence similarity, we need to align the sequences. We define the alignment process as an optimization problem and discuss efficient algorithms based on dynamic programming for these problems, both considering global and local alignments. The implementation of these algorithms in Python is put forward and discussed. We also address *BioPython* functions to align pairs of sequences.

6.1 Introduction: Comparing Sequences and Sequence Alignment

One of the major contributions that Bioinformatics can bring to biological research is the possibility to help in elucidating the function of given genes or the proteins they encode. A common approach to address these tasks is to generate hypotheses regarding their biological function, based on the similarity of their sequences to others that have a determined function, preferably with an experimental validation.

Indeed, what we are trying to achieve here is to infer *homology*, i.e. the existence of shared ancestry, based on sequence *similarity*. It is important to emphasize that these concepts are not equivalent. However, in practice, sequences that show a high degree of similarity have a high probability of being homologous and sharing similar functions. This probability grows with the increase in the similarity degree. When the similarity reaches high levels (the definition of a proper threshold is unfortunately not obvious), we can infer function with some credibility, although the ideal is always to verify experimentally.

So, in this scenario, to be able to compare pairs of sequences is of foremost importance. At a first sight, this seems a straightforward problem. Indeed, we can intuitively define similar sequences as those who share the same symbols in the same order. Thus, this seems like a trivial computational task, but this is not the case.

We need to consider the sources of variability between sequences that share a common ancestor: *mutations*. Biological mutations can lead to the changes in one or more nucleotides, but there are also insertions and deletions of nucleotides. These make the comparison of sequences more complex, since we can not compare sequences position by position.

To overcome this difficulty, we need to resort to sequence *alignment*, the computational procedure of seeking shared individual characters in both sequences, keeping the same order. The

Bioinformatics Algorithms. DOI: 10.1016/B978-0-12-812520-5.00006-7

Protein / **DNA**

Global

```
Protein                 DNA
LGPSSGCASRIWTKSA        -CAGTGCATG-ACATA
 ||| |  | |||||          ||| || |  ||| |
TGPS-G--S-IWSKSG        TCAG-GC-TCTACAGA
```

Local

```
Protein                 DNA
LGPSSGCASRIWTKSA        -CAGTGCATGTACAGA
     |||||||                    |||||||
TWNR-GCASRIWMRDW        TTCG-TC-TGTACAGT
```

Figure 6.1: Examples of different global and local alignments of DNA and protein sequences. Gaps are represented with the hyphen symbol.

process of sequence alignment will consist on introducing spacing characters, known as *gaps*, to maximize shared characters between the two sequences.

In this case, similar sequences would be those that have a large number of equal (or identical) characters, after a proper process of sequence alignment. Although this provides the general idea of sequence alignment and similarity, we clearly need this to be more formally defined to be able to design proper algorithms, and this will be the aim of the next section.

It is important to define also which biological sequences are we considering in these tasks. The most frequent targets for sequence alignment are DNA (or in more rare cases RNA) and proteins, defined in the same way as given in the last two chapters. Protein sequence alignment is a very important tool mainly for sequence annotation, i.e. providing functions to sequences (and parts of sequences), while DNA sequence alignment, apart from annotation, is also used in other applications related for instance with phylogenetic analysis.

Also, there are two broad classes of sequence alignment: global and local. In the first case, we are interested in aligning the complete sequences, while in the latter the aim is to find good alignments of adjacent parts of both sequences, ignoring the remaining. Fig. 6.1 provides a couple of examples for local and global alignments, considering both DNA and protein sequences.

6.2 Visual Alignments: Dot Plots

Before delving further into the algorithms for sequence alignment, in this section, we will cover tools that can be used to have a visual perspective on the most similar parts of two sequences. In this regard, the main tools are dot plots, graphical matrix representations of two sequences, representing one sequence over the rows and the other over the columns, highlighting within the matrix regions of the sequences where the similarity is high.

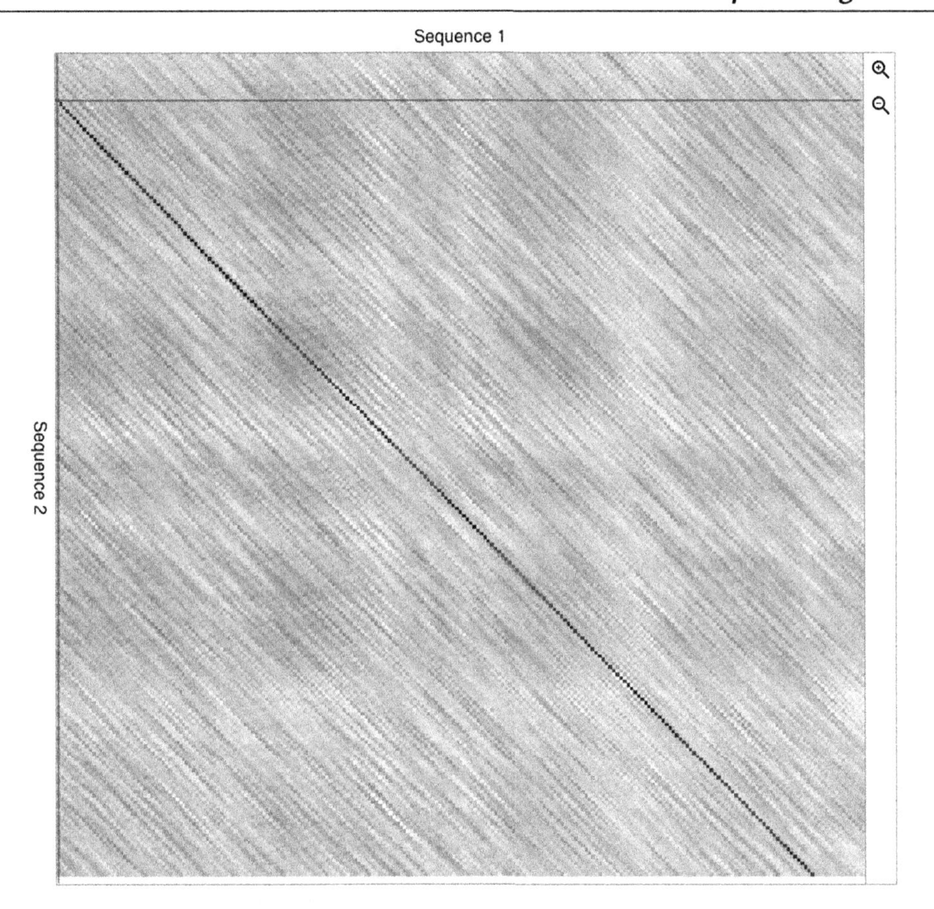

Figure 6.2: Example of a dot plot produced by the tool *dotlet*.

An example of the output of these tools is given in Fig. 6.2, created by the *dotlet* web-based application (`https://dotlet.vital-it.ch`), considering two protein sequences. The output shows that these sequences have a region of very high similarity, shown as a darker diagonal, from the beginning of sequence 1 (over the columns) and starting around position 15 of sequence 2 (along the rows), that goes until the end of sequence 2 (but not of sequence 1).

One important application of dot plots is the possibility of seeking for repeats in a single sequence. In this case, we put the same sequence over rows and columns of the matrix. Repeated parts of the sequence may be visually identified by looking for dark diagonals, outside of the main diagonal, while vertical/horizontal rows represent repeats of a single character.

In their simplest form, these matrices place dots in the cells where the characters in both sequences coincide. The algorithm to build these matrices can be implemented in a very simple function, which returns a binary matrix, i.e. containing the values 0 or 1. In this case, the values of 1 represent the dots in the matrix. This function can be implemented as follows.

```
def create_mat(nrows, ncols):
    mat = []
    for i in range(nrows):
        mat.append([])
        for j in range(ncols):
            mat[i].append(0)
    return mat

def dotplot(seq1, seq2):
    mat = create_mat(len(seq1), len(seq2))
    for i in range(len(seq1)):
        for j in range(len(seq2)):
            if seq1[i] == seq2[j]:
                mat[i][j] = 1
    return mat
```

In the previous code, the function **create_mat** is used to create a matrix filled with zeros. The function **dotplot** fills the matrix, placing ones in the appropriate places, i.e. when the characters in the corresponding positions of the sequences are equal.

The function may be tested by printing its content. In the following code, the asterisk symbol is used to print the dots, while positions with zero are printed as white spaces.

```
def print_dotplot(mat, s1, s2):
    import sys
    sys.stdout.write(" " + s2+"\n")
    for i in range(len(mat)):
        sys.stdout.write(s1[i])
        for j in range(len(mat[i])):
            if mat[i][j] >= 1:
                sys.stdout.write("*")
            else:
                sys.stdout.write(" ")
        sys.stdout.write("\n")

def test():
    s1 = "CGATATAG"
```

```
    s2 = "TATATATT"
    mat1 = dotplot(s1, s2)
    print_dotplot(mat1, s1, s2)

test()
```

Since, in many cases, this simple algorithm leads to much noise, it is common to filter the results. One of the possible strategies is to consider a window around each position, and count the number of matching characters for each sequence in such neighborhood. In this case, we only fill a given cell if, within this neighborhood, the number of matching characters exceeds a given parameter, typically named as *stringency*. A function implementing this strategy is given below.

```
def extended_dotplot (seq1, seq2, window, stringency):
    mat = create_mat(len(seq1), len(seq2))
    start = int(window/2)
    for i in range(start,len(seq1)-start):
        for j in range(start, len(seq2)-start):
            matches = 0
            l = j - start
            for k in range(i-start, i+start+1):
                if seq1[k] == seq2[l]: matches += 1
                l += 1
                if matches >= stringency: mat[i][j] = 1
    return mat

def test():
    s1 = "CGATATAGATT"
    s2 = "TATATAGTAT"
    mat2 = extended_dotplot(s1, s2, 5, 4)
    print_dotplot(mat2, s1, s2)

test()
```

The function **extended_dotplot** receives, together with both sequences, the size of the window and the *stringency* parameter. In the example, dots will only be placed when, for a given position, 4 out of 5 characters match in the window, which in this case contains the character in that position, the two previous and the two following ones (window size of 5) in each sequence.

6.3 Sequence Alignment as an Optimization Problem

6.3.1 Problem Definition and Complexity

To define the process of sequence alignment in a more formal way, we first need to understand that this is actually an optimization problem, i.e. a task where there are several possible solutions and we need to select which is the best one. The different solutions for this problem arise from the different possible combinations of the positions where we put the gaps in each of the sequences. Note that the order of the sequences' characters can not be altered.

To be able to look at the different solutions and select the best one, we need to define a proper objective function for the optimization problem. Having that defined, the *pairwise sequence alignment problem* can be defined as follows:

Inputs: two sequences in a well defined alphabet; an objective function to evaluate each alignment solution;
Output: the optimal pairing of the characters in each sequence, placing gaps in appropriate positions, to maximize the objective function.

Before looking in more detail to the specificities of the objective function, let us try to understand the potential complexity of this problem by estimating the number of possible solutions. For that purpose, we will consider a simplification stating that both sequences have the same size n. This constraint will only serve to get our expressions simpler, and will suffice to provide an idea of the complexity. Note that this constraint will not apply to our algorithms, as we will be able to align sequences of different sizes.

Since, for our problem, solutions vary in the way gaps are placed in the sequences, we can easily figure out that the maximum size of the whole alignment will be $2n$, for the case where we place n gaps, since it does not make sense to have a gap in both sequences.

Thus, the total number of solutions will be provided by the possible combinations of size n from $2n$ elements, given by:

$$\binom{2n}{n} = \frac{(2n)!}{n!^2} \tag{6.1}$$

To have an idea of the value represented, for $n = 20$ (a very small value for real biological sequences), this amounts to about 120 billion solutions. This provides an idea that the problem is quite complex and that it is impossible to address by using a brute-force approach, where all possible solutions would be attempted. We will see in the next section that there are algorithms that can be efficient for this task, provided the objective function has some features. So let us first define the objective functions to use.

6.3.2 Objective Function: Substitution Matrices and Gap Penalties

The objective function, in any optimization problem, has the purpose of evaluating all feasible solutions for the problem, assigning to each a numerical value, being the aim of the optimization process to find the solution(s) that minimize (or maximize) this value, i.e. the one(s) with the lowest (or highest) value for the defined objective function.

In this case, assuming a maximization problem, an adequate objective function should be one that has high values for alignments of similar sequences, and has low values for less similar (or more distant) sequences. In this scenario, the objective function also takes the designation of *score*.

A very simple solution, which can be used in some cases, is to simply count the number of equal symbols in the aligned sequences. In this case, we are assigning a value of 1 for columns with equal characters (matches), and 0 both to the case of distinct characters (mismatches) and columns where gaps occur, thus not penalizing those columns.

This is an example of an *additive* objective function, where the contributions of each column are independent, being summed up to obtain the final value of the objective function. This is not an assumption we can safely make regarding biological sequences, since many mutations affect several adjacent positions, while there are other interactions between nucleotides/aminoacids in nearby positions. However, in most cases, we will still consider this as a valid assumption, since it allows the development of efficient algorithms, as we will see in the next sections. Indeed, to consider interactions among the different columns leads to more complex optimization problems, more difficult to decompose.

Variants of the previous objective function arise by giving different score values for the three cases: matches, mismatches, and gaps. While the first is typically a positive value, the two latter are negative. In the case of DNA (or RNA) sequence alignments, typical values for match scores can be $+1$ or $+2$, while for mismatches and gaps values as -3 or -2 are common.

The simple objective functions described above may be adequate in many areas, including DNA/RNA sequence alignment, but are not widely adopted in the alignment of protein sequences. Indeed, when defining objective functions, we will see that there are different typical options for nucleotide and aminoacid sequences, which derive in large part from the size of the alphabets (4 vs 20), but also from biochemical features of the molecules represented by each type of sequence.

Indeed, for protein sequences, there are 20 possible aminoacids, and these can be grouped according to their biochemical properties. So, in the case of mismatches, there can be cases where the two aminoacids are very different in terms of their properties, while in other cases the two aminoacids are more similar. This leads to the necessity of assigning different scores to each case.

Given this context, objective functions for biological sequence alignment are typically defined in a more general and flexible way, considering two components:

- a *substitution matrix*, which includes a score value for all cases where there are no gaps in the columns, assigning a value for each pair of possible symbols in the alphabet (e.g. each aminoacid or nucleotide pair);
- a *gap penalty* function, which defines how to penalize a gap (or a sequence of gaps depending on the model).

For protein sequence alignments, substitution matrices are calculated based on the probabilities of finding a given pair of aligned aminoacids in a *good alignment*, i.e. an alignment of homologous sequences, as compared to the probability of obtaining such a pairing by chance. To reach the final matrix, a database with good alignments needs to be used. These databases typically include protein sequences (or fragments) with high quality alignments with no gaps.

Based on one of these databases, the probability of occurrence of a given pair of aminoacids is estimated by its relative frequency in the alignments. Also, for each pair of aminoacids, the expected probability is estimated as if pairs were obtained by chance. This probability may be calculated, assuming independence of the occurrence of each aminoacid, by the multiplication of the probability of occurrence of each individual aminoacid, estimated by their relative frequency of occurrence in the whole database.

The score s for a given aminoacid pair a, b is thus given by the expression:

$$s(a,b) = round(2 \times log_2 \frac{P_{(a,b)}}{p_a p_b}) \tag{6.2}$$

where $P_{(a,b)}$ is the estimated probability of occurrence of the pair of aminoacids (a, b) in the database alignments, and p_a and p_b are the estimated probabilities for the occurrence of aminoacids a and b, respectively.

Note that the application of the logarithm guarantees that pairs of aminoacids appearing more frequently than expected will bring a positive score (ratio is larger than 1), while the other ones will lead to a negative contribution to the objective function. This is normally denoted as the *log odd ratio* of an observed over an expected event. Note that the result is rounded to reach integer matrices.

As a simple example, let us assume the database contains 1000 aminoacid pairs in total. Here, SS occurs 40 times in the alignments columns and the relative frequency of S is 10%. Applying the previous expression we would have: $score(S, S) = round(2 \times log_2 \frac{40/1000}{0.1 \times 0.1}) = 4$. In

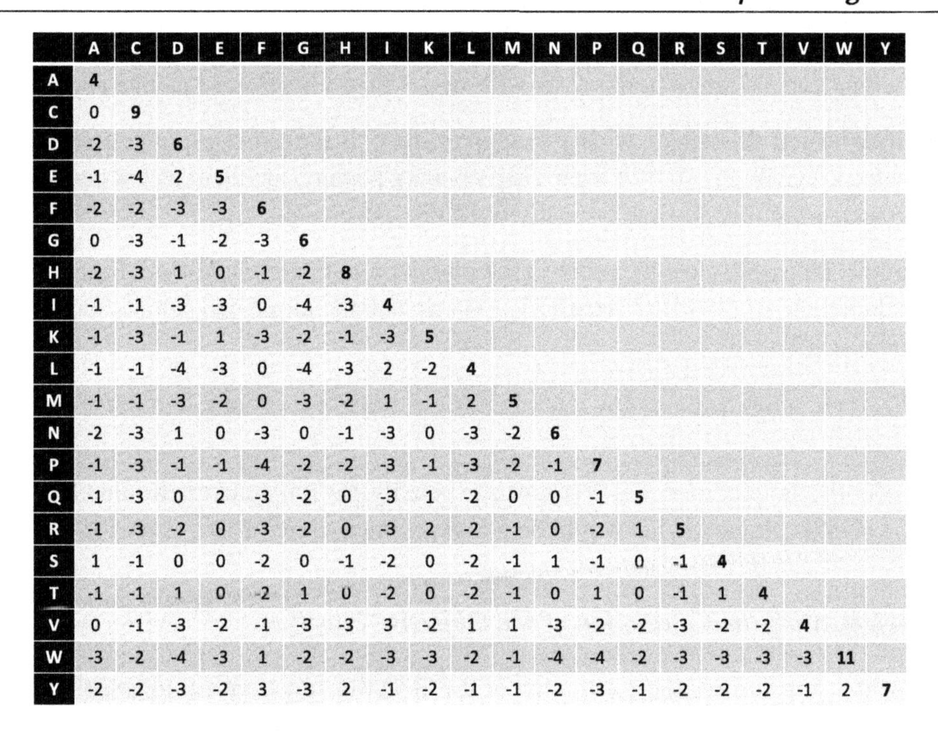

	A	C	D	E	F	G	H	I	K	L	M	N	P	Q	R	S	T	V	W	Y
A	4																			
C	0	9																		
D	-2	-3	6																	
E	-1	-4	2	5																
F	-2	-2	-3	-3	6															
G	0	-3	-1	-2	-3	6														
H	-2	-3	1	0	-1	-2	8													
I	-1	-1	-3	-3	0	-4	-3	4												
K	-1	-3	-1	1	-3	-2	-1	-3	5											
L	-1	-1	-4	-3	0	-4	-3	2	-2	4										
M	-1	-1	-3	-2	0	-3	-2	1	-1	2	5									
N	-2	-3	1	0	-3	0	-1	-3	0	-3	-2	6								
P	-1	-3	-1	-1	-4	-2	-2	-3	-1	-3	-2	-1	7							
Q	-1	-3	0	2	-3	-2	0	-3	1	-2	0	0	-1	5						
R	-1	-3	-2	0	-3	-2	0	-3	2	-2	-1	0	-2	1	5					
S	1	-1	0	0	-2	0	-1	-2	0	-2	-1	1	-1	0	-1	4				
T	-1	-1	1	0	-2	1	0	-2	0	-2	-1	0	1	0	-1	1	4			
V	0	-1	-3	-2	-1	-3	-3	3	-2	1	1	-3	-2	-2	-3	-2	-2	4		
W	-3	-2	-4	-3	1	-2	-2	-3	-3	-2	-1	-4	-4	-2	-3	-3	-3	-3	11	
Y	-2	-2	-3	-2	3	-3	2	-1	-2	-1	-1	-2	-3	-1	-2	-2	-2	-1	2	7

Figure 6.3: *BLOSUM62* substitution matrix used for protein sequence alignment. Rows and columns represent aminoacids in a single letter alphabet.

the same context, if SL occurs 9 times, considering the relative frequency of L to be 15%, we would have: $score(S, L) = round(2 \times log_2 \frac{9/1000}{0.1 \times 0.15}) = -1$.

In practice, the most used matrices are the ones from the *BLOSUM (BLOcks of Amino Acid SUbstitution Matrix)* family, which use as its core the conserved local alignments from the *Blocks* database. There are several of these matrices that vary in the set of local alignments used, being those selected based on their level of similarity. For instance, the matrix *BLOSUM62*, shown in Fig. 6.3, is built using alignments with a similarity level of over 62%.

An alternative, also forming a quite popular family, are the *PAM (Percent Accepted Mutations)* matrices, which are derived from closely related families of proteins with a known evolutionary distance. Unlike the previous, numbers in the designations of PAM matrices stand for evolutionary distances, and, therefore, lower numbers should be used to compare closer (more similar) sequences.

The other important component of the objective function is the model used to compute the *gap penalties*. In this regard, the simplest alternative is to use a single parameter (g) that defines a constant penalty for each column where a gap occurs. Typical values for g in protein sequence alignment can range from -7 to -12, or -2 to -3 in DNA sequence alignment.

A more sophisticated alternative, widely used in practice, is the *affine gap penalty* model, which penalizes heavily the start of a gap (gap opening penalty g), but does not penalize so much a gap extension (penalty can typically be of $r = -1$ or -2).

As an example, let us calculate the score of the global protein alignment from Fig. 6.1 (top left), considering the *BLOSUM62* substitution matrix and the affine gap penalty model with $g = -8$ and $r = -2$:

$$score = -1 + 6 + 7 + 4 - 8 + 6 - 8 - 2 + 4 - 8 + 4 + 11 + 1 + 5 + 4 + 0 = 25 \quad (6.3)$$

If we consider the simpler model of constant gap penalty, the only change would be on the score of the eighth column, which would be -8, reaching a score of 19.

6.3.3 Implementing the Calculation of the Objective Function

To implement, using Python, the calculation of the objective function, i.e. to be able to assign scores to alignments, we will need to take into account both components mentioned above, the substitution matrix and the gap penalties.

Starting with the first, we will implement substitution matrices by assigning, to each possible pair of symbols from the desired alphabet, a numerical value. In this way, we allow the most flexibility in the definition of the scores using a substitution matrix, or considering simpler methods that use scores for matches and mismatches. Note that, in this last case, all pairs with the same symbol will take the match score (diagonal of the matrix), and all the remaining will take the mismatch score.

The data structure chosen to keep all this information was a Python dictionary, where we will define keys as strings with two characters (the symbols in the alignment from each sequence), and values as the scores assigned to each pair of symbols. These matrices may be defined filling the dictionary by hand with the desired scores, but will normally be filled using functions that implement different strategies. The following code shows a function to create the substitution matrix based on the values of a match and a mismatch score, also receiving as input the desired alphabet.

```
def create_submat (match, mismatch, alphabet):
    sm = {}
    for c1 in alphabet:
        for c2 in alphabet:
```

```
            if (c1 == c2):
                sm[c1+c2] = match
            else:
                sm[c1+c2] = mismatch
    return sm

def test_DNA():
    sm = create_submat(1,0,"ACGT")
    print(sm)

test_DNA()
```

For more complex substitution matrices, as it is the case with those used in protein alignments, as the *BLOSUM62* shown above, the most used approach to load these matrices is to read them from files. The next code block shows a function to read a substitution matrix from file, where the scores arc separated by tabs and the first row is used to define the alphabet used.

```
def read_submat_file (filename):
    sm = {}
    f = open(filename, "r")
    line = f.readline()
    tokens = line.split("\t")
    ns = len(tokens)
    alphabet = []
    for i in range(0, ns):
        alphabet.append(tokens[i][0])
    for i in range(0,ns):
        line = f.readline();
        tokens = line.split("\t");
        for j in range(0, len(tokens)):
            k = alphabet[i]+alphabet[j]
            sm[k] = int(tokens[j])
    return sm

def test_prot():
    sm = read_submat_file("blosum62.mat")
    print(sm)

test_prot()
```

Having defined the substitution matrices, we can proceed to the calculation of alignment scores, which will also include the definition of different gap penalty models. Let us start by assuming the simpler model of penalizing each gap equally by a constant value. Note that, in this case, the score of the alignment may be calculated summing the scores of individual columns, which are all independent.

The following functions show how the score of an alignment, represented by a list with two sequences, possibly containing gaps, can be calculated. The first calculates the score of a column of the alignment, while the latter sums these values to reach the final score.

```
def score_pos (c1, c2, sm, g):
    if c1 == "-" or c2=="-":
        return g
    else:
        return sm[c1+c2]

def score_align (seq1, seq2, sm, g):
    res = 0;
    for i in range(len(seq1)):
        res += score_pos (seq1[i], seq2[i], sm, g)
    return res
```

Using these functions, we can calculate the score of the global alignments from Fig. 6.1 (top), as it is shown in the next code chunk. In the example, the DNA alignment is evaluated considering a match score of 2, mismatch score of -2 and gap penalty (constant) of -3. Regarding the protein alignment, the *BLOSUM62* substitution matrix is used and gaps are penalized with a constant of $g = -8$.

```
def test_DNA():
    sm = create_submat(2,-2,"ACGT")
    seq1 = "-CAGTGCATG-ACATA"
    seq2 = "TCAG-GC-TCTACAGA"
    g = -3
    print(score_align(seq1, seq2, sm, g))

def test_prot():
    sm = read_submat_file("blosum62.mat")
    seq1 = "LGPSSGCASRIWTKSA"
    seq2 = "TGPS-G--S-IWSKSG"
    g = -8
```

```
    print(score_align(seq1, seq2, sm, g))

test_DNA()
test_prot()
```

A more complex scenario emerges when the affine gap penalty model is used, since in this case the value of each column can be dependent of previous columns, in the case of multiple consecutive gaps. The following function handles this case, by using two flags which are set to `True` when inside gap sequences. The function is applied to the previous example of a protein sequence alignment.

```
def score_affinegap (seq1, seq2, sm, g, r):
    res = 0
    ingap1 = False
    ingap2 = False
    for i in range(len(seq1)):
        if seq1[i]=="-":
            if ingap1: res += r
            else:
                ingap1 = True
                res += g
        elif seq2[i]=="-":
            if ingap2: res += r
            else:
                ingap2 = True
                res += g
        else:
            if ingap1: ingap1 = False
            if ingap2: ingap2 = False
            res += sm[seq1[i]+seq2[i]]
    return res

def test_prot():
    sm = read_submat_file("blosum62.mat")
    seq1 = "LGPSSGCASRIWTKSA"
    seq2 = "TGPS-G--S-IWSKSG"
    g = -8
    r = -2
    print(score_affinegap(seq1, seq2, sm, g, r))
```

```
test_prot()
```

6.4 Dynamic Programming Algorithms for Global Alignment

6.4.1 The Needleman-Wunsch Algorithm

We have already realized that the pairwise sequence alignment problem is quite complex with a huge number of possible solutions, even for moderate sized sequences. Since we do not have the possibility of trying all possible solutions, we need to find more intelligent approaches for the problem.

The main idea driving these algorithms will be a divide-and-conquer approach, quite common to address complex problems by sub-dividing those into simpler problems and combining their solutions. This is made possible by the additive nature of the scoring functions we have presented in the previous section, which allows to decompose a solution (an alignment) on its different columns, being the overall score the sum of the scores of the columns.

Notice that the affine gap penalty assumption does not obey this principle, and, therefore, the algorithms we will study in these sections will consider the simpler gap penalty based on a constant per gap (column). It is still possible to define similar algorithms for the affine gap penalty model, but these are more complex and will not be covered in this text.

Based on these observations, in the early days of Bioinformatics, different researchers have proposed the use of *dynamic programming* (DP) algorithms to address biological sequence alignment. DP is a general-purpose class of optimization algorithms based on a divide-and-conquer approach, where optimal solutions for sub-problems and their scores are re-used (and not recomputed) when solving larger problems. This idea has been applied to sequence alignment considering that the alignment of two sequences can be composed from alignments of sub-sequences of those sequences.

We will next detail these algorithms, starting with the *Needleman-Wunsch* (NW) algorithm for global sequence alignment, whose name is derived from the two authors that proposed it. Consider we want to align two sequences A and B of sizes n and m, respectively. The individual symbols in A and B can be accessed by subscripting, using indexes, and thus we can represent $A = a_1a_2 \dots a_n$ and $B = b_1b_2 \dots b_m$.

When executing the NW algorithm, we will build a matrix S, where the elements of sequence A will be placed to index rows and the elements of sequence B will be placed to index columns. We will add an extra column and an extra row in the beginning to represent alignments with gaps in all positions of one of the sequences.

Sequence B

			b_1	b_2	b_3	b_4	b_5
		gap	H	G	W	A	G
	gap	0	-8	-16	-24	-32	-40
a_1	P	-8	-2	-10	-18	-25	-33
Sequence A a_2	H	-16	0	-4	-12	-20	-27
a_3	S	-24	-8	0	-7	-11	-19
a_4	W	-32	-16	-8	11	3	-5
a_5	G	-40	-24	-10	3	11	9

Example:

$S_{1,1} = max\,(S_{0,0} + sm(\text{"H"},\text{"P"}),\ S_{0,1} + g,\ S_{1,0} + g) = max(0\text{-}2, -8\text{-}8, -8\text{-}8) = -2$

Figure 6.4: Example of an S matrix for the *Needleman-Wunsch* algorithm.

One of the purposes of the algorithm is to fill this matrix S considering that, in each position, the element will indicate the score of the optimal alignment of the sub-sequences of A and B, considering all symbols from the beginning until the character represented in the respective row from A and column from B. Thus, the element $S_{i,j}$ of the matrix will indicate the optimal score for the alignment of the sub-sequence of A with the first i characters ($a_1 ... a_i$) and of the sub-sequence of B with the first j characters ($b_1 ... b_j$). Notice that we consider the matrix S to be indexed in rows and columns starting from 0 in both cases, being the row and column with index 0 the ones for gaps.

The structure of the S matrix can be visualized in Fig. 6.4 for a given example, considering the alignment of two sequences of aminoacids: $A = PWSHG$ and $B = HGWAG$. Notice that the matrix S has dimensions of $n + 1$ rows and $m + 1$ columns.

The main idea of this algorithm is that this matrix S can be filled cell by cell, using adjacent cells to reach the value of the target cell. This will mean using the previously calculated scores of smaller alignments to get the optimal score for the current alignment through composition. To fill S, we can use the following recurrence relation:

$$S_{i,j} = max(S_{i-1,j-1} + sm(a_i, b_j), S_{i-1,j} + g, S_{i,j-1} + g), \forall 0 < i \leq n, 0 < j \leq m \quad (6.4)$$

where $sm(c_1, c_2)$ gives the value of the substitution matrix for symbols c_1 and c_2, while g provides the penalty value for a gap, considering a constant penalty per gap position (with $g < 0$).

Notice that this recurrence expression defines possible orders for the calculation of all values in S, since to calculate $S_{i,j}$, the values of $S_{i-1,j-1}$, $S_{i-1,j}$ and $S_{i,j-1}$ need to be previously known. In practice, we fill the matrix by rows or by columns going from left to right and from top to bottom, i.e. increasing the indexes.

This recurrence relation states that an alignment may always be obtained by the composition of other alignments. In particular, it builds an alignment by adding an extra column to a previous alignment and computing the new score by summing the previous score with the one from the added column.

There are three ways to add a column to a previous alignment: to add a column with the next symbols from each sequence, to add the next symbol from the first sequence and a gap, or to add the next symbol from the second sequence and a gap. These are the three hypotheses given in the recurrence from Eq. (6.4). In the first hypothesis, the score of the new column is calculated by checking the respective entry in the substitution matrix, while in the other two the score is given by g. The new scores are added to the previous ones and the maximum score is taken for the new alignment and used to fill S in that position.

To fill S using this recurrence relation, we need to define how to initialize the matrix, filling the first row and the first column. In both cases, the idea will be to multiply the number of columns in the alignment by g, since these alignments are composed of only gaps in one of the sequences. Thus, $S_{i,0} = i * g, \forall 0 < i \leq n$ and $S_{0,j} = j * g, \forall 0 < j \leq m$.

Fig. 6.4 provides an example of the application of the algorithm to fill S. In this case, the substitution matrix used is the *BLOSUM62* provided in Fig. 6.3 and $g = -8$.

Notice that, in example from the figure, the gaps' row and column are shaded gray being calculated as stated above. For the remaining cells, the recurrence relation from Eq. (6.4) is applied. An example is provided in the figure for the cell $S_{1,1}$.

From the definition put forward for the matrix S, it is easy to conclude that the overall score of the best alignment, for the whole sequences, is given in the lower right corner of the matrix (cell highlighted with dots in the figure).

The process explained above allows to reach the score for the optimal alignment (indeed for all optimal alignments of sub-sequences from the original sequences starting at the first position), but it is still not sufficient to obtain the best alignment itself. To be able to reconstruct the best alignment, we need to "memorize" the decisions took every time the recurrence relation is applied. Indeed, we need to store the information of which was the hypothesis (from the three possible ones) that provided the highest score, and consequently which was the previous alignment.

This is normally achieved by encoding these choices using a 3 symbol alphabet and building a *trace-back matrix*, typically denoted as T, with the same dimensions as the S matrix. Fig. 6.5

	gap	H	G	W	A	G
gap	0	-8	-16	-24	-32	-40
P	-8	-2	-10	-18	-25	-33
H	-16	0	-4	-12	-20	-27
S	-24	-8	0	-7	-11	-19
W	-32	-16	-8	11	3	-5
G	-40	-24	-10	3	11	9

Figure 6.5: Example of the trace-back matrix for the *Needleman-Wunsch* algorithm in the example provided.

provides the T matrix for the previous example in a graphical way, by using arrows which indicate if the best previous alignment came from a diagonal in the matrix (first hypothesis of the recurrence with no gaps), or if it came from the previous row (vertical) or column (horizontal), cases where there was a gap in one of the sequences.

Having the trace-back information available allows to recover the best alignment by traversing the matrix. The process starts in the lower right corner of the matrix, follows the directions (arrows) from the T matrix, and terminates in the upper left corner (row and column with index zero). When moving in the matrix, the alignment is built in the reverse order (from the last column to the first). If the move is diagonal, the column consists of the characters in the row and column from the original cell (first hypothesis of the recurrence with no gaps). If the move is vertical, we are only moving in the first sequence, with a gap in the second, while in a horizontal move the reverse happens.

Fig. 6.6 provides an illustration of this process for the previous example. The path over the matrix is highlighted and the final alignment is provided. As a validation, the optimal score is recalculated and we can check it matches the one in the corner of the matrix.

6.4.2 Implementing the Needleman-Wunsch Algorithm

The algorithm explained in the previous section will be implemented in a set of Python functions provided below. The first function, given in the next block, provides the implementation of the core algorithm filling the matrices S and T, applying the recurrence defined in the previous section, as well as the initialization of the gaps' row and column. In this case, we used integer values in the trace-back matrix representation, where the value 1 is used for diagonal, 2 is used for vertical and 3 for horizontal. The auxiliary function **max3t** provides the integer to fill T using this encoding scheme.

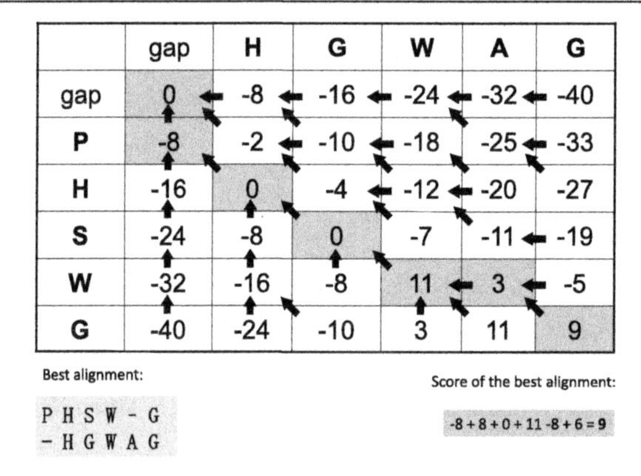

Figure 6.6: Example of the process of recovering the optimal alignment from the trace-back information in the *Needleman-Wunsch algorithm*, for the example provided above.

Note that, in this function, the matrices S and T are represented as lists of lists, and these lists are filled when the values are being calculated through the use of the **append** function. The function first initializes the gaps' row and column, and then fills the remaining of the matrices applying the recurrence relation. The result of the function is a tuple, providing in the first position the matrix S and in the second the matrix T.

```
def needleman_Wunsch (seq1, seq2, sm, g):
    S = [[0]]
    T = [[0]]
    ## initialize gaps' row
    for j in range(1, len(seq2)+1):
        S[0].append(g * j)
        T[0].append(3)
    ## initialize gaps' column
    for i in range(1, len(seq1)+1):
        S.append([g * i])
        T.append([2])
    ## apply the recurrence relation to fill the remaining of the
   matrix
    for i in range(0, len(seq1)):
        for j in range(len(seq2)):
            s1 = S[i][j] + score_pos (seq1[i], seq2[j], sm, g);
            s2 = S[i][j+1] + g
            s3 = S[i+1][j] + g
            S[i+1].append(max(s1, s2, s3))
```

```
            T[i+1].append(max3t(s1, s2, s3))
    return (S, T)

def max3t (v1, v2, v3):
    if v1 > v2:
        if v1 > v3: return 1
        else: return 3
    else:
        if v2 > v3: return 2
        else: return 3
```

The next function takes the trace-back (T) matrix built from the last function, together with the two sequences, and implements the process of recovering the optimal alignment. The algorithm starts in the bottom right corner of the matrix and uses the information in T to update the position and gather the alignment. A vertical (horizontal) cell in T leads to moving to the previous row (column), and adds to the alignment a column with a symbol from the sequence A (B) in the respective row (column) and a gap in the other. A diagonal cell in T leads to moving to the previous row and column, and adds to the alignment a column with a symbol from sequence A and another from sequence B, in the respective row and column. The algorithm stops when the top left corner of the matrix is reached. Notice that the alignment is represented by two strings of the same size, and is built in this function from the end to the beginning, i.e. new columns are added always in the beginning.

```
def recover_align (T, seq1, seq2):
    res = ["", ""]
    i = len(seq1)
    j = len(seq2)
    while i>0 or j>0:
        if T[i][j]==1:
            res[0] = seq1[i-1] + res[0]
            res[1] = seq2[j-1] + res[1]
            i -= 1
            j -= 1
        elif T[i][j] == 3:
            res[0] = "-" + res[0]
            res[1] = seq2[j-1] + res[1]
            j -= 1
        else:
            res[0] = seq1[i-1] + res[0]
```

```
            res[1] = "-" + res[1]
            i -= 1
    return res
```

The following code block defines a testing function for the previous code, using the example of a protein sequence alignment from the previous section, aligning sequences "PHSWG" and "HGWAG", using the *BLOSUM62* substitution matrix and $g = -8$.

```
def print_mat (mat):
    for i in range(0, len(mat)):
        print(mat[i])

def test_global_alig():
    sm = read_submat_file("blosum62.mat")
    seq1 = "PHSWG"
    seq2 = "HGWAG"

    res = needleman_Wunsch(seq1, seq2, sm, -8)
    S = res[0]
    T = res[1]
    print("Score of optimal alignment:", S[len(seq1)][len(seq2)])
    print_mat(S)
    print_mat(T)
    alig = recover_align(T, seq1, seq2)
    print(alig[0])
    print(alig[1])

test_global_alig()
```

6.5 Dynamic Programming Algorithms for Local Alignment

6.5.1 The Smith-Waterman Algorithm

In the last section, we covered a dynamic programming algorithm for the global alignment of biological sequences, working with an additive score that takes into account a substitution matrix and constant gap penalties. Here, we will discuss the changes that we need to consider in the algorithm, for the case of local alignments.

First, we need to address the relevant changes in the problem definition. When performing a local alignment, we will be interested in the best partial alignment of sub-sequences from the

	gap	H	G	W	A	G
gap	0	0	0	0	0	0
P	0	0	0	0	0	0
H	0	8	0	0	0	0
S	0	0	8	0	1	0
W	0	0	0	19	11	3
G	0	0	6	11	19	17

Examples:

$S_{1,1} = max\,(S_{0,0} + sm(\text{"H"},\text{"P"}),\ S_{0,1} + g,\ S_{1,0} + g,\ 0) = max(0\text{-}2, -8\text{-}8, -8\text{-}8, 0) = 0$

$S_{4,3} = max\,(S_{3,2} + sm(\text{"H"},\text{"P"}),\ S_{3,3} + g,\ S_{4,2} + g,\ 0) = max(8+11, 0\text{-}8, 0\text{-}8, 0) = 19$

Figure 6.7: Example of an S matrix for the *Smith-Waterman* algorithm.

two sequences, which maximize the objective function (score). Notice that the sub-sequences need to consider all characters in a sequence from a starting til an ending position. There are no changes in the way scores are calculated, but now alignments covering only parts of the sequences are acceptable.

This definition imposes a number of changes in the algorithm from the previous section, leading to the *Smith-Waterman* algorithm, that again takes its name from the original authors. The major change in the algorithm is the reformulation of the recurrence relation used in the DP. In this case, if the best alternative from the three considered before leads to a negative score, this means that in that position the best is to restart the alignment from this position onward, ignoring the previous parts of the sequences. This will imply to add 0 as an alternative in the recurrence relation that becomes:

$$S_{i,j} = max(S_{i-1,j-1} + sm(a_i, b_j), S_{i-1,j} + g, S_{i,j-1} + g, 0), \forall 0 < i \leq n, 0 < j \leq m \quad (6.5)$$

Also, the initialization of the matrix (first row and column) will be done filling with zeros: $S_{i,0} = 0, \forall 0 < i \leq n$ and $S_{0,j} = 0, \forall 0 < j \leq m$, as these will be cases where the best local alignment will ignore columns with gaps only that would reduce the score.

Fig. 6.7 shows the S matrix for an example, where the sequences are the same that were used in the previous section, but now the aim is to perform a local alignment. Two examples are shown for the application of the previous recurrence relation.

Note that in this case, the best alignment can occur in any cell of the matrix, corresponding to the highest score value, i.e. the maximum value in the S matrix. In the example, there are two alternative best alignments, both with a score of 19.

As before, to be able to recover the optimal alignment we need to keep trace-back information (the T matrix). In this case, this matrix can have four possible values, the three as before

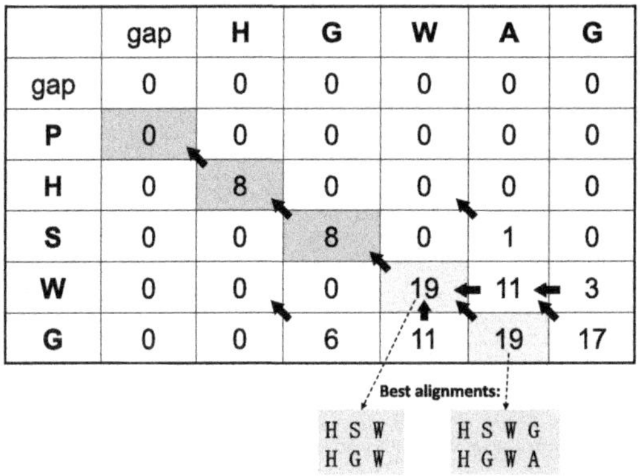

	gap	H	G	W	A	G
gap	0	0	0	0	0	0
P	0	0	0	0	0	0
H	0	8	0	0	0	0
S	0	0	8	0	1	0
W	0	0	0	19	11	3
G	0	0	6	11	19	17

Best alignments:

```
H S W
H G W
```

```
H S W G
H G W A
```

Figure 6.8: Example of the process of gathering the trace-back information and recovering the optimal alignment for the *Smith-Waterman* algorithm, for the example provided above.

(diagonal, horizontal, and vertical) and an extra hypothesis corresponding to the case where the alignment is terminated (where the S matrix is filled with 0). The T matrix is graphically shown for the example in Fig. 6.8, as before by overlapping arrows. The cells where the alignment is terminated have no overlapping arrows.

Also in the figure, we can check the process of recovering the alignment. In this case, this process starts in the cell with the highest score. Then, it proceeds as before following the arrows, until the upper left corner is reached (meaning the sequences terminate) or a cell with the indication that the alignment should terminate is reached (no arrow provided). In any of those cases the alignment is terminated. This process is shown by the highlighted cells in gray, for the two alternative optimal alignments.

6.5.2 Implementing the Smith-Waterman Algorithm

The *Smith-Waterman* algorithm was implemented using Python functions as before. The first code block shows the core function that builds the S and T matrices, also returning the maximum score. The implementation is similar to the one of global alignments, with the changes explained above. In this case, a tuple with the S and T matrices is also returned, but an extra element is added to the tuple with the score of the optimal alignment.

```
def smith_Waterman (seq1, seq2, sm, g):
    S = [[0]]
    T = [[0]]
    maxscore = 0
    for j in range(1, len(seq2)+1):
```

```
        S[0].append(0)
        T[0].append(0)
    for i in range(1, len(seq1)+1):
        S.append([0])
        T.append([0])
    for i in range(0, len(seq1)):
        for j in range(len(seq2)):
            s1 = S[i][j] + score_pos (seq1[i], seq2[j], sm, g);
            s2 = S[i][j+1] + g
            s3 = S[i+1][j] + g
            b = max(s1, s2, s3)
            if b <= 0:
                S[i+1].append(0)
                T[i+1].append(0)
            else:
                S[i+1].append(b)
                T[i+1].append(max3t(s1, s2, s3))
                if b > maxscore:
                    maxscore = b
    return (S, T, maxscore)
```

The following code shows the function used to perform the recovery of the optimal alignment, given the *S* and *T* matrices computed using the previous function. The algorithm implemented first finds the cell in the matrix with the highest score (using auxiliary function **max_mat**), and that is the starting point. The process of moving through the matrix and building the alignment are similar to the ones implemented in function **recover_align**. The main difference lies in the termination criterion, which is, in this case, to find a cell where the *T* matrix has a value of 0.

Notice that this function only handles one optimal alignment, and therefore when multiple alignments exist with the same score (as it is the case in the example from the previous section) only one is returned.

```
def recover_align_local (S, T, seq1, seq2):
    res = ["", ""]
    i, j = max_mat(S)
    while T[i][j]>0:
        if T[i][j]==1:
            res[0] = seq1[i-1] + res[0]
            res[1] = seq2[j-1] + res[1]
```

```
            i -= 1
            j -= 1
        elif T[i][j] == 3:
            res[0] = "-" + res[0];
            res[1] = seq2[j-1] + res[1]
            j -= 1
        elif T[i][j] == 2:
            res[0] = seq1[i-1] + res[0]
            res[1] = "-" + res[1]
            i -= 1
    return res

def max_mat(mat):
    maxval = mat[0][0]
    maxrow = 0
    maxcol = 0
    for i in range(0,len(mat)):
        for j in range(0, len(mat[i])):
            if mat[i][j] > maxval:
                maxval = mat[i][j]
                maxrow = i
                maxcol = j
    return (maxrow,maxcol)
```

An example of the use of these functions to reach the alignment of the protein sequences given in the example is provided next.

```
def test_local_alig():
    sm = read_submat_file("blosum62.mat")
    seq1 = "HGWAG"
    seq2 = "PHSWG"
    res = smith_Waterman(seq1, seq2, sm, -8)
    S = res[0]
    T = res[1]
    print("Score of optimal alignment:", res[2])
    print_mat(S)
    print_mat(T)
    alinL= recover_align_local(S, T, seq1, seq2)
    print(alinL[0])
```

```
    print(alinL[1])

test_local_alig()
```

6.6 Special Cases of Sequence Alignment

In the previous sections, we have proposed a general purpose objective function for sequence alignment problems. Depending on the application, the objective function may be defined to achieve different aims, using the previous algorithms.

One obvious application is the calculation of the identity between two sequences, as a value in the range [0, 1]. This can achieved by obtaining the identical characters in both sequences, doing a global alignment that scores matches as 1, and mismatches and gaps as 0. The score obtained may be divided by the sequence length to obtain a normalized identity value (if the sequences have a different size, the largest will be used).

The following function implements this process, receiving as input the two sequences and the set of allowed characters (the *alphabet* parameter, set by default to consider DNA sequences).

```
def identity(seq1, seq2, alphabet = "ACGT"):
    sm = create_submat(1,0,alphabet)
    S,_ = needleman_Wunsch(seq1, seq2, sm, 0)
    equal = S[len(seq1)][len(seq2)]
    return equal / max(len(seq1), len(seq2))
```

Another example is the calculation of the *edit distance* between two sequences, which is defined as the minimum number of operations required to transform one string into the other, considering as allowable operations the insertion, deletion, or substitution of a single character. The following function implements this definition, using the previously defined NW algorithm.

```
def edit_distance(seq1, seq2, alphabet = "ACTG"):
    sm = create_submat(0, -1, alphabet)
    S = needleman_Wunsch(seq1, seq2,sm,-1)[0]
    res = -1*S[len(seq1)][len(seq2)]
    return res
```

In this case, we consider matches to have a score of 0, while gaps and mismatches are scored with −1. Note that gaps correspond to insertions or deletions, while mismatches correspond to substitutions. Thus, the score will count the total number of these operations, if taken its absolute value. The overall score corresponds to the negative of the edit distance, and thus maximizing the score will correspond to minimizing the edit distance.

Another special case of alignment is the determination of the *longest common sub-sequence* to two sequences. Note that, in this case, the sub-sequences are not required to be of consecutive positions within the original sequences. The following function provides a solution to the problem, using global alignments provided by the NW algorithm.

```
def longest_common_subseq (seq1, seq2, alphabet = "ACGT"):
    sm = create_submat(1, 0, alphabet)
    _,T = needleman_Wunsch(seq1, seq2, sm, 0)
    alin = recover_align(T, seq1, seq2)

    sizeal = len(alin[0])
    lcs = ""
    for i in range(sizeal):
        if alin[0][i] == alin[1][i]:
            lcs += alin[0][i]
    return lcs
```

In this case, matches will score 1, while gaps and mismatches have a score of 0. Thus, the NW will return the algorithm that maximizes the identical characters in both sequences as in the first function of this section. In this case, we will get the alignment, being necessary to perform a post-processing to identify the columns with matches, gathering the character matching in those positions.

A variant of this problem, sometimes named as the *longest sub-string problem*, is similar to the previous, but considering that the sub-string cannot include insertions or deletions, i.e. should be exactly the same in both sequences.

Although the problem definition seems quite similar, the approach for this case will be different. Indeed, in this case, we can use the SW algorithm for local alignments, but need to make sure we do not have mismatches or gaps within the optimal alignment. One way to assure this condition is to impose a heavy penalty to both, that will make all solutions containing any of those cases score less than solutions containing only matches. This can be done by setting the gap and mismatch penalties to a negative value larger than the size of the sequences, in absolute terms.

The following function implements this solution.

```
def longest_common_string (seq1, seq2, alphabet = "ACGT"):
    m = max(len(seq1), len(seq2))
    pen = -1 * (m+1)
    sm = create_submat(1, pen, alphabet)
    S,T,_ = smith_Waterman(seq1, seq2, sm, pen)
    alinL= recover_align_local(S, T, seq1, seq2)
    return alinL[0]
```

6.7 Pairwise Sequence Alignment in BioPython

The BioPython framework includes a specific module, named **Bio.pairwise2** that provides an implementation of dynamic programming algorithms, similar to those put forward in this chapter. We will provide here a few examples of the use of this module, while the full documentation may be found in `http://biopython.org/DIST/docs/api/Bio.pairwise2-module.html`.

When performing alignments using this module, several functions are available, which start by either "global" or "local", depending on the type of alignment. The name of the function terminates with a code of two characters which indicates the parameters it takes: the first indicates the parameters for matches/mismatches or a substitution matrix, and the second indicates the parameters for gap penalties.

Regarding the first character, if it takes the value of "x" the alignment considers the match score to be 1 and the mismatch score to be 0. If "m" is provided, the function allows to define a match and a mismatch score using appropriate parameters. On the other hand, if "d" is selected, a dictionary may be passed to the function defining a full substitution matrix.

Regarding the second character, "x" is used when no gap penalties are imposed ($g = 0$), while "s" is used to define an affine gap penalty model allowing to define different penalties for gap opening and extension (or constant gap penalties by setting these values to be the same).

The following example shows how to perform an alignment of two DNA sequences, using a match score of 1, while the mismatch score and gap penalties are both 0. The code prints the number of alternative optimal alignments and the alignments themselves with the scores.

```
from Bio import pairwise2
from Bio.pairwise2 import format_alignment

alignments = pairwise2.align.globalxx("ATAGAGAATAG", "ATGGCAGATAGA")
print (len(alignments))
```

```
for a in alignments:
    print(format_alignment(*a))
```

The next example shows how to align two protein sequences, using the *BLOSUM62* substitution matrix, an opening gap penalty of −4 and an extension penalty of −1.

```
from Bio import pairwise2
from Bio.pairwise2 import format_alignment
from Bio.SubsMat import MatrixInfo

matrix = MatrixInfo.blosum62
for a in pairwise2.align.globalds("KEVLA", "EVSAW", matrix, -4, -1):
    print(format_alignment(*a))
```

Finally, the last example shows how to perform local alignments: in the first case of DNA sequences using a match score of 3, mismatch score of −2 and constant gap penalty g of −3. The second shows a local alignment of the same protein sequences from the last example, also using the same set of parameters.

```
from Bio import pairwise2
from Bio.pairwise2 import format_alignment
from Bio.SubsMat import MatrixInfo

matrix = MatrixInfo.blosum62

local_dna = pairwise2.align.localms("ATAGAGAATAG", "GGGAGAATC",
   3,-2,-3,-3)
for a in local_dna: print(format_alignment(*a))

local_prot = pairwise2.align.localds("KEVLA", "EVSAW", matrix, -4,
   -1)
for a in local_prot: print(format_alignment(*a))
```

Bibliographical Notes and Further Reading

BLOSUM matrices have been proposed by Henikoff and Henikoff in [74], while Margaret Dayhoff firstly proposed PAM matrices [43]. The algorithms of *Needleman-Wunsch* and *Smith-Waterman* take the names of theirs authors and have been proposed in [118] and [141], respectively. Edit distance was introduced by Levenshtein [97].

Exercises and Programming Projects

Exercises

1. a. Consider the application of the *Smith-Waterman* algorithm to the sequences: S1: ANDDR; S2: AARRD. The alignment parameters should be the *BLOSUM62* substitution matrix and the value of $g = -8$. Calculate (by hand); (i) the S matrix with the best scores; (ii) the trace-back matrix; (iii) the optimal alignment and its score. Check if there are any alternative optimal alignments.
   b. Write a program in Python, using the functions defined in this chapter, that allows to confirm the results you obtained in the previous exercise.
2. a. Consider the application of the *Needleman-Wunsch* algorithm to the following DNA sequences: S1: TACT; S2: ACTA. The used parameters are the following: gap penalty (g): −3, match (equal characters): 3, mismatch: −1. Calculate (by hand); (i) the S matrix with the best scores; (ii) the trace-back matrix; (iii) the optimal alignment and its score. Check if there are any alternative optimal alignments.
   b. Write a program in Python, using the functions defined in this chapter, that allows to confirm the results you obtained in the previous exercise.
3. Write and test a function that, given a binary matrix (with elements 0 or 1), coming from a function that creates dotplot matrices, identifies the largest diagonal containing ones (it can be the main diagonal or any other diagonal in the matrix). The result should be a tuple with: the size of the diagonal, the row where it begins, the column where it begins.
4. Consider the functions to calculate pairwise global alignments. Note that, in the case there are distinct alignments with the same optimal score, the functions only return one of them. Notice that these ties arise in the cases where, in the recurrence relation of the DP algorithm, there are at least two alternatives that return the same score.
   a. Define a function **needleman_Wunsch_with_ties**, which is able to return a *trace-back matrix* (T) with each cell being a list of optimal alternatives and not a single one.
   b. Define a function **recover_align_with_ties**, which taking a *trace-back matrix* created by the previous function, can return a list with the multiple optimal alignments.
5. Considering the functions to calculate pairwise local alignments, define similar functions to the previous exercise for the case of multiple optimal alignments. Note that, in this case, ties may also arise due to multiple equal scores in the S matrix (check the example from Figs. 6.7 and 6.8).
6. Write and test a function that, given two lists of sequences ($l1$ and $l2$), searches for each sequence in the $l1$ the most similar sequence in $l2$ (considering similarity based on identity, as defined above). The result will be a list with the size $l1$, indicating in each position i the index in $l2$ of the most similar sequence to the i-th sequence in $l1$.

7. Write and test a function that, given two DNA sequences *s1* and *s2*, searches for the best possible local alignment between a putative protein encoded by a reading frame from *s1* and a putative protein encoded by a reading frame from *s2* (check Section 4.4 for the details on reading frame calculations). The result will be a tuple with the best alignment and its score. The parameters of the alignment should be passed as arguments to the function.

Programming Projects

1. Using the object-oriented functionalities of Python, develop a set of classes to represent alignments and implement the algorithms described in this chapter. These could be integrated within classes to handle biological sequences, as proposed in previous chapters. As a suggestion, the following classes may be defined:
   - A class to keep substitution matrices, implementing their creation in different ways (e.g. with match/mismatch scores or loading from files) and access to scores for a pair of symbols;
   - A class to keep alignments, allowing to create them, to access their columns, and calculate scores given the alignment parameters;
   - A class to implement the alignment algorithms based on DP, which should keep alignment parameters as attributes.
2. Implement a class to represent dot plots, implementing methods to create, print and analyze these matrices.
   a. Start by considering binary matrices and the algorithms defined in this chapter.
   b. Consider more complex approaches, by implementing matrices with numerical values in the range [0, 1]. These may be calculated with filters or using substitution matrices of pairs of characters, or neighborhood windows.

CHAPTER 7

Searching Similar Sequences in Databases

In this chapter, we discuss the problem of finding similar sequences to a target sequence in databases of a large dimension. We discuss how the need to repeat pairwise comparisons a large number of times changes the requirements of the problem and demands more efficient solutions. We discuss existing heuristic algorithms and tools for this task, and proceed to implement a simplified version of the most popular one (BLAST). We finish the chapter by looking at *BioPython* and checking how we can run and analyze the results from BLAST using Python scripts.

7.1 Introduction

In the last chapter, we have discussed how to design and implement efficient algorithms for pairwise sequence alignment, which may be used to compare pairs of sequences calculating their similarity. We have discussed different parameters of scoring functions that can be used to measure the similarity between these sequences.

In many situations in biological research, we come across DNA or protein sequences (e.g. coming from DNA sequencing projects) for which we do not have any information. One common request for a bioinformatician is to provide some annotation to this sequence, i.e. to provide hypotheses for the functions of the sequence.

Trying to infer the function from similar sequences is a way to address this task, where we try to identify sequences that are similar to the target (query) sequence which are already annotated with a given function. In this task, we need to scan databases containing such sequences and compare our query sequence against all sequences in the database, seeking to find the ones that show higher degree of similarity.

Here, the use of the algorithms from the last chapter seems an obvious choice. Indeed, all we need to do is to run the pairwise sequence alignment algorithms to compare the query with each of the sequences in the database, and select the one(s) with the highest score.

Note that in this process we need to select a number of parameters: the type of alignment (local or global), the gap penalty model and parameters, and a proper substitution matrix (or match/mismatch scores). The selection of these parameters deeply influences the final results, i.e. the set of sequences showing similarity.

Bioinformatics Algorithms. DOI: 10.1016/B978-0-12-812520-5.00007-9

One example of this approach is provided by the following function, which uses the functions provided in the previous chapter. This function takes as input a query sequence, a list of sequences (e.g. those in our database), the substitution matrix and the gap penalty (assumed to be fixed), and returns the best local alignment of the query with a sequence in the database (*ls*).

```
def align_query (query, ls, sm, g):
    bestScore = -1
    bestSeq = None
    bestAl = None
    for seq in ls:
        al = smith_Waterman(query, seq, sm, g)
        if al[2] > bestScore:
            bestScore = al[2]
            bestSeq = seq
            bestAl = al
    bestAlin = recover_align_local(bestAl[0], bestAl[1], query,
   bestSeq)
    return bestAlin, bestScore
```

While we seem to have solved the problem of this chapter, there is an important hurdle to the approach described above. Indeed, while dynamic programming algorithms for sequence alignment may be considered efficient algorithms when run once (or a limited number of times), they become less appropriate when we need to run them a large number of times (possibly millions of times).

Looking at the complexity of dynamic programming algorithms, we see that they have a quadratic complexity, since we need to fill two matrices (with scores and trace-back) which have as their dimensions (rows and columns) the sizes of the sequences. When running this algorithm for each sequence in the database, we will multiply this by the size of the database. Knowing that the databases where we conduct these searches are typically of a large dimension, with several millions of sequences, this is a problem in most practical situations.

The most popular solution for this problem has been the development of some algorithms (and programs) of a heuristic nature that can be up to 100 times faster than the dynamic programming algorithms. Of course, this comes also with a disadvantage, since we loose the exact nature of the algorithm, i.e. these algorithms do not guarantee to find the best possible solution (alignment). This may bring problems in the application of these algorithms in some scenarios, especially in cases where the similarity of the sequences is low, i.e. the query does not have sequences in the database that are highly similar.

A few heuristic algorithms have been proposed by the Bioinformatics community over the last years, to address this task. The most popular ones are *FASTA*, that appeared in the early days of Bioinformatics, and *BLAST (Basic Local Alignment Search Tool)* that is now a widely used program available to be run by the community in a large number of servers, including the main research institutes, as the NCBI or the EBI.

In the next section, we will use *BLAST* as an example and explore some of its algorithmic features. The following section will contain a simplified Python version to exemplify the implementation of some of these features. The chapter closes with some examples of the use of the *BLAST* servers and programs using *BioPython* functions.

7.2 BLAST Algorithm and Programs

7.2.1 Overview of the BLAST Algorithm

The *BLAST* algorithm is currently the most used to perform searches over DNA or protein sequence databases. The aim is to search for good alignments between a query sequence and the sequences of a defined database. The basic idea of the algorithms is to use short "words" (i.e. sub-sequences) and search for matches of high similarity (with no gaps) of these words in the query and in the sequences from the database. These matches will form a basis that can be further extended, in both directions, to obtain high-quality alignments of larger dimension.

The main steps of the *BLAST* algorithm may be summarized in the following:

1. Remove regions of low complexity (e.g. sequence repeats) from the sequence that may compromise the quality of the alignment;
2. Obtain all possible "words" of size w (a parameter of the algorithm), i.e. sub-sequences of length w occurring in the query sequence;
3. For each word from the previous step, compile the list of all possible words of size w that can be defined in the allowable alphabet, whose alignment score (with no gaps) is higher than a threshold T (parameter of the algorithm);
4. Search in all sequences from the database, all occurrences of the words collected in the last step, which represent matches (hits) of size w between the query and one of the database sequences;
5. Extend all hits from the last step, in both directions, while the score follows a given criterion (typically, the criterion is dependent on the size of the extension);
6. Select the alignments in the previous step with highest scores, normalized for its size (these are named the high-scoring pairs-HSPs).

In the most recent version of *BLAST*, from 1997, the criterion to extend alignments is more demanding, requiring two near hits with scores above T separated by a distance smaller than a given parameter. This change leads to less extensions and a more efficient algorithm, while the obtained results are shown to be in a similar level. With this strategy, it is also possible to include gaps within the two hits, and as a result BLAST can return gapped alignments.

Notice that the parameters w and T are quite important in the set of returned results and algorithm's efficiency. Choosing a smaller value of w or T increases sensitivity, i.e. the number of interesting sequences found, but also increases the processing time.

7.2.2 BLAST Programs

The described algorithm gave rise to a number of programs specific for different sequence types, which can be locally installed or run over distinct servers available in the web. The most important of these servers is maintained by NCBI and can be accessed in the URL `https://www.ncbi.nlm.nih.gov/BLAST`.

The main programs are *BLASTN* and *BLASTP*, for nucleotide and protein searches. *BLASTN* can be used to search different nucleotide databases for similar sequences, being the default the *nr/nt*, a non-redundant collection of nucleotide sequences, while many other alternatives exist, including to filter by a given species genome, to use the *RefSeq* database, to consider only ribosomal RNA sequences (16S), etc.

BLASTN allows to optimize the search for highly similar sequences, allowing a faster search for longer sequences (using the *megablast* program). In *BLASTN*, there are a number of parameters that can be set including the word size w whose default value is 11, and the ones defining the scoring function, the match/mismatch scores (default 2 and -3) and the gap penalties for opening and extension (default values of -5 and -2).

On the other hand, *BLASTP* can be used to search for protein sequence databases, such as the *nr*, the non-redundant set of protein sequences, *RefSeq*, *UniProt* (curated sequences from the *SwissProt* database) or sequences from the PDB database. The set of adjustable parameters are similar to *BLASTN*, with different default values: w is set to 6, while the scoring function uses a substitution matrix (*BLOSUM62* by default), and gap penalties for opening and extension (default values of -11 and -1). For protein alignments, there are alternative programs which are not covered in this book, such as *PSI-BLAST*, *PHI-BLAST* and *DELTA-BLAST*.

Also, there are three other programs in the *BLAST* suite which may be used to search for sequences of a different type:

- *BLASTX* – takes a DNA sequence as query, but searches over protein sequences, and thus can be used to find potential protein products encoded by a nucleotide sequence; the DNA sequence is translated considering all 6 reading frames (relevant definitions and algorithms may be found in Chapter 4);

- *TBLASTN* – takes a protein sequence as query, but searches over DNA sequence databases, thus trying to identify database sequences encoding proteins similar to the one in the query; sequences from the databases are translated considering the 6 reading frames, prior to the alignments;
- *TBLASTX* – takes a DNA sequence as input, searching over DNA sequence databases, but in both cases the sequences are translated considering the 6 reading frames and the matches are searched over protein sequences (this leads to 36 comparisons); this method can be used to identify nucleotide sequences similar to the query based on their coding potential.

7.2.3 Significance of the Alignments

One important aspect to discuss relates with the statistical significance of the alignments obtained. Although a detailed discussion on this subject is outside the scope of this book, we will briefly explain the underlying problems and present the main metrics returned by these programs to address this issue.

The scoring functions used by *BLAST* and dynamic programming, based on substitution matrices and gap penalties, are relative to the size of the sequences. So, they are useful to compare different alignments for the same pair of sequences, but cannot be used to compare alignments of different sequences with distinct sizes. So, when comparing hits of a query with other sequences of distinct sizes, there is the need to compute normalized scores that can be used to compare alignments of different sizes. *BLAST* computes normalized scores using measures from information theory, and as such the normalized scores are provided in bits.

However, even these scores normalized for sequence size are not enough to compare any type of alignment, and thus can not provide an answer to the question: is the similarity found between the query and the sequence, for a given alignment, statistically significant? Or, in other words, how probable is that this similarity occurs by pure chance? These are important questions to be able to infer possible homology from sequence similarity.

The most popular metric to evaluate the significance of the alignment, provided as a result for each HSP in *BLAST*, is the E value. This value indicates the number of expected alignments with a score at least as high as the one provided by the current HSP. Although we will not discuss details, this value is calculated taking into account the score of the alignment, the size of the database, and length of the sequences included, as well as the parameters of the alignment.

The lower the value of E is, i.e. the closer it is to zero, the more significant the considered match is. It is difficult to define a threshold for what is considered significant, being used values from 10^{-5} up to 0.05 in different scenarios and by different authors.

When checking if an alignment indicates homology or not, it is also important not to rely only on the E value, but look also at other results. Since *BLAST* returns local alignments, it is of foremost importance to look at the coverage of the alignment in the query (and also in the sequence found), i.e. looking at which parts of the overall sequences are included in the alignment proposed. This will help in understanding if there might be a global homology or just similarity in parts of the sequences (e.g. protein domains or other local patterns).

7.3 Implementing Our Own BLAST

In this section, we will build our own version of a very simplified version of *BLAST*, which we will name *MyBlast*. Although much simpler than *BLAST* itself, it will allow to get an idea of some of the algorithmic constructions used to build these tools.

The first step in our *MyBlast* program will be to create a database. We will assume here the simplest hypothesis: the database will be a list of sequences (strings), and we will define below a function to load this database from a text file, where each sequence will be in a separate line.

```
def read_database (filename):
    f = open (filename)
    db = []
    for line in f:
        db.append(line.rstrip())
    f.close()
    return db
```

The next step will be to pre-process the query, creating a map that will identify all words of size w (a parameter) occurring in it, keeping for each word the positions where it occurs. This technique is commonly named *hashing* and if, in a first look, it seems a waste of time, it is very useful if one needs to repeatedly find occurrences of these words in the query. The data structure used to keep these results will be a Python dictionary, since it allows a very efficient access of the values associated to its keys.

Thus, the next function creates a dictionary (map) from the query sequence following this idea. The dictionary returned has the words in the query as keys, and lists of positions where these occur as values.

```
def build_map (query, w):
    res = {}
    for i in range(len(query)-w+1):
```

```
        subseq = query[i:i+w]
        if subseq in res:
            res[subseq].append(i)
        else:
            res[subseq] = [i]
    return res
```

In our simplified version of *MyBlast*, we will not use a substitution matrix, rather considering a match score of 1 and a mismatch score of 0, not considering gaps. Thus, we only consider perfect hits, i.e. our threshold will be a score equal to w.

The next code chunk shows a function that, given a sequence, finds all matches of words from this sequence with the query. The result will be a list of hits, where each hit is a tuple with: (index of the match in the query, index of the match in the sequence). Notice that we use the map created from the query using the previous function, instead of the query itself, increasing the efficiency of the search.

```
def get_hits (seq, m, w):
    res = []  # list of tuples
    for i in range(len(seq)-w+1):
        subseq = seq[i:i+w]
        if subseq in m:
            l = m[subseq]
            for ind in l:
                res.append( (ind,i) )
    return res
```

The next step will be to extend the hits that were found by the previous function. Again, here, we will greatly simplify the process by considering that the hit will be extended, in both directions, while the contribution to the increase in the score is larger or equal to half of the positions in the extension. The result is provided as a tuple with the following fields: starting index of the alignment on the query, the starting index of the alignment on the sequence, the size of the alignment, and the score (i.e. the number of matching characters).

```
def extends_hit (seq, hit, query, w):
    stq, sts = hit[0], hit[1]
    ## move forward
    matfw = 0
    k=0
    bestk = 0
```

```
    while 2*matfw >= k and stq+w+k < len(query) and sts+w+k < len(seq
   ):
        if query[stq+w+k] == seq[sts+w+k]:
            matfw+=1
            bestk = k+1
        k += 1
    size = w + bestk
    ## move backwards
    k = 0
    matbw = 0
    bestk = 0
    while 2*matbw >= k and stq > k and sts > k:
        if query[stq-k-1] == seq[sts-k-1]:
            matbw+=1
            bestk = k+1
        k+=1
    size += bestk

    return (stq-bestk, sts-bestk, size, w+matfw+matbw)
```

The next function will identify the best alignment between the query and a given sequence, using the previous ones. We will identify all hits of size w and extend all those hits. The one with the best overall score (highest number of matches) will be selected. The result is provided as a tuple with the format returned by the last function.

```
def hit_best_score(seq, query, m, w):
    hits = get_hits(seq, m, w)
    bestScore = -1.0
    best = ()
    for h in hits:
        ext = extends_hit(seq, h, query, w)
        score = ext[3]
        if score > bestScore or (score== bestScore and ext[2] < best
   [2]):
            bestScore = score
            best = ext
    return best
```

The final step is to apply the previous functions to compare a query with all the sequences in the database. In this case, we will find the best alignment of the query with each sequence in the database, and find the best overall alignment of the query with a given sequence. The result will be a tuple similar to the ones described above, adding in the last position the index of the sequence with the best alignment.

```
def best_alignment (db, query, w):
    m = build_map(query, w)
    bestScore = -1.0
    res = (0,0,0,0,0)
    for k in range(0,len(db)):
        bestSeq = hit_best_score(db[k], query, m, w)
        if bestSeq != ():
            score = bestSeq[3]
            if score > bestScore or (score== bestScore and bestSeq[2]
    < res[2]):
                bestScore = score
                res = bestSeq[0], bestSeq[1], bestSeq[2], bestSeq[3],
    k
    if bestScore < 0: return ()
    else: return res
```

7.4 Using BLAST Through BioPython

The *BioPython* package that we have covered in a few of the previous chapters has interfaces that allow to run the *BLAST* programs, both when these are locally installed and through remote access to the servers. The functions provided by *BioPython* allow to prepare queries for *BLAST*, defining the query sequences and relevant parameters, execute queries, recover the results, and handle their processing. These functionalities are quite useful to automate these procedures, allowing their setup and execution in large scale.

In the following, we will address functions that allow the remote call to *BLAST* queries through the NCBI servers. We will not cover here the use of locally installed *BLAST* programs, although many of the steps are similar.

The most relevant *BioPython* module for remote calls is *Bio.Blast.NCBIWWW*, being the core function **qblast**. This function receives as parameters: the program to use (a string from the set: "blastn", "blastp", "blastx", "tblastn" or "tblastx"), the database to search (a string, for instance "nr", "nt" or "swissprot") and the query sequence (a string, the sequence in *FASTA* format or an identifier from NCBI such as GI). There are also several optional parameters that allow to define the type of output (XML by default), the threshold for the E value, the substitution matrix to use, the gap penalties, among others.

The next example shows how to run a *BLASTN* search in the Python shell, over the non-redundant nucleotide database, considering the sequence in the "example_blast.fasta" file (*FASTA* format).

```
>>> from Bio.Blast import NCBIWWW
>>> from Bio import SeqIO
>>> record = SeqIO.read(open("example_blast.fasta"), format="fasta")
>>> result_handle = NCBIWWW.qblast("blastn", "nt", record.format("
    fasta"))
```

Notice that the third line reads the sequence from the *FASTA* file (check Section 4.8 for the details on the *SeqIO* module), while the last line executes the *BLAST* search with the given parameters (this can take a while since it is made over an Internet connection). The resulting variable is a handle for the results that may be used to save the results to file, as shown in the next code block, or serve as an input to analysis functions that will be shown afterwards. The next code chunk saves the *BLAST* results into an XML file.

```
>>> save_file = open("my_blast.xml", "w")
>>> save_file.write(result_handle.read())
>>> save_file.close()
>>> result_handle.close()
```

BioPython provides parsers for XML result files from *BLAST* that can be used to load files saved as shown in the previous examples, or other files coming from running *BLAST* queries directly in the web servers.

If you have a result saved in an XML file "my_blast.xml", the following code reads the file and returns a handle to be used in the analysis.

```
>>> result_handle = open("my_blast.xml")
```

The handle returned in this example can be used as an input to the **read** function, when the file only has the results of a single *BLAST* search, or to the **parse** function, when there are multiple results. The result is an object of the class **BlastRecord** in the first case, or an iterator over objects of this class, in the second, that can be accessed using a cycle `for` or using the function **next**.

```
>>> from Bio.Blast import NCBIXML
>>> blast_record = NCBIXML.read(result_handle)

>>> blast_records = NCBIXML.parse(result_handle)
>>> for blast_record in blast_records:
```

An object of the class **BlastRecord** contains all information available from the result of a BLAST search, including also the parameters used in the process. The results are organized in a hierarchical way, with three different levels of information:

- At the level of the **BlastRecord**, there is the list of *alignments* with respective information, but also some of the general parameters used in the search: *matrix*, that keeps the substitution matrix; *gap_penalties*, which keeps the penalties used for gaps; *database*, keeping the target database.
- At the level of each *alignment*, we can find the set of HSPs (notice that a single sequence can align with the query in different locations, giving rise to different HSPs) in the field *hsps*, but also some information related to the full alignment of the query with a given sequence, such as: *title, accession, hit_id, hit_def*, providing the description, accession number, the identifier and the definition of the sequence, or *alignment.length*, that provides the full length of the alignment.
- At the level of the HSP, we can find information specific for the local alignment, such as *expect, score, query_start, sbjct_start, align_length, sbjct. match*, which contain the *E* value, the normalized score, the index in the query where the alignment starts, the index in the sequence where the alignment starts, the HSP length, the part of the query sequence part in the HSP alignment, the part of the sequence in the alignment, and the matches of both.

A more complete reference of the contents at each level may be found in Section 7.4 of the *BioPython*'s tutorial [8]. The next example shows how to navigate this information, printing some of the fields in the different levels of information.

```
>>> E_threshold = 0.001
>>> for blast_record in blast_records:
        for alignment in blast_record.alignments:
...         for hsp in alignment.hsps:
...             if hsp.expect < E_threshold :
...                 print ("****Alignment****")
...                 print ("sequence:", alignment.title)
...                 print ("length:", alignment.length)
...                 print ("e value:", hsp.expect)
...                 print (hsp.query[0:75] + "..." )
...                 print( hsp.match[0:75] + "...")
...                 print( hsp.sbjct[0:75] + "...")
```

To illustrate further the use of these functions, let us consider a simple example and write a script that can show the different functionalities of these functions. We will use a *FASTA* file ("interl10.fasta") containing a human protein, the interleukin-10 precursor, available at the URL: https://www.ncbi.nlm.nih.gov/protein/10835141.

The first part of our script, shown below, will load the *FASTA* file, check the number of aminoacids in the sequence (should be 178), run the *BLASTP* query using the non-redundant protein sequence database and default parameters. Then, we will save the *BLASTP* results to an XML file, so we can recover them later to parse the results.

Note that this step is not strictly needed since we can use the handle (variable *result_handle*) directly later. However, since we are running remote *BLAST*, which can take a while, in this way we only need to run this once, and the part of interpreting the results can be run without further access to the Internet and avoiding additional delay.

```
from Bio.Blast import NCBIXML
from Bio.Blast import NCBIWWW
from Bio import SeqIO

record = SeqIO.read(open("interl10.fasta"), format="fasta")
print (len(record.seq))
result_handle = NCBIWWW.qblast("blastp", "nr", record.format("fasta")
   )
save_file = open("interl-blast.xml", "w")
save_file.write(result_handle.read())
save_file.close()
result_handle.close()
```

Next, we will show how to open the XML file, recover the results from the *BLAST* search, and print some of the parameters used in the search.

```
result_handle = open("interl-blast.xml")
blast_record = NCBIXML.read(result_handle)

print ("PARAMETERS:")
print ("Database: " + blast_record.database)
print ("Matrix: " + blast_record.matrix)
print ("Gap penalties: ", blast_record.gap_penalties)
```

The next code chunk shows how we can access the first alignment (the one with lowest E value), checking which sequence matches (showing accession number, identifier and definition), the alignment length and the number of HSPs (in this case a single one with the full alignment).

```
first_alignment = blast_record.alignments[0]
print ("FIRST ALIGNMENT:")
print ("Accession: " + first_alignment.accession)
print ("Hit id: " + first_alignment.hit_id)
print ("Definition: " + first_alignment.hit_def)
print ("Alignment length: ", first_alignment.length)
print ("Number of HSPs: ", len(first_alignment.hsps))
```

Since there is a single HSP, we can check the results in more detail using the following code:

```
hsp = first_alignment.hsps[0]
print ("E-value: ", hsp.expect)
print ("Score: ", hsp.score)
print ("Length: ", hsp.align_length)
print ("Identities: ", hsp.identities)
print ("Alignment of the HSP:")
print (hsp.query)
print (hsp.match)
print (hsp.sbjct)
```

We may also wish to gather some relevant information about the top 10 alignments, getting the sequences that matched and the E value. This is done in the following code:

```
print ("Top 10 alignments:")
for i in range(10):
    alignment = blast_record.alignments[i]
    print ("Accession: " + alignment.accession)
    print ("Definition: " + alignment.hit_def)
    for hsp in alignment.hsps:
        print ("E-value: ", hsp.expect)
    print()
```

As an example of a more specific task, we may be interested in checking which organisms are present in the top 20 alignments. This is done in the next code block, where regular expressions (check Section 5.5) are used to filter the *hit_def* field to obtain the relevant information.

```
import re
specs = []
for i in range(20):
    alignment = blast_record.alignments[i]
```

```
    definition = alignment.hit_def
    x = re.search("\[(.*?)\]", definition).group(1)
    specs.append(x)

print ("Organisms:")
for s in specs: print(s)
```

Bibliographical Notes and Further Reading

BLAST has been proposed in 1990 in a paper by Altshul et al. [11], while the improved version with gapped alignments was proposed in 1997 in [12].

FASTA provides an alternative suite to *BLAST*. It has been firstly presented in [123], and can be found in several servers including `http://www.ebi.ac.uk/Tools/sss/fasta`.

Exercises and Programming Projects

Exercises

1. Consider the function **get_hits** above. Create a variant that allows at most 1 character to be different between the sequence and the query words.
2. a. Write a function that given two sequences of the same length, determines if they have at most d mismatches (d is an argument of the function). The function returns `True` if the number of mismatches is less or equal to d, and `False` otherwise.
   b. Using the previous function find all approximate matches of a pattern p in a sequence. An approximate match of the pattern can have at most d characters that do not match (d is an argument of the function).
3. Search in the *UniProt* database the record for the human protein APAF (O14727). Save it in the *FASTA* format. Using *BioPython* perform the following operations:
   a. Load the file and check that the protein contains 1248 aminoacids.
   b. Using *BLASTP*, search for sequences with high similarity to this sequence, in the "swissprot" database.
   c. Check which the global parameters were used in the search: the database, the substitution matrix, and the gap penalties.
   d. List the best alignments returned, showing the accession numbers of the sequences, the E value of the alignments, and the alignment length.
   e. Repeat the search restricting the target sequences to the organism *S. cerevisiae* (suggestion: use the argument *entrez_query* in the **qblast** function).

f. Check the results from the last operation, listing the best alignments, and checking carefully in each the start position of the alignment in the query and in the sequence.
g. What do you conclude about the existence of homologous genes in the yeast for the human protein APAF ?

Programming Projects

1. Improve the implementation of *MyBlast* by changing it to a class, which allows further configuration and considering the following variants:
   - Allow the database to be provided by an FASTA file, keeping in a dictionary the sequences and their descriptions.
   - Consider extending only those hits that have other hits in a neighborhood of less than n positions (n should be a parameter of the function). In the extensions, you can use dynamic programming to achieve the best alignment of the characters between the hits.
   - Add the option to use substitution matrices in the calculation of the matches. In this case, you should pre-compute all possible words of size w that score over T (T is a parameter) against words in the query, after mapping it. Adjust the hit extension for this scenario, creating criteria for stopping extensions based on scores and extension length.
   - Adapt the functions to allow returning a ranking of the best alignments, and not only the one that scores highest.

CHAPTER 8

Multiple Sequence Alignment

In this chapter, we generalize the alignment problem to consider multiple sequences and discuss the impacts on the problem complexity and available algorithms. We review the main classes of optimization algorithms for this problem and discuss their main advantages and limitations. Then, we focus on progressive alignment and implement a simplified version of one of these algorithms in Python. We close the chapter reviewing how to handle alignments in *BioPython*.

8.1 Introduction: Problem Definition and Complexity

In Chapter 6, we have defined pairwise sequence alignment and discussed its need when checking for similarity between two biological sequences. We have looked at the problem complexity and have been able to find efficient algorithms to address its solution.

However, in many situations, we are faced with the need to align more than two sequences. This is of particular relevance, for instance, when determining the function of proteins, where the combination of different sequences with potential homology with our target, from distinct organisms, can bring additional confidence in generating hypotheses for functional annotation, and provides a more reliable way to identify conserved regions of those proteins that might have a relevant functional role. Protein alignments can also greatly help in determining their secondary or tertiary structures.

Also, in epidemiology, aligning DNA sequences of related organisms or strains is of foremost importance to understand their evolution and regions of relevance for virulence factors. It is also important to mention that there is a close relationship between multiple sequence alignment and phylogenetic analysis which will be covered in the next chapter.

The problem of *multiple sequence alignment* (MSA) generalizes the pairwise alignment, considering N ($N > 2$) sequences to be aligned. The aim of the alignment is the same as before: to introduce gaps in the sequences as a way to highlight the similar characters among them, placing similar characters in the same columns.

Note that MSAs can be global or local, as it happens with their pairwise counterparts. We will here deal predominantly with glocal MSA, as the concept of local MSA is strongly related to motif discovery, a topic that will be addressed in Chapters 10 and 11.

Bioinformatics Algorithms. DOI: 10.1016/B978-0-12-812520-5.00008-0

Although the definition of the problem is very similar, the complexity increases when the number of sequences grows. Indeed, the problem is considered NP-hard in terms of complexity. Thus, there are no efficient algorithms to solve the problem, when N becomes large.

One of the changes when handling MSA is the definition of the objective function in the formulation of the optimization problem. The objective function we used for pairwise sequence alignment was of an additive nature, summing the scores of the different columns to reach the final score. We will use the same strategy when scoring MSAs.

Also, the score of a column in pairwise alignment depends on the substitution matrix and the gap penalties defined. In MSA, these two components need also to be considered. Still, we also need to deal with the fact that, in MSA, each column contains more than two characters. We, thus, need to define how to combine the multiple characters of a column and calculate a score.

One common approach to address this is the *sum of pairs* (SP) method. In this case, for each column, we consider all possible pairs of the characters in this column and sum the respective scores. We may choose to adapt the gap penalties to avoid over-penalizing columns with a large number of gaps.

Let us provide an example. Consider the following alignment:

```
-RADNS
ARCD-A
AR-D-A
```

Using the *BLOSUM62* matrix (check Fig. 6.3) and a penalty of $g = -8$ for each gap, the score SP of this alignment is the following:

$$\begin{aligned} Score &= (sm(A, A) + g) + 3 \times sm(R, R) + (sm(A, C) - g) + 3 \times sm(D, D) \\ &\quad + (2 \times g) + (2 \times sm(S, A) + sm(A, A)) \\ &= (4 - 8) + 3 \times 5 + (0 - 8) + 3 \times 6 + (-8 \times 2) + (1 + 1 + 4) \\ &= 11 \end{aligned}$$

8.2 Classes of Optimization Algorithms for Multiple Sequence Alignment

8.2.1 Dynamic Programming

For pairwise alignment, we have already seen that there are efficient algorithms, based on Dynamic Programming (DP), to solve those optimization problems, assuring optimal solutions,

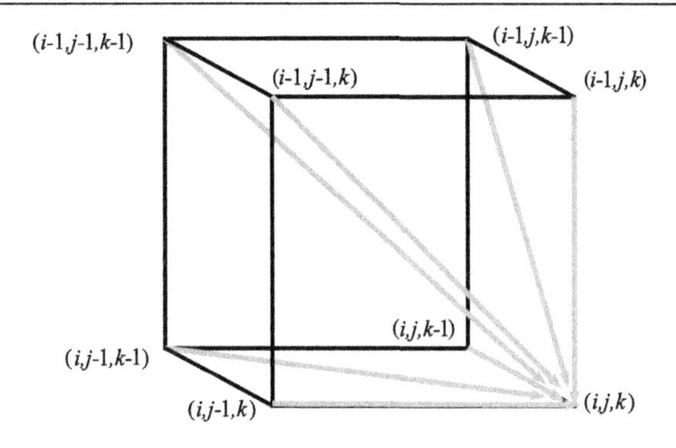

Figure 8.1: Example of the recurrence rule for a dynamic programming algorithm for multiple sequence alignment.

considering the defined objective functions. So, an immediate question would be: can we generalize these algorithms for N sequences?

The answer is "Yes, but ...". Indeed, we can generalize DP methods for MSA, but these are not efficient when the number of sequences to align grows. In terms of complexity, the pairwise alignment case has, as we saw in the previous chapter, a DP algorithm with a quadratic complexity, with the average size of the sequences to align as the basis. When generalizing for MSA, the complexity becomes exponential, with N being the exponent, and thus it will only be possible to use this method for a small number of sequences, below the typical number required in biological research.

Indeed, if, in the pairwise case, we need to fill 2-dimensional matrices (of scores and trace-back), in MSA we would to fill N-dimensional structures (hypercubes) to assure an optimal solution. We will show an example for the case where $N = 3$. In this case, the S and T matrices we need to fill in the DP algorithm (revisit Chapter 6 for the details) will be 3-dimensional. You can imagine filling a large cube with smaller cubes in it, while the arrows for the trace-back would connect vertexes along faces and edges or crossing through the smaller cubes (Fig. 8.1).

An example for the recurrence relation for the DP, in this case, will be given by the following expression:

$$
\begin{aligned}
S_{i,j,k} = max(&S_{i-1,j-1,k-1} + sm(a_i, b_j) + sm(a_i, c_k) + sm(b_j, c_k),\\
&S_{i-1,j-1,k} + sm(a_i, b_j) + g,\\
&S_{i-1,j,k-1} + sm(a_i, c_k) + g,\\
&S_{i,j-1,k-1} + sm(b_j, c_k) + g,\\
&S_{i-1,j,k} + 2g,
\end{aligned}
$$

$$S_{i,j-1,k} + 2g,$$
$$S_{i,j,k-1} + 2g), \forall 0 < i \leq n, 0 < j \leq m, 0 < k \leq p$$

In the previous expression, we follow the nomenclature for Chapter 6, namely in the definition of the recurrence relation for the *Needleman-Wunsch* algorithm, and will consider the objective function to be given by the sum of pairs method. The sequences to align will be $A = a_1a_2 \dots a_n$, $B = b_1b_2 \dots b_m$ and $C = c_1c_2 \dots c_p$, and the gaps will be penalized by g (constant penalty for each gap in the column).

There have been some attempts to reduce the computational cost of DP algorithms for MSA, by avoiding the need to try all possible paths in the N-dimensional hypercube. These approaches are based on the fact that the optimal MSA imposes a given pairwise alignment for each pair of sequences. This can be seen as a projected path over the 2-dimensional space of the pairwise alignment matrices.

For an optimal MSA, it is possible to calculate an upper bound for the cost of the projected path, for a given pair of sequences. This limits the set of possible paths to consider in the pairwise alignment, which in turn can be used to limit the set of hypotheses to explore when searching for the MSA. Although this has brought significant improvements, the number of sequences that could be aligned was still around 10, thus not serving most of the practical purposes.

8.2.2 Heuristic Algorithms

An alternative to DP algorithms, which may be used when we need to scale MSA for a larger number of sequences, which is the most common practical scenario, is to use heuristic (also called approximation) algorithms, which are not able to assure optimal solutions, but have an acceptable complexity in terms of their running times.

Broadly, heuristic algorithms for MSA can be classified as:

- **Progressive** – start by aligning two sequences and then iteratively add the remaining sequences to the alignment;
- **Iterative** – consider an initial alignment and then try to improve this alignment by moving, adding or removing gaps;
- **Hybrid** – can combine different strategies, and use complementary information (e.g. protein structural information, libraries of good local alignments).

The general idea of progressive algorithms is to create an initial alignment of a core of the two most related sequences, adding, in subsequent iterations, increasingly distant sequences, until the final alignment with all sequences is attained.

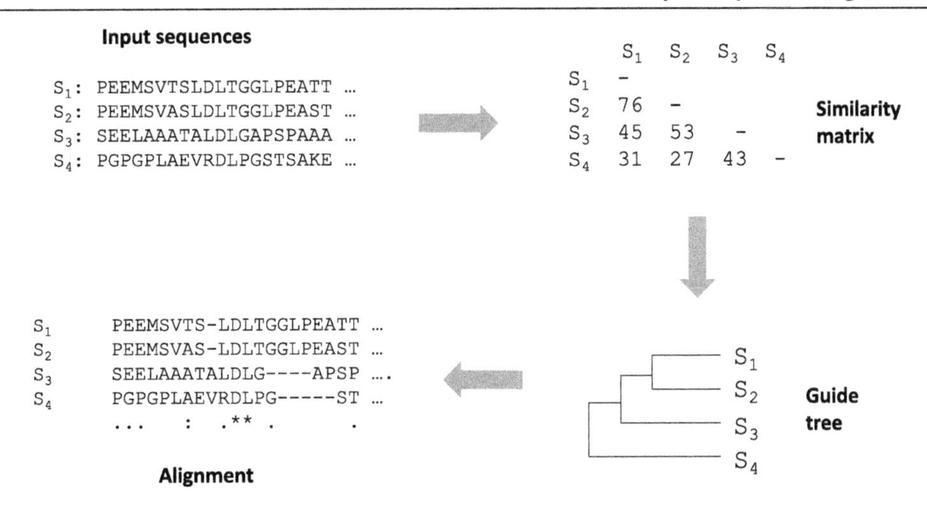

Figure 8.2: Overall workflow of the CLUSTAL, representing progressive MSA algorithms.

Although there are several variants of these algorithms, we will here detail the main steps of the *CLUSTAL* algorithm, a classical MSA method, that although being now discontinued, has given rise to a successful family of algorithms, namely *CLUSTALW* and, more recently, *Clustal Omega*, currently widely used. An overview of the main steps of *CLUSTAL* is provided in Fig. 8.2.

The first step of the algorithm is to calculate the pairwise alignments of all pairs of sequences in the set. From these alignments, a matrix is created containing the similarities between each pair of sequences, which will serve as input to an algorithm that creates a guide tree from this matrix. The methods to create this tree will be covered in more detail in the next chapter.

The selection of the two first sequences is of foremost importance in these methods, as it will define the core of the alignment. From the previous matrix, and guide tree, these sequences will typically be the ones with higher similarity, with their common ancestor nearer the leaves of the tree. The order of the remaining sequences to consider by the algorithm will be defined by the guide tree (in the figure, the first two will be S1 and S2, followed by S3 and S4).

One important step of these algorithms is the way further sequences are added to an existing alignment. The usual way to address this step is to create, from the existing alignment, a summary of the content of each column. In the simplest version, one could represent each column by the most common character (consensus), but this would be overly simplistic. Instead, it is common to represent the relative frequencies of the different characters (this is normally called a profile and will be covered in more detail in Chapter 11).

Thus, in each iteration, we need to align a sequence with a previous alignment, represented as a profile. This implies following an algorithm similar to the ones used in pairwise alignment,

```
S1      PEEMSVTS-LDLTGGLPEATT ...
S2      PEEMSVAS-LDLTGGLPEAST ...
S3      SEELAAATALDLG----APSP ....

S4:     PGPGPLAEVRDLPGSTSAKE ...
```

Score =
(SM(V,L)* 2 + SM(A,L)) / 3

Figure 8.3: Example of the calculation of the score of a column when combining an alignment with a sequence. The selected columns are highlighted. SM stands for the substitution matrix used.

but in this case adjusted to align a sequence to a profile, which implies changing the way the score for a column is computed.

An example of this process in *CLUSTAL* is shown in Fig. 8.3. In this case, to define the match score (diagonal move in the S/T matrices in the DP algorithm), all possible characters appearing in the profile, weighted by their frequency, are combined with the symbol in the sequence to add. Thus, the score will be a weighted average of the scores of the possible pairs.

In the step of adding new sequences to an alignment, the moto "once a gap, always a gap" prevails, since when a gap appears in an alignment it will be maintained throughout the next steps.

Notice that, in the more general case, we may need to join alignments from two profiles, corresponding to different branches of the guide tree, to build a new alignment. The process is easily thought as a generalization of the method described above.

Within the *CLUSTAL* family, the *CLUSTALW* algorithm was quite popular for many years. Regarding the original version, it added a number of improvements, namely the adaptation of the parameters, both the substitution matrix used and the gap penalties, according to both the stage of the alignment and the residues present in the sequences in given regions.

The specific rules for these adaptations were taken from observation of good and bad alignments, taking into account also protein structural information. As an example of such a rule, stretches of hydrophilic residues usually indicate loop or random coil structural regions, leading to a reduction of gap opening penalties in these regions.

On the other hand, regarding substitution matrices, these are varied along the alignment process, depending on the estimated divergence of the sequences being added to the alignment at each stage. Also, when calculating scores in the alignment of profiles, sequences are down-weighted if they are very similar to others in the set, and up-weighted if they are distant from the others.

Recently, the *CLUSTAL* family has switched to the use of the more recent *CLUSTAL Omega*. This algorithm greatly increased the efficiency of the algorithm used to create the guide tree that determines the order of the sequences in the alignments. Thus, it allows the algorithm to be usable with a larger number of sequences (in the site at the EBI, they currently allow 4000, but report testing with over 100,000).

Although very popular and quite used in practice, progressive alignments have some problems, mostly related to their heuristic greedy nature. Indeed, wrong decisions made in early steps of the alignment will not be corrected afterwards. The worst results of these algorithms occur when most of the sequences to align have low similarity.

An approach developed to counteract such problems has led to consistency based methods, from which the most well known is probably *T-coffee*. The idea is to maximize the agreement of pairwise alignments and, thus, try to avoid errors in the progressive algorithms. *T-coffee* builds libraries of global and local pairwise alignments, which are used to calculate weights for each position pair in each pairwise alignment. These weights are then used in the progressive alignment as scores in the DP algorithm when adding new sequences to the alignment. This method provides MSA solutions that are typically more accurate, but it suffers from some computational efficiency problems, not being suited for a large number of sequences, when compared for instance with *ClustalOmega*.

Iterative algorithms can be an alternative. They start by generating an alignment using a given method, which may be a progressive alignment or any other method. Then, they try to improve this alignment by making selected changes and evaluating the effect of those changes over the objective function.

One alternative to improve the alignments is to repeatedly realign sub-groups of sequences, or sub-groups of columns, by changing the position of gaps or adding/removing gaps. More evolved optimization meta-heuristics, such as genetic algorithms, have also been tried, with some success.

Hybrid algorithms include recent proposals such as *MUSCLE*, which combine progressive alignments based on profiles, with techniques from iterative methods, in this case working by refining the guide tree to improve the resulting alignment. It is one of the more efficient methods currently available.

On the other hand, the *MAFFT* algorithm brings the application of Fast Fourier Transforms (FFT) to MSA algorithms, being the sequences of aminoacid symbols converted into sequences of volumes and polarities. FFTs allow to rapidly identify homologous regions, allowing fast algorithms for MSA, which also combine progressive and iterative methods. Recent versions allow a lot of flexibility in the configuration with different trade-offs of quality and speed.

8.3 Implementing Progressive Alignments in Python

We will implement, in this section, a simplified version of a progressive algorithm. The implementation proposed for this algorithm will be object-oriented, thus defining the basis to implement further algorithms. We will thus define a few core classes that will be described firstly, before addressing the implementation of the algorithm itself.

The starting point of this implementation will be the class **MySeq** proposed in Section 4.6, which implements the biological sequences that will be aligned (DNA, RNA, or proteins). The remaining classes will be explained further in the next subsections.

8.3.1 Representing Alignments: Class MyAlign

We will start by developing a class to represent alignments. This class, although quite simple, will allow more modularity, defining a natural result for pairwise or multiple sequence alignment algorithms.

The class **MyAlign** will define alignments using two variables: the alignment type (DNA, RNA, protein) and the list of sequences included in the alignment that will be defined as strings, including the symbol "-" to represent gaps. The constructor and some core methods to access the information kept in the alignment are given below:

```
class MyAlign:

    def __init__(self, lseqs, al_type = "protein"):
        self.listseqs = lseqs
        self.al_type = al_type

    def __len__(self): # number of columns
        return len(self.listseqs[0])

    def __getitem__(self, n):
        if type(n) is tuple and len(n) ==2:
            i, j = n
            return self.listseqs[i][j]
        elif type(n) is int: return self.listseqs[n]
        return None

    def __str__(self):
        res = ""
```

```
        for seq in self.listseqs:
            res += "\n" + seq
        return res

    def num_seqs(self):
        return len(self.listseqs)

    def column (self, indice):
        res = []
        for k in range(len(self.listseqs)):
            res.append(self.listseqs[k][indice])
        return res

if __name__ == "__main__":
    alig = MyAlign(["ATGA-A","AA-AT-"], "dna")
    print(alig)
    print(len(alig))
    print(alig.column(2))
    print(alig[1,1])
    print(alig[0])
```

Apart from the constructor, the three other special methods define what is the length of an alignment (defined as the number of columns), how to access elements of the alignment (both using a single integer value, returning a sequence, and using two indexes for rows – sequences – and columns), and how to print an alignment as a string. The remaining two methods allow to easily access the number of sequences and a column in the alignment. The last part shows a very simple example of the use of these methods.

Another important method in this class, which will be used to implement our MSA algorithm, is the calculation of the consensus of the alignment. A consensus is defined as the sequence composed of the most frequent character in each column of the alignment, ignoring gaps. This definition is implemented by the following method, included in the **MyAlign** class:

```
class MyAlign:

(...)

    def consensus (self):
        cons = ""
        for i in range(len(self)):
```

```
            cont = {}
            for k in range(len(self.listseqs)):
                c = self.listseqs[k][i]
                if c in cont:
                    cont[c] = cont[c] + 1
                else:
                    cont[c] = 1
            maximum = 0
            cmax = None
            for ke in cont.keys():
                if ke != "-" and cont[ke] > maximum:
                    maximum = cont[ke]
                    cmax = ke
            cons = cons + cmax
        return cons

if __name__ == "__main__":
    alig = MyAlign(["ATGA-A","AA-AT-"], "dna")
    print(alig.consensus())
```

This method collects the frequencies of the characters in each column, using a dictionary, and selects the most common. Notice that, in this way, ties are broken arbitrarily, as dictionary keys have no inherent order.

8.3.2 *Pairwise Alignment: Class* AlignSeq

The algorithms for pairwise sequence alignment based on Dynamic Programming were previously implemented in Chapter 6, using Python functions. In this case, we will use an object-oriented implementation.

The first class defined for this purpose, **SubstMatrix**, implements substitution matrices in a very similar way to the ones we implemented before, being used the same dictionary based representation (see Section 6.3.3 for the details). The class has an attribute keeping the alphabet, and a second one keeping the dictionary with the scores for pairs of characters.

We will not provide here the full content of this class, given the similarity to the code presented before (the full code is available in the book's website). Below, we just provide the constructor, a few access methods, and the headers of the remaining methods, where in most cases the names are similar to the ones used in Chapter 6.

```
class SubstMatrix:

    def __init__(self):
        self.alphabet = ""
        self.sm = {}

    def __getitem__(self, ij):
        i, j = ij
        return self.score_pair(i, j)

    def score_pair(self, c1, c2):
        if c1 not in self.alphabet or c2 not in self.alphabet:
            return None
        return self.sm[c1+c2]

    def read_submat_file(self, filename, sep):
        (...)

    def create_submat(self, match, mismatch, alphabet):
        (...)
```

Note that, as before, we can create substitution matrices reading the full set of scores from file (method **read_submat_file**) or defining two scores for matches and mismatches (method **create_submat**).

The class that allows to execute the alignments is named **PairwiseAlignment**. This class will have a set of variables keeping the parameters of the alignment (substitution matrix and gap penalty), the sequences to align, and the S and T matrices from the DP algorithms. As before, we do not show here the full code (available in the site).

```
from MyAlign import MyAlign
from MySeq import MySeq
from SubstMatrix import SubstMatrix

class PairwiseAlignment:

    def __init__(self, sm, g):
        self.g = g
        self.sm = sm
        self.S = None
```

```
        self.T = None
        self.seq1 = None
        self.seq2 = None

    def score_pos (self, c1, c2):
        (...)

    def score_alin (self, alin):
        (...)

    def needleman_Wunsch (self, seq1, seq2):
        if (seq1.seq_type != seq2.seq_type): return None
        (...)
        return self.S[len(seq1)][len(seq2)]

    def recover_align (self):
        (...)
        return MyAlign(res, self.seq1.seq_type)

    def smith_Waterman (self, seq1, seq2):
        if (seq1.seq_type != seq2.seq_type): return None
        (...)
        return maxscore

    def recover_align_local (self):
        (...)
        return MyAlign(res, self.seq1.seq_type)
```

Notice that, in this implementation, the methods that run the algorithms return the optimal scores, and alter the matrices S and T defined as class variables. The methods to recover the alignments use the content of those matrices to return the optimal alignments, as an object of the class **MyAlign** described in the previous section.

8.3.3 Implementing Multiple Sequence Alignment: Class MultipleAlign

In this section, we provide the implementation of the simplified progressive MSA algorithm. This algorithm will receive as input a set of sequences, provided in a given order, as well as the parameters that define the objective function of the alignment (substitution matrix and gap penalties), returning the best possible alignment of the sequences.

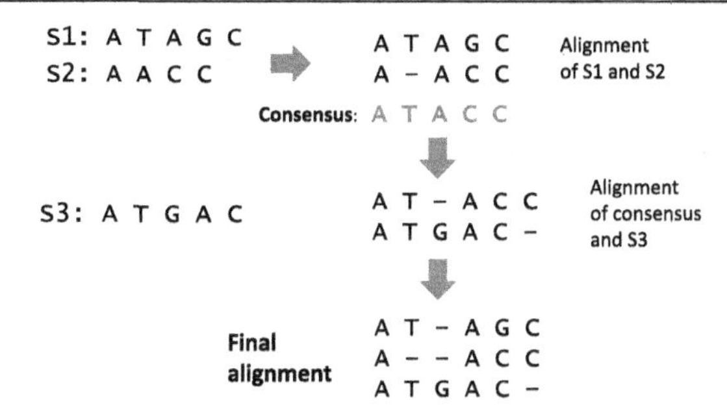

Figure 8.4: Example of the application of the MSA algorithm to be implemented in this chapter.

In our implementation, the *Needleman-Wunsch* algorithm, described in Chapter 6, will be used to provide for the pairwise alignments. The MSA will start by using this algorithm to align the first two sequences, according to the defined parameters. Afterwards, in each iteration, the following sequence will be added to the alignment. This process first takes the current alignment and calculates its consensus. Then, this consensus is aligned with the new sequence using the *Needleman-Wunsch* algorithm. The resulting alignment is reconstructed based on the columns of the consensus, filling the column with gaps if the alignment puts a gap in that position.

An example of an alignment with 3 sequences is shown in Fig. 8.4, where the alignment parameters used were: match score: 1, mismatch score: -1, gap penalty: -1. Note that in the calculation of the consensus, in the column where there is a tie, the consensus selected the first character in lexicographical order (in this case, C).

We will implement this algorithm resorting to the class **MultipleAlignment**, built over the basis created by the set of classes proposed in the previous sections. This class will have two variables: one containing the sequences to be aligned (objects of class **MySeq**), and the other the alignment parameters in the form of an object of the class **PairwiseAlignment**, which will be used to perform the pairwise alignments using the *Needleman-Wunsch* algorithm. The implementation of this class is provided below.

```
from PairwiseAlignment import PairwiseAlignment
from MyAlign import MyAlign
from MySeq import MySeq
from SubstMatrix import SubstMatrix

class MultipleAlignment():

    def __init__(self, seqs, alignseq):
```

```
        self.seqs = seqs
        self.alignpars = alignseq

    def add_seq_alignment (self, alignment, seq):
        res = []
        for i in range(len(alignment.listseqs)+1):
            res.append("")
        cons = MySeq(alignment.consensus(),alignment.al_type)
        self.alignpars.needleman_Wunsch(cons, seq)
        align2 = self.alignpars.recover_align()
        orig = 0
        for i in range(len(align2)):
            if align2[0,i]== '-':
                for k in range(len(alignment.listseqs)):
                    res[k] += "-"
            else:
                for k in range(len(alignment.listseqs)):
                    res[k] += alignment[k,orig]
                orig+=1
        res[len(alignment.listseqs)] = align2.listseqs[1]
        return MyAlign(res, alignment.al_type)

    def align_consensus(self):
        self.alignpars.needleman_Wunsch(self.seqs[0], self.seqs[1])
        res = self.alignpars.recover_align()

        for i in range(2, len(self.seqs)):
            res = self.add_seq_alignment(res, self.seqs[i])
        return res

def test():
    s1 = MySeq("ATAGC")
    s2 = MySeq("AACC")
    s3 = MySeq("ATGAC")
    sm = SubstMatrix()
    sm.create_submat(1,-1,"ACGT")
    aseq = PairwiseAlignment(sm,-1)
    ma = MultipleAlignment([s1,s2,s3], aseq)
    al = ma.align_consensus()
```

```
    print(al)

if __name__ == "__main__":
    test()
```

In the code, the function **align_consensus** provides the implementation of the MSA algorithm, returning an object of the class **MyAlign**. The function **add_seq_alignment** is used as an auxiliary function, that adds new sequences to existing alignments, being called in each iteration of the main algorithm to align the consensus of the previous alignment with the new sequence, returning the new alignment. The **test** function defines a simple example, with the scenario previously depicted in Fig. 8.4.

8.4 Handling Alignments in BioPython

The *BioPython* project, which we have been studying in different chapters along this book, provides features that allow to manipulate alignments, being able to read and write alignments in different formats. Note that at this stage, this project does not provide the implementation of algorithms for MSA, although it provides some wrappers for local installation of some tools (e.g. *CLUSTALW*), which we will not cover here.

The core class for handling alignments is **MultipleSeqAlignment**, where instances of this class keep pairwise or multiple sequence alignments (in this case, an alignment with 2 sequences is just a particular case). Objects of this class may be created directly by the programmer defining sets of **Seq** and **SeqRecord** objects (see Sections 4.7 and 4.8 for the details of these classes). An example is shown in the code block below.

```
from Bio import Alphabet
from Bio.SeqRecord import SeqRecord
from Bio.Align import MultipleSeqAlignment
from Bio.Alphabet import IUPAC
from Bio.Seq import Seq

seq1 = "MHQAIFIYQIGYPLKSGYIQSIRSPEYDNW"
seq2 = "MH--IFIYQIGYALKSGYIQSIRSPEY-NW"
seq3 = "MHQAIFI-QIGYALKSGY-QSIRSPEYDNW"

seqr1 = SeqRecord(Seq(seq1,Alphabet.Gapped(IUPAC.protein)),id="seq1")
seqr2 = SeqRecord(Seq(seq2,Alphabet.Gapped(IUPAC.protein)),id="seq2")
seqr3 = SeqRecord(Seq(seq3,Alphabet.Gapped(IUPAC.protein)),id="seq3")
```

```
alin = MultipleSeqAlignment([seqr1, seqr2, seqr3])
print(alin)
```

The alignments represented as objects of the class **MultipleSeqAlignment**, may be accessed through indexing in different ways. Using a single index, within squared brackets, we can index rows (sequences), while two indexes, separated by a colon, will index rows and columns, respectively. In both indexes, we can use slicing to get intervals of sequences and/or columns. The following examples show some of the possibilities.

```
print(alin[1]) # 2nd sequence
print(alin[:,2]) # 3rd column
print(alin[:,3:7])  # 4th to 7th columns (all sequences)
print(alin[0].seq[:3]) # first 3 columns of seq1
print(alin[1:3,5:12]) # sequences 2 and 3; 4th to 10th column
```

Also, and similarly to what happens to sequences, *BioPython* also provides a class to allow input/output operations of alignments in different formats, the class **AlignIO**. This includes the methods **read** and **parse**, to read a single or multiple alignments, respectively, and the method **write** that allows to write alignments to file. On the other hand, the method **convert** may be used to directly convert between different alignment formats.

The following code reads a file with an alignment in the "clustal" format. The remaining of the code gets the number of columns in the alignment and prints its rows and identifiers of the sequences.

```
from Bio import AlignIO
alin2 = AlignIO.read("PF05371_seed.aln", "clustal")

print("Size:", alin2.get_alignment_length())
for record in alin2:
    print (record.seq, record.id)
```

Finally, in the next code block, we show two alternative ways of writing the previous alignment to an FASTA file, by using the functions **write** and **convert**, respectively.

```
AlignIO.write(alin2, "example_alin.fasta", "fasta")

AlignIO.convert("PF05371_seed.aln", "clustal", "example_alin.fasta",
    "fasta")
```

Bibliographical Notes and Further Reading

Dynamic programming for multiple sequence alignment was proposed by Lipman et al. in a program named *MSA* [103], including some of the improvements that could increase efficiency, which were firstly proposed by Carrillo et al. in a previous paper [34].

Progressive alignments for MSA were firstly proposed by Feng and Doolitle in 1987 [63]. The *CLUSTAL* algorithm was firstly proposed in 1988 [77], and improved in 1994 with *CLUSTALW* [147]. The most recent member of the family, *Clustal Omega* was proposed in 2014 [139].

T-coffee was firstly published in 2000 [119]. On the other hand, the *MUSCLE* algorithm was proposed in [58], while *MAFFT* has been firstly described in [87].

Exercises and Programming Projects

Exercises

1. Consider the following four sequences of DNA:

   ```
   S1: ACATATCAT
   S2: AACAGATCT
   S3: AGATATTAG
   S4: GCATCGATT
   ```

   Write a Python script, using the code developed in this chapter, to generate a multiple alignment of these sequences using the progressive algorithm implemented in class **MultipleAlignment**. Consider the parameters to be: match score = 1, mismatch = −1, gap penalty $g = -1$.
2. Implement a method calculating the score sum of pairs (SP) of a given alignment. The method should be included in the **MultipleAlignment** class, taking as input an alignment (object of class **MyAlign**). Notice that the parameters for the score are given in an internal variable of the class (*alignpars*).
3. a. Consider the application of the *Needleman-Wunsch* algorithm to the protein sequences: S1: APSC; S2: TAPT, using the *BLOSUM62* matrix and $g = -4$. Calculate the optimal alignment.
   b. Based on the result of the previous exercise, could the following alignment be provided by the progressive algorithm implemented in this chapter?

   ```
   -AP-SC
   TAPT--
   TAT-S-
   ```

c. Calculate the SP score of the previous alignment.
d. Write a Python script, using the code developed in this chapter to confirm your results.

4. Write a method to add to the class **MyAlign** that, given an alignment (*self*), returns the list of columns (indexes) in the alignment that are rich in polar basic aminoacids (R, H, or K). To be considered rich, the column needs to include at least half of the aminoacids in this group.
5. Write a method to add to the class **MyAlign** that, given an alignment (*self*), returns a string with a symbol for each column of the alignment, following these rules: '*', if the column is fully conserved (it has all symbols equal and no gaps); ':', if the column has at least half of the symbols equal; '.': if the column does not match any of the previous, but has no gaps; ' ', in all other cases.
6. Write a method to add to the class **MultipleAlign** that, given an alignment (object of class **MyAlign**), identifies the columns where this alignment has high quality. In this case, a column is considered of high quality if the score of the alignment, calculated by the SP method is larger than 0. The method should return a list of indexes of the selected columns.
7. Consider the following sequence of aminoacids:

```
MEEPQSDPSVEPPLSQETFSDLWKLLPENNVLSPLPSQAMDDLMLSPDDIEQWFTE
DPGPDEAPRMPEAAPPVAPAPAAPTPAAPAPAPSWPLSSSVPSQKTYQGSYGFRLG
FLHSGTAKSVTCTYSPALNKMFCQLAKTCPVQLWVDSTPPPGTRVRAMAIYKQSQH
MTEVVRRCPHHERCSDSDGLAPPQHLIRVEGNLRVEYLDDRNTFRHSVVVPYEPPE
VGSDCTTIHYNYMCNSSCMGGMNRRPILTIITLEDSSGNLLGRNSFEVRVCACPGR
DRRTEEENLRKKGEPHHELPPGSTKRALPNNTSSSPQPKKKPLDGEYFTLQIRGRE
RFEMFRELNEALELKDAQAGKEPGGSRAHSSHLKSKKGQSTSRHKKLMFKTEGPDSD
```

a. Through the NCBI site, use the *BLASTP* application to search for similar sequences (alternatively use *BioPython* interface for this purpose) – see Chapter 7 for details.
b. From the result, select 10 to 12 matching sequences. Try to select different species and avoid sequences marked as "PREDICTED". Save those into a file in the FASTA format. Keep as sequence identifiers (after the ">" in the first line) the species name, without spaces.
c. In the EBI site, use the *Clustal Omega* application to get a multiple sequence alignment using the previous set of sequences. Save the alignment (in the "clustal" format) and the guide tree in two different files.
d. Load the previous alignment using *BioPython*.
e. Calculate the consensus of the alignment. Suggestion: check the class **AlignInfo** in the *BioPython* documentation or implement a function yourself.

f. Calculate a list of positions where the alignment is conserved (i.e. all sequences have the same aminoacid and there are no gaps). Calculate the percentage of these positions in the whole alignment.
g. From the previous information, get the region of the alignment more conserved, i.e. the longest sequence of consecutive conserved positions from the list above.

Programming Projects

1. Complete the definition of the classes provided in this chapter (**MySeq**, **MyAlign**, **SubstMatrix**, **PairwiseAlignment**, **MultipleAlignment**) with suitable methods coming from the concepts and algorithms from the previous chapters (including exercises), written in an object-oriented paradigm.
2. Add methods to the **MultipleAlignment** class that allow to consider the alignment of profiles with sequences. Redefine the MSA algorithm, allowing the user to input the guide tree of the alignment and use it to build the MSA.
3. Considering the class **MultipleAlign** developed above, implement an iterative algorithm. This should first create an alignment by randomly adding gaps to each sequence (the number of gaps per sequence may be defined by a user parameter). Implement methods to allow adding and removing gaps from the alignment. Using these methods, implement an iterative algorithm that tries changes over the alignment, keeping the ones improving the alignment. Try to implement more intelligent changes over the alignments.
4. Implement a Dynamic Programming algorithm for MSA considering 3 sequences, according to what was defined above. Adapt the algorithm for different scoring models, namely in the gap penalties.

CHAPTER 9

Phylogenetic Analysis

In this chapter, we review phylogenetic analysis problems and related algorithms, i.e. those addressing the construction of phylogenetic trees from sequences. We discuss the main classes of algorithms to address the problem, focusing on distance-based approaches, and providing a Python implementation for one of the simplest algorithms. We also check some of the Bio-Python's resources to handle phylogenetic trees.

9.1 Introduction: Problem Definition and Relevance

Phylogenetics is the branch of Biology which studies the evolutionary history and relationships among individuals or species. The inference of philogenies, represented as phylogenetic trees that explain the evolutionary relationship among these individuals or species, has traditionally been achieved based on heritable traits, mainly morphological. The availability of DNA sequences has allowed to reshape this field, making it much more rigorous, since it is based on the direct product of evolution, given that mutations occur directly over DNA.

Thus, in the context of Bioinformatics, phylogenetic analysis is concerned with the determination of how a given set of sequences has evolved from a common ancestor through a process of natural evolution. This evolutionary process is depicted in the form of an evolutionary tree, where bifurcations represent events of mutation from a common ancestor, that give rise to two (or more) different branches.

A phylogenetic tree will represent known sequences (DNA, RNA, protein) in its leaves (typically from different species or other taxonomic categories), while internal tree nodes will represent common ancestors of the sequences below (see Fig. 9.1). Note that we will here discuss mainly rooted trees, with a unique node, the root, that will correspond to the common ancestral of all sequences (taxa).

The structure of a rooted tree may be represented by the clusters (sets of sequences) covered by each internal node. For instance, the tree in Fig. 9.1 may be represented by the sets: $\{1, 2\}$, $\{3, 4\}$, $\{1, 2, 3, 4\}$, $\{1, 2, 3, 4, 5\}$, not being here represented the trivial sets with the individual sequences. The height of the nodes in the tree represent a measure of time, and traveling from the root to the leaves represents the passage of time.

There are also unrooted trees, which are mainly used to illustrate the relationships among the leaves without explicitly representing or inferring the common ancestral. These will not be

Bioinformatics Algorithms. DOI: 10.1016/B978-0-12-812520-5.00009-2

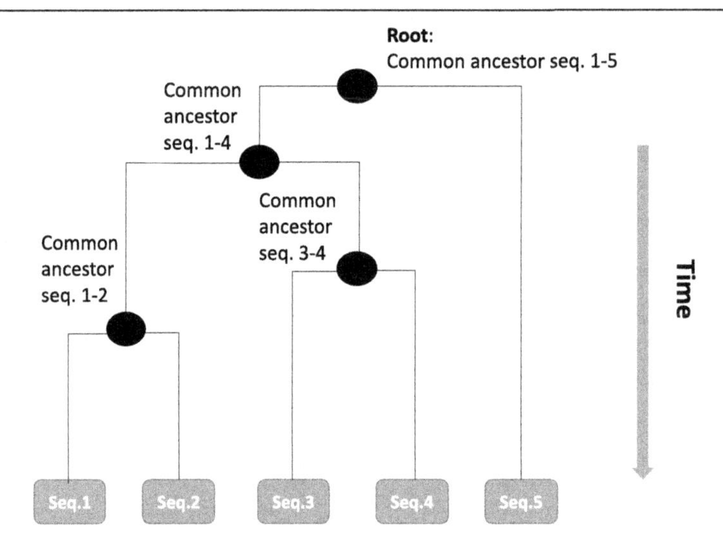

Figure 9.1: Example of a tree representing a philogeny with five sequences.

covered in this book, but can be easily obtained by "removing" the root node from a rooted tree and creating the resulting graph (graphs will be the subject of some of the next chapters, namely Chapter 13). In the case of the figure, it would be a graph with the 5 sequence nodes and 3 internal nodes.

There are different times of philogenies, depending on the entities represented in the leaves. Interspecific philogenies typically represent species or other taxonomic categories in the leaves, being the trees representative of ancestral species and branches represent events where mutations lead to the appearance of novel species. These phylogenies are typically inferred from DNA sequences of highly conserved sequences, mainly sequences encoding ribosomal RNA genes (e.g. 16S rRNA). Some interesting projects attempt to create global phylogenies of the whole set of living beings on Earth, such as *The tree of life* (`http://tolweb.org/tree/`).

Another type of relevant phylogenies include intraspecific phylogenies that can be used to highlight variants between different species, such as differences between strains or individuals.

Phylogenetic analysis has a number of quite relevant applications. From what has been said, one of these applications is in helping to classify new species and to provide rigorous support for the definition of taxonomic categories. It also has important applications on areas such as forensics, for instance in fatherhood determination, food contamination or identification of individuals committing crimes, epidemiology, on learning more about pathogens' outbreaks and acquired mutations, and conservation policies for endangered species.

On the level of functional annotation, phylogenies of protein sequences can be of importance, together with sequence alignment (as studied in the previous chapters), to better understand the functions of specific sequences, and relate conserved domains to their functional roles.

The computational problem of inferring a phylogeny can be defined as the one of outputting the best possible evolutionary tree, given as the input the set of sequences assumed to be evolutionary related and, thus, deriving from a common ancestor. This is clearly an optimization problem, since for a given set of sequences there are a number of possible alternative trees that could explain their evolution from a common ancestor.

The underlying optimization task is likely to be very complex since the number of possible trees grows very rapidly, even for a moderate number of sequences in the input. For example, for a scenario of 30 sequences to create a tree, we would have around 10^{40} different trees, a number that clearly shows the complexity of the problem.

Before proceeding, and to define a proper optimization problem, we need to define suitable objective functions that could be prone to mathematical optimization. This is a hard task that has been addressed in different ways by the numerous distinct algorithms and tools that have been proposed for this problem. We will cover those in the next section, together with the main classes of available algorithms.

9.2 Classes of Algorithms for Phylogenetic Analysis

The algorithms for phylogenetic analysis can be broadly classified into three main classes, which mainly differ on the strategy used to compute the objective function, while also proposing alternative approaches to search over the huge solution space for this problem. The main classes are the following:

- **Distance-based algorithms**: includes methods based on the previous computation of a matrix of pairwise distances among the sequences (based on the sequence alignment), seeking to find the trees where the distances are consistent with the ones in the input matrix;
- **Maximum parsimony**: includes methods that search for the trees where the number of needed mutations (in internal nodes of the tree) to explain the variability of the sequences is minimized;
- **Statistical/Bayesian methods**: defines probabilistic models for the occurrence of different types of mutations, and uses those models to score trees based on their probability (or likelihood), searching for the most likely trees that explain the sequences according to the assumed model.

In this book, we will give an emphasis to the methods on the first group, while trying to provide an overview over the remaining classes. This will be done in the next subsections.

9.2.1 Distance-Based Methods

Distance-based methods for phylogeny inference rely on objective functions that measure the consistency of the distances between the leaves (representing sequences) in the tree, with the ones obtained from sequence similarity (through sequence alignment). Thus, the methods will try to adjust the structure of the tree and the length of the edges connecting the nodes to mimic the pairwise distances between sequences.

The first step in distance-based methods is the definition of a matrix of distances between the sequences. In this case, distance will be the reverse of similarity, and the methods used to calculate these distances will consist on aligning the sequences and calculating the distances based on this alignment.

One simple way of achieving this task is to calculate the percentage of columns where the alignment contains mismatches or gaps. Of course, more sophisticated distance functions may be defined using the set of parameters for pairwise alignment discussed in Chapter 6.

Given the matrix of distances, the objective function may be defined as an error function, that tries to minimize the differences between the distances in the tree and the distances in the matrix. One common approach is to consider the sum of squared errors:

$$score(T) = \sum_{i,j \in S} (d_{ij}(T) - D_{ij})^2 \tag{9.1}$$

where S is the set of input sequences, T represents the tree (solution to be scored), $d_{ij}(T)$ represents the distance of the leaves representing sequences i and j in the tree T, and D_{ij} represents the distance between sequences i and j in the input matrix D given from sequence alignment.

Note that the distance between two nodes (u and v) in a rooted phylogenetic tree T is related to the vertical distances traveled from the origin u to the destination v. If w is the nearest common ancestor of u and v, the distance between u and v is given by the sum of $d_{uw} + d_{wv}$. In the tree from Fig. 9.1, for instance, to travel between the leaves representing sequences 1 and 4 would imply going from sequence 1 to the common ancestor of sequences 1 to 4, and traveling back down to sequence 4. Both these distances are computed by the difference of the heights of the nodes: $d_{uw} = h(w) - h(u)$ and $d_{wv} = h(w) - h(v)$, where $h(x)$ denotes the height of node x (note that $d(w)$ is larger than $d(u)$ and $d(v)$, since w is a common ancestor of u and v).

If we assume the molecular clock hypothesis, that states that the mutation rate in all branches of the tree is uniform, this implies the distance between all leaves and the root is the same (as it is the case with the tree in the figure). In this case, the tree is *ultrametric* and the height of each leaf can be defined as 0. Therefore, in the previous expression, we would have $d_{uv} = 2 \times h(w)$.

Given an objective function based on these general principles, a number of methods have been proposed. Unfortunately, when the number of sequences grows, the solution space grows very rapidly, and the problem has been proven to be NP-hard. As such, there are no algorithms that can efficiently provide guaranteed optimal solutions for reasonable dimensions (number of sequences).

So, most algorithms used in practice are heuristic, providing reasonable solutions in practice for most of the problem instances. One of the simplest and most popular methods in *Unweighted Pair Group Method Using Arithmetic Averages* (UPGMA), which is based on agglomerative hierarchical clustering algorithms.

This algorithm firstly considers each sequence (tree leaf) to be in its own cluster, which are put at height zero in the tree. The algorithm starts by considering the pair of closest sequences/clusters (i.e. the minimum value in the matrix D), and joining these sequences creating an internal node, with height in the tree equal to half of the distance between these sequences.

Those sequences will act, in the next iteration as a single cluster (merging the two clusters with each sequence), being the distances to the remaining sequences/clusters calculated by the average of the distances, and leading to an update of D where the columns and rows of the connected sequences disappear and a new row and new column appear for the new cluster.

In subsequent iterations, the algorithm proceeds by finding the pair of clusters with minimum distances and repeating the same steps: joining the clusters, adding an internal node to the tree with the given height and updating the distance matrix D. The algorithm stops when all sequences are within a single cluster, which corresponds to the root node of the tree.

An example of the application of the UPGMA algorithm is shown in Fig. 9.2 to a set of five sequences. The input distance matrix is shown in the upper left corner. In the first step, clusters representing sequences 1 and 2 are merged, being the respective internal node in the tree placed at height $h = 1$. The matrix D is updated removing the rows and columns for sequences 1 and 2, and creating new ones for the new cluster. The distances of this cluster to the remaining are calculated by the mean of the distances of the sequences in the cluster (sequences 1 and 2, in this case) to the remaining.

In the following step, clusters representing sequences 4 and 5 are merged, and the algorithm proceeds in a similar manner as for the previous ones. The third step merges the cluster of sequences 1 and 2 with the one of sequence 3, while the final step merges the two remaining clusters.

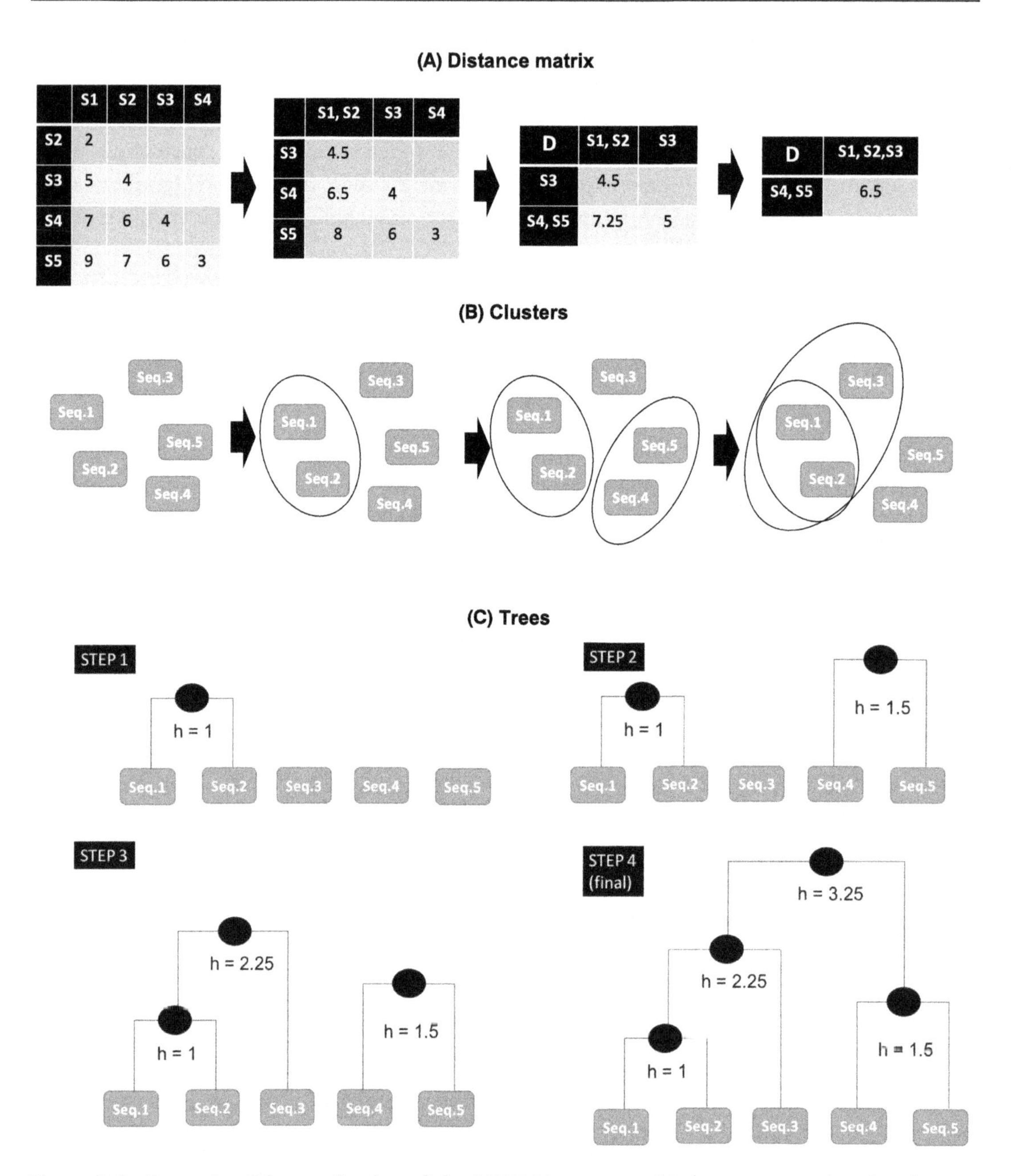

Figure 9.2: Example of the application of the UPGMA to a set of five sequences, showing the different steps of the algorithm concerning the state of: (A) the distance matrices D; (B) the clusters created; (C) the evolutionary tree.

As a general rule, in UPGMA, the distance between two clusters A and B is calculated averaging the distances of all pairs made from elements of both sets:

$$\frac{1}{|A|.|B|}\sum_{i\in A}\sum_{j\in B} D_{ij} \tag{9.2}$$

Within the execution of the algorithm, clusters are merged. If, in a given iteration, clusters A and B are joined to obtain a new cluster $(A \cup B)$, the distances from the new cluster to each cluster X can be calculated by a weighted average of the distances already calculated in the matrix:

$$D(A \cup B, X) = \frac{|A|.D(A, X) + |B|.D(B, X)}{|A| + |B|} \tag{9.3}$$

where $D(X, Y)$ here represents the distance of clusters X and Y in D.

An alternative is to use the algorithm WPGMA (*Weighted Pair Group Method with Arithmetic Mean*) where the distances of new clusters to existing ones are calculated as the arithmetic mean of the distances of the joined clusters:

$$D(A \cup B, X) = \frac{D(A, X) + D(B, X)}{2} \tag{9.4}$$

The UPGMA algorithm is quite simple and has gained popularity in the community, but suffers from many limitations. One of the assumptions of the algorithm is that the mutation rate is uniform and, thus, the trees are ultrametric. If the input distance matrix is ultrametric, the algorithm returns the optimal solution. However, this is an assumption that is rarely true in practice and which leads to erroneous results in many situations.

An alternative method is *Neighbor Joining* (NJ), originally created to infer unrooted trees, and that does not assume the constant mutation rates across the different lineages. Note that, although it creates unrooted trees, the result of the NJ algorithm is in some cases given as a rooted tree, by adding a root. There are different ways to do so, being the most used to place the root at the midpoint of the longest distance connecting two leaves in the tree.

The main difference of NJ, when compared to UPGMA, is that the criterion to select the clusters to merge in each step does not only take into account the distances between the clusters, but also seeks to select cluster pairs where the nodes are apart from the other ones. To achieve this purpose, the original distance matrix D is pre-processed to create a new matrix Q, where each cell is calculated as:

$$Q_{ij} = (n-2)D_{ij} - \sum_{k=1}^{n} D_{ik} - \sum_{k=1}^{n} D_{jk} \tag{9.5}$$

It will be this matrix Q that will be used to select the nearest clusters in each step, as it is the case in UPGMA. Notice that this implies that the clusters to be merged (or the nodes that will be joined by edges) are selected based on a trade-off, seeking pairs of nodes that have a shortest distance among themselves (first term of the previous expression) and that have large distances from other clusters (measured by the last two sums in the previous expression).

In each step of the algorithm, the matrix D is recalculated. Taking u as the new cluster (new node in the tree) created joining nodes a and b, the distances from the other clusters/nodes (represented as i) are calculated as:

$$D_{ui} = \frac{1}{2}(D_{ai} + D_{bi} - D_{ab}) \tag{9.6}$$

From this new matrix D, the new matrix Q is recalculated applying the expression given above, and the algorithm proceeds by choosing the minimum value to select the clusters to merge (nodes to join).

9.2.2 Maximum Parsimony

The maximum parsimony methods define the objective functions of the phylogenetic inference problem by estimating the number of mutations that are implied by the tree to explain the input sequences. Thus, in principle, shorter trees that explain the data are preferred, following the Occam's razor principle that states a preference for simpler models.

These methods are typically based on a previously defined multiple sequence alignment of the input sequences, which allows to identify a set of columns in the alignment that are considered informative about the possible philogeny, namely columns where there are mutations (gaps or mismatches) and thus variations of the different sequences. From this information, this method tries to identify the tree that requires the minimum number of mutations to explain the variations in the sequences.

The simplest way of identifying the most parsimonious tree is the enumeration of all possible solutions and their cost. However, this is again an NP-hard problem, which implies that this is a non-viable strategy in practice, being only possible to achieve for a small number of sequences (typically less than 10).

An alternative which has been proposed for this problem is the use of *branch and bound* methods (we will cover those in more detail in the context of motif discovery in Chapter 10). In this case, an exact solution will still be found, but without having to test all possible trees, since some areas of the search space may be discarded since we have guaranteed that these do not contain the optimal solution. In practice, this allows to increase the number of sequences in the input, still guaranteeing optimal solutions, up to about 20 sequences.

In other cases, there is the need to resort to the development of heuristic methods, which may allow to consider a larger number of sequences. Among these heuristics, one may find those based on nearest-neighbor interchanges, sub-tree pruning, and regrafting, tree bisection and reconnection, among other tree rearrangement methods, which also include meta-heuristics as genetic algorithms or simulated annealing.

Overall, these methods have the advantage of making easier to identify relationships between tree branches and sequence mutations. However, they show more limitations when the sequences are distant and require deeper philogenies.

9.2.3 Statistical Methods

Maximum likelihood methods rely on a statistical model of the occurrence of mutations in the DNA. This model is used to estimate the likelihood of possible trees, by multiplying the estimated probability of each mutation event implied by the tree. Models of DNA evolution include, among others, Kimura's two parameter model, the Jukes-Cantor or the Tamura-Nei model.

This class of methods is typically quite intensive computationally, being also NP-hard in terms of complexity. The *pruning algorithm*, a variant of Dynamic Programming can be used to reduce the complexity, by computing the likelihood of sub-trees, thus decomposing the overall problem in smaller sub-problems.

Maximum likelihood methods offer a great flexibility given the plethora of possible mutation models that can be considered. One advantage is that they allow statistical flexibility by permitting varying rates of evolution across sites and lineages.

Bayesian inference methods are an alternative that offer some similarity with the previous ones. Bayesian methods work based on a prior probability distribution applied to the possible phylogenetic trees. The search methods generally use variants of Markov chain Monte Carlo algorithms, including different sets of local move operators which apply changes over the trees.

9.3 Implementing Distance-Based Algorithms in Python

In this section, we will provide an implementation of the UPGMA algorithm, explained in detail in Section 9.2.1. This will be based on the core set of classes that we started developing in the previous chapter (see Section 8.3), namely those implementing sequences, alignment parameters and algorithms.

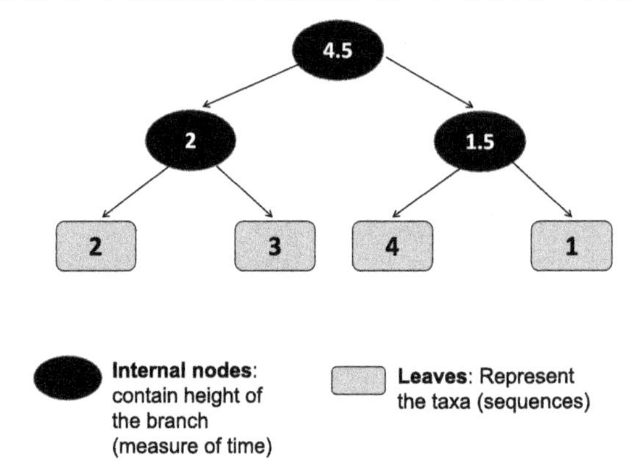

Figure 9.3: Example of a binary tree representing a phylogenetic tree in our Python implementation.

In the next subsections, we will start by presenting a class to represent evolutionary trees (as binary trees), followed by a class implementing the agglomerative hierarchical clustering algorithm, and, finally, a class to run the UPGMA algorithm with biological sequences as input.

9.3.1 Implementing Binary Trees

To represent evolutionary trees we will use a well-known data structure in Computer Science: binary trees. Note that, in this way, we assume all internal nodes have two branches going out of them, and therefore all mutations give rise to two alternative lineages.

Binary trees are recursive data structures, since every tree is composed of a node (possibly containing some information), a left sub-tree and a right sub-tree, which may be null or each represent another tree with the same structure.

There are normally two types of trees of interest: leaves, that contain some information in the nodes, but where both sub-trees are null, and trees representing internal nodes (including the root), also containing information, and having right and left sub-trees.

In the case of phylogenetic trees, we will represent the input sequences in the leaves, and the internal nodes will be bifurcation points representing mutation events. Thus, the trees considered in our implementation will be of the form shown in Fig. 9.3. Notice that each internal node has, as its information content, its height (a numerical value), while leaves contain the index of the sequence (an integer).

We will represent binary trees recursively using a class named **BinaryTree**, where each instance will have the following set of attributes:

- *value* – integer to keep the index of the sequence to represent in the leaves, being −1 in internal nodes;
- *distance* – numerical value to keep the height of the node (will be 0 in the leaves);
- *left* and *right* – left and right sub-trees; in the leaves will be None.

The following code shows the constructor of this class, together with a method to print the tree, implemented based on a recursive structure that tries to represent depth of the nodes by padding the respective lines with tabs. The **test** function creates the tree in Fig. 9.3 and prints it.

```
class BinaryTree:

    def __init__(self, val, dist=0, left = None, right = None):
        self.value = val
        self.distance = dist
        self.left = left
        self.right = right

    def print_tree(self):
        self.printtreerec(0, "Root")

    def print_tree_rec (self, level, side):
        tabs = ""
        for i in range(level): tabs += "\t"
        if self.value >= 0:
            print(tabs, side, " - value:", self.value)
        else:
            print(tabs, side, "- Dist.: ", self.distance)
            if (self.left != None):
                self.left.print_tree_rec(level+1, "Left")
            if (self.right != None):
                self.right.print_tree_rec(level+1, "Right")

def test():
    a = BinaryTree(1)
    b = BinaryTree(2)
    c = BinaryTree(3)
    d = BinaryTree(4)
    e = BinaryTree(-1, 2.0, b, c)
```

```
    f = BinaryTree(−1, 1.5, d, a)
    g = BinaryTree(−1, 4.5, e, f)
    g.print_tree()

if __name__ == '__main__':
    test()
```

One important method to implement in this class will allow to determine which leaves (sequences) are below a given node, or in other words, which is the cluster that corresponds to a given tree. This method, named **get_cluster**, is implemented in the code below and follows a general structure when designing algorithms over binary trees: if it is an internal node, first call the method recursively for the left sub-tree, and then call the method for the right tree, consolidating the results (in this case, merging the two sets of results); it is a leaf, terminate the recursion returning the result (in this case, a set with a single value).

```
class BinaryTree:

    def get_cluster(self):
        res = []
        if self.value >= 0:
            res.append(self.value)
        else:
            if (self.left != None):
                res.extend(self.left.get_cluster())
            if (self.right != None):
                res.extend(self.right.get_cluster())
        return res

def test():
    a = BinaryTree(1)
    (...)
    print(f.get_cluster())
    print(g.get_cluster())

if __name__ == '__main__':
    test()
```

9.3.2 Implementing the UPGMA Algorithm

The UPGMA algorithm has been described in some detail in Section 9.2.1, being an example provided in Fig. 9.2. Since the input for this algorithm is a distance matrix, we will here firstly implement a class **NumMatrix** that will allow to keep and manipulate these matrices.

The code for this class is provided below, including a set of methods to access information (number of rows/columns, accessing/setting values from row and column indexes), printing the matrix, adding or removing rows/columns and providing a copy of the matrix.

There is also an important method (**min_dist_indexes**) that returns the row and column of the matrix with the minimum value (ignoring the zeros). Note that the matrices defined in this class are triangular and, thus, only cells where the row index is larger than the column index are provided (the remaining values are filled with zeros). This implies ignoring those in several of the methods.

```
class NumMatrix:

    def __init__(self, rows, cols):
        self.mat = []
        for i in range(rows):
            self.mat.append([])
            for j in range(cols):
                self.mat[i].append(0.0)

    def __getitem__(self, n):
        return self.mat[n]

    def num_rows (self):
        return len(self.mat)

    def num_cols (self):
        return len(self.mat[0])

    def get_value (self, i, j):
        if i>j: return self.mat[i][j]
        else: return self.mat[j][i]

    def set_value(self, i, j, value):
        if i>j: self.mat[i][j] = value
```

```
        else: self.mat[j][i] = value

    def print_mat(self):
        for r in self.mat: print(r)
        print()

    def min_dist_indexes (self):
        m = self.mat[1][0]
        res= (1,0)
        for i in range(1,self.num_rows()):
            for j in range(i):
                if self.mat[i][j] < m:
                    m = self.mat[i][j]
                    res = (i, j)
        return res

    def add_row(self, newrow):
        self.mat.append(newrow)

    def add_col(self, newcol):
        for r in range(self.num_rows()):
            self.mat[r].append(newcol[r])

    def remove_row(self, ind):
        del self.mat[ind]

    def remove_col(self, ind):
        for r in range(self.num_rows()):
            del self.mat[r][ind]

    def copy(self):
        newm = NumMatrix(self.num_rows(), self.num_cols())
        for i in range(self.num_rows()):
            for j in range(self.num_cols()):
                newm.mat[i][j] = self.mat[i][j]
        return newm
```

Next, a general purpose agglomerative hierarchical clustering algorithm will be implemented. We will implement this algorithm within a new class, named **HierarchicalClustering**. The only attribute in this class will consist of a distance matrix, the input of the method.

The code of the class is provided below, being the **execute_clustering** the main method, which runs the algorithm and returns a binary tree (an instance from the class described in the last section) as the result.

The method starts by initializing the set of trees, creating the leaf nodes, and the distance matrix by copying the original input matrix. The main `for` cycle firstly identifies the indexes of the minimum distance in the matrix to identify clusters to join. A new tree is created joining the branches relative to these two clusters. If it is the last iteration, this tree is returned as the final result.

On the other hand, if this is not the case, the algorithm proceeds with the following processes: (i) the joined branches are removed from the list of trees to handle, (ii) the distance matrix is updated by removing the rows and columns of the joined clusters and adding a new one for the new cluster (notice the new distances are calculated based on Eq. (9.3)), and, (iii) the new tree is added to the set of active trees to handle in posterior iterations.

```
from BinaryTree import BinaryTree
from NumMatrix import NumMatrix

class HierarchicalClustering:

    def __init__(self, matdists):
        self.matdists = matdists

    def execute_clustering(self):
            ## initialization of the tree leaves and matrix
        trees = []
        for i in range(self.matdists.num_rows()):
            t = BinaryTree(i)
            trees.append(t)
        tableDist = self.matdists.copy()
        ## iterations
        for k in range(self.matdists.num_rows(), 1, -1):
            mins = tableDist.min_dist_indexes() ## minimum distance
 in D
            i,j = mins[0], mins[1]
            ## create new tree joining clusters
            n = BinaryTree(-1, tableDist.get_value(i, j)/2.0, trees[i
 ], trees[j])
            if k>2:
```

```
                ## remove trees being joined from the list
                ti = trees.pop(i)
                tj = trees.pop(j)
                ## calculating distances for new cluster
                dists = []
                for x in range(tableDist.num_rows()):
                    if x != i and x != j:
                        si = len(ti.get_cluster())
                        sj = len(tj.get_cluster())
                        d = (si*tableDist.get_value(i,x) + sj*
    tableDist.get_value(j,x)) / (si+sj)
                        dists.append(d)
                ## updating the matrix
                tableDist.remove_row(i)
                tableDist.remove_row(j)
                tableDist.remove_col(i)
                tableDist.remove_col(j)
                tableDist.add_row(dists)
                tableDist.add_col([0] * (len(dists)+1))
                ## add the new tree to the set to handle
                trees.append(n)
            else: return n

def test():
    m = NumMatrix(5,5)
    m.set_value(0, 1, 2)
    m.set_value(0, 2, 5)
    m.set_value(0, 3, 7)
    m.set_value(0, 4, 9)
    m.set_value(1, 2, 4)
    m.set_value(1, 3, 6)
    m.set_value(1, 4, 7)
    m.set_value(2, 3, 4)
    m.set_value(2, 4, 6)
    m.set_value(3, 4, 3)
    hc = HierarchicalClustering(m)
    arv = hc.execute_clustering()
    arv.print_tree()
```

```
if __name__ == '__main__':
    test()
```

Finally, we will define the class **UPGMA** which will apply the generic hierarchical clustering algorithm defined above to biological sequences. This class will have attributes to keep a set of sequences that will be the leaves of the tree (objects of the class **MySeq** defined earlier), the alignment parameters (object of the class **PairwiseAlignment**) and the distance matrix (object of the class **NumMatrix**).

In the code below, we show the implementation of this class, considering a distance metric consisting of the number of distinct characters between the two sequences, after a global alignment (with the *Needleman-Wunsch* method). This is calculated in the method **create_mat_dist** that fills the class variable *matdist*. Note that we can easily create other distance metrics by changing or replacing this function.

The **run** method is used to create an object of the class **HierarchicalClustering** and execute the clustering algorithm, returning the resulting tree.

```
from NumMatrix import NumMatrix
from HierarchicalClustering import HierarchicalClustering
from MySeq import MySeq
from PairwiseAlignment import PairwiseAlignment
from SubstMatrix import SubstMatrix

class UPGMA:

    def __init__(self, seqs, alseq):
        self.seqs = seqs
        self.alseq = alseq
        self.create_mat_dist()

    def create_mat_dist(self):
        self.matdist = NumMatrix(len(self.seqs), len(self.seqs))
        for i in range(len(self.seqs)):
            for j in range(i, len(self.seqs)):
                s1 = self.seqs[i]
                s2 = self.seqs[j]
                self.alseq.needleman_Wunsch(s1, s2)
                alin = self.alseq.recover_align()
```

```
                ncd = 0
                for k in range(len(alin)):
                    col = alin.column(k)
                    if (col[0] != col[1]): ncd += 1
                self.matdist.set_value(i, j, ncd)

    def run(self):
        ch = HierarchicalClustering(self.matdist)
        t = ch.execute_clustering()
        return t

def test():
    seq1 = MySeq("ATAGCGAT")
    seq2 = MySeq("ATAGGCCT")
    seq3 = MySeq("CTAGGCCC")
    seq4 = MySeq("CTAGGCCT")
    sm = SubstMatrix()
    sm.create_submat(1, -1, "ACGT")
    alseq = PairwiseAlignment(sm, -2)
    up  = UPGMA([seq1, seq2, seq3, seq4], alseq)
    arv = up.run()
    arv.print_tree()

if __name__ == '__main__':
    test()
```

9.4 BioPython Functions for Phylogenetic Analysis

To finalize this chapter, we will take a brief look at some of the *BioPython* features for phylogenetic analysis, implemented mainly in the module *Bio.Phylo*. This module mainly implements data structures to represent phylogenetic trees and methods to load, save, and explore these trees. The methods for tree inference need to be run externally, while there are a few wrappers in the module to facilitate its use, which we will not cover here.

To understand the main data structures of this module let us work with a simple example. Save a text file with the name "simple.dnd" and the following content:

```
(((A,B),(C,D)),(E,F,G));
```

This simple line will represent a tree, in the Newick format. The following code can be used to read this file, using the **read** function, and print the tree using two different functions, being the first a print of the tree's content and the second a simple graphical representation of the tree.

```
tree = Phylo.read("simple.dnd", "newick")
print(tree)
Phylo.draw_ascii(tree)
```

This module supports functions for input/output of phylogenetic trees in a large list of formats. The functions **read** and **parse** can be used to read a single or several trees from a file, respectively, while the function **write** can be used to write trees to file, and the function **convert** can directly convert file formats.

The following example shows a simple use of these functions, where a file is read and converted to a new format. This new file is then read and the trees it contains are printed. The example file may be found in the book's website and in the *BioPython* tutorial.

```
tree2 = Phylo.read("int_node_labels.nwk", "newick")
Phylo.draw_ascii(tree2)

Phylo.convert("int_node_labels.nwk", "newick", "tree.xml", "phyloxml"
    )
trees = Phylo.parse("tree.xml", "phyloxml")
for t in trees: print(t)
```

Looking at the result of the **print** method, we can understand the structure of the tree representation in this implementation. The **Tree** object is unique for each complete phylogenetic tree and contains global information about the tree, such as whether it is rooted or unrooted. The recursive representation of the tree is provided by the **Clade** objects. The **Tree** object has one root **Clade**, and under that, it's composed of nested lists of clades all the way down to the leaves. Notice that unlike our own implementation, trees in this module are n-ary, meaning that each node may have more than two sub-trees, being represented by a list of clades.

The module has a large number of methods that allow to retrieve information from the tree, as it is the case with: searching for an element traversing the tree, listing all leaves and internal nodes of a tree, finding the common ancestor of a set of leaves, calculating the distance between two leaves, among many others. Also, it is possible to modify the tree and its clades, for instance by pruning the tree, removing nodes, or splitting nodes into new branches. Details of these methods are provided in Section 13.4 of the *BioPython* tutorial.

This module also allows to color the tree branches with different colors as it is shown in the next example, where we take the example tree defined above and we color the branch including leaves E, F and G, and their common ancestor in salmon, and the branch with leaves C and D in blue.

```
from Bio.Phylo.PhyloXML import Phylogeny

treep = Phylogeny.from_tree(tree)
Phylo.draw(treep)

treep.root.color = "gray"
mrca = treep.common_ancestor({"name": "E"}, {"name": "F"})
mrca.color = "salmon"
treep.clade[0, 1].color = "blue"
Phylo.draw(treep)
```

Bibliographical Notes and Further Reading

The agglomerative hierarchical clustering algorithm used by UPGMA is generally attributed to Sokal and Michener [142]. The *Neighbor-Joining* algorithm has been proposed by Saitou and Nei [135]. The book by Felsenstein [62] contains a thorough explanation on phylogenetics inference algorithms, covering the three classes presented in this chapter.

The *Bio.Phylo* module included in *BioPython* has been described in an article by Talevich et al. [146].

The MEGA free software application (`http://www.megasoftware.net`) can be used to run the methods described in this chapter, applying them to generate evolutionary trees, given sets of user defined sequences. It includes representative methods from all classes described in Section 9.2.

Alternative software tools are PAML (*Phylogenetic Analysis by Maximum Likelihood*) (`http://abacus.gene.ucl.ac.uk/software/paml.html`) and PhyML (`http://www.atgc-montpellier.fr/phyml/`) [69]. Both these tools have available wrappers in *BioPython*.

Exercises and Programming Projects

Exercises

1. Consider the sequences of the first exercise of the previous chapter. Assume that the multiple sequence alignment obtained was the following:

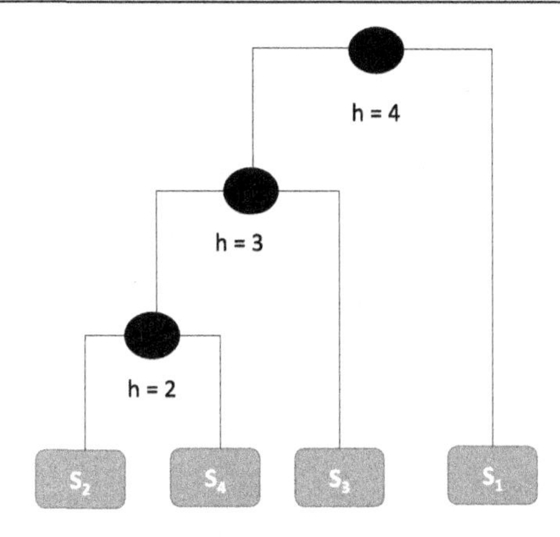

Figure 9.4: Example of a binary tree representing a phylogenetic tree for the exercise.

```
S1: A-CATATC-AT-
S2: A-GATATT-AG-
S3: AACAGATC-T--
S4: G-CAT--CGATT
```

a. Assuming the metric distance to be the number of distinct characters in pairwise alignment, and taking the pairwise alignments imposed by the multiple alignment above, calculate the distance matrix.

b. Apply the algorithm UPGMA to build the tree for these sequences.

c. Write a Python script that allows you to check your results.

2. a. Consider the phylogenetic tree represented in Fig. 9.4. Assume it was built by the UPGMA algorithm as implemented in our Python code, from 4 sequences (S_1, S_2, S_3, S_4). Using the notation D_{ij} to represent the distance between sequences S_i and S_j, which of the following expressions are true?
   - $D_{24} = 2$,
   - $D_{12} > 4$,
   - $D_{23} + D_{34} = 12$,
   - $D_{32} > 8$.

   b. Considering our Python implementation, write a script that creates and prints the tree in the figure.

3. Considering the class **BinaryTree** implemented in this chapter, add methods that:
   a. Return the size of the tree, which will be given by a tuple with two values: the number of internal nodes of the tree, the number of leaves.
   b. Search if there is a leaf that contains a given value passed as a parameter of the method. The result should be a Boolean value (`True` if the value exists; `False`, otherwise).

c. Return the common ancestor of two sequences/taxa (identified as integer values), i.e. will return the simplest tree (with less height) that contains the leaves with those values.
d. Generalize the previous function to a set of sequences as input.
e. Return the distance between two leaves identified by their integer values.
f. Return the distance between the two leaves (identified by their integer values) that are nearest in the tree (i.e. have their common ancestor at the smallest height).

4. Implement the WPGMA variant of the UPGMA algorithm, changing the way the distance between clusters is calculated (as described above). Compare the results of both approaches.
5. Consider the last exercise of the previous chapter. Read the tree obtained from *Clustal Omega*. Draw the tree with the *Bio.Phylo* module. Explore the tree using the available functions.

Programming Projects

1. Implement the *Neighbor-Joining* method applied to rooted trees, as described above in Section 9.2.1.
2. Extend the class UPGMA, by considering other distance metrics for the sequences (e.g. edit distance). Take into account the sequence type (nucleotide vs protein) to define meaningful metrics.
3. Implement a simple maximum parsimony method, by firstly creating a multiple sequence alignment of the sequences. Add a function to identify informative columns (where there are mutations). Implement a cost function for a tree, taking those columns into account.

CHAPTER 10

Motif Discovery Algorithms

In this chapter, we revisit and formalize the definition of deterministic motif. We introduce the concept of search and solution space and formally define the problem of deterministic motif search in a set of biologically related sequences. We start by describing brute-force algorithms based on an exhaustive search and then present more efficient approaches for motif discovery.

10.1 Introduction: Problem Definition and Relevance

Within the context of biological sequence mining, the term *motif* refers to a non-trivial sequential pattern that is shared across multiple sequences. With non-trivial, we mean that the motif has a minimum relevant length and captures a combination of symbols that differs from the underlying symbol distribution. However, its main feature is its recurrence, i.e. the fact that it occurs in several of the analyzed sequences.

The relevance of a motif arises when the set of sequences where it occurs are associated to genomic elements, such as genes, that share a certain biological property or are under the same regulatory control. The hypothesis will then be that the motif will play a role in the associated biological phenomenon.

For instance, in DNA sequences, motifs may occur in the promoter region of genes and indicate the presence of binding sites of single proteins or protein complexes that have a regulatory role in gene transcription. In protein sequences, motifs may indicate the existence of conserved domains, i.e. parts of the protein that play specific biological functions, such as enzyme binding sites for substrates or other molecules.

We can distinguish two main classes of motifs: *deterministic* and *probabilistic*. Deterministic motifs are often captured by enhanced regular expression syntax, and as it name indicates are either present or absent in the input sequences. These were already addressed in the context of Chapter 5.

On the other hand, probabilistic motifs form a loose model that captures the underlying variability of the motif occurrences. When a given segment of a sequence is presented to the model, the probability of being part of the motif can be derived. As we will see in the next chapter, position weight matrices (PWMs) can be used as a way to represent probabilistic motifs. In this chapter, we will discuss the algorithmic framework to define deterministic motifs and present motif finding algorithms with increasing efficiency level.

Bioinformatics Algorithms. DOI: 10.1016/B978-0-12-812520-5.00010-9

We will start by introducing the concepts that will help us to better define the problem of efficiently discovering the sequence motifs conserved among a set of related biological sequences. Note that we will use the terms *sequence pattern* and *motif* with equivalent meaning throughout this chapter.

As we have seen in previous chapters, biological sequences are defined as strings of symbols, which for DNA consist of four letters $\Sigma_{DNA} = \{A, C, G, T\}$ and for protein sequences of twenty letters, $\Sigma_{Protein} = \{A, C, D, E, F, G, H, I, K, L, M, N, P, Q, R, S, T, V, W, Y\}$. The characteristics of nucleotide sequences and proteins, such as the size of the alphabet, the typical length of the sequence or the underlying symbol distribution raises different algorithmic challenges. Therefore, motif discovery algorithms need to be optimized according to the nature of sequences to which they are applied.

As we mentioned before, a deterministic motif can be captured by regular expression syntax. In some cases, the motif is composed by a continuous combination of symbols, while in other more complex cases the motif may contain certain parts, possibly of variable length, corresponding to highly variable symbols, which apart from their length do not add any relevant information. These two cases can be differentiated using the terms, *sub-string* and *sub-sequence*. As defined in [70], a sub-string corresponds to a consecutive part of a string, while a sub-sequence corresponds to a new sequence obtained by eventually dropping some symbols from the original sequence and preserving the respective order. The later is a more general case than the former. For instance, for a sequence $S = abcdef$, $acdf$ is a sub-sequence of S and bcd is a sub-string of S.

If $|S|$ denotes the length of the sequence S and $S = xyz$, $|x| \geq 0$ and $|z| \geq 0$ then x is said to be a *prefix* of S and y is *suffix* of S. Alternatively, S is a *superstring* or *supersequence* of x.

Generally, a deterministic motif M can be defined as a non-null string of symbols from an enhanced alphabet Σ'. The alphabet Σ' results from an extension of Σ with an additional set of symbols E, $\Sigma' = \Sigma \cap E$. E adds greater expressiveness to the patterns and may contain symbols that express the existence of gaps or symbols that reflect positions that may be occupied by more than one symbol. The set of all the possible regular expressions that result from the substitution of symbols from Σ' in M defines a *pattern language*: $L(M)$. A motif M is said to *match* or *hit* a sequence S, if S contains some string y that occurs in $L(M)$.

We have seen in Chapter 5 that the restriction enzyme *EcoRI* cuts the DNA when it finds the sequence pattern $GAATTC$. In this case, we have a very clear pattern represented by a unique string. However, it is often the case that the binding sequences change slightly. This leads to more subtle patterns that need to comprise more than one sequence match. This is, for instance, the case of *HindII* [140], a type II restriction enzyme that cuts the DNA in specific sequences that are six nucleotides long. When the DNA is read from the 5' to 3' sense, the

```
gccatcgtttatcgtcaacattaaaaccgctcaagttaataacggccgatcacgttaaat
atggtcgacacaagaaaaggtctttatgggctattactatatctctcgacaaaatgaaaa
ctagtgtacgtcagttgtggggcgcagaagttaacaattgaacagttaaaaagagcgtgt
aatgttcatgaaaagatcttttgttgacttttctatcaatacactactgttgtgacaag
gggagtcaacaataactttattgccattttcctgaaattattcggtactcgagaacaaa
```

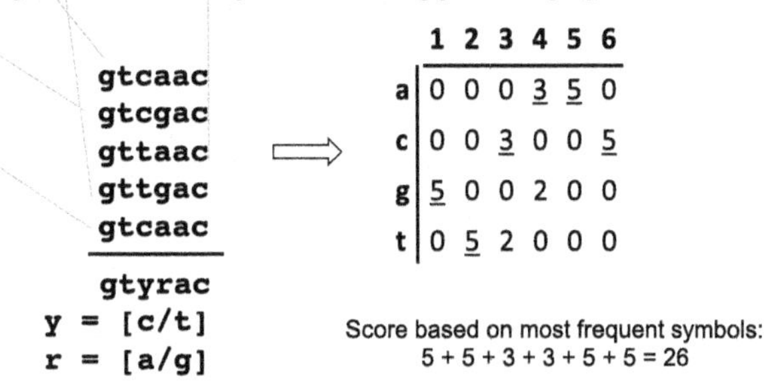

Figure 10.1: Multiple representations of a motif with matches in five DNA sequences.

cutting points are formed by the sequences: GT followed by T or C, then followed by A or G and AC. In this case, we have a degenerate motif but still highly conserved.

When considering for instance transcription factors (TFs), their motifs become more complex since these proteins often bind to sequences that are significantly degenerated. This degeneracy highly increases the complexity of *de novo* motif discovery.

The *EcoRI* can be described using only the Σ_{DNA} alphabet, while *HindII* is a good candidate to be described an extended version of the alphabet Σ'_{DNA}. Fig. 10.1 depicts the occurrence of the strings along the input sequence and their alignment. The second and third position of the motif can be captured by the symbol Y and R that respectively represent C or T and A or G. The IUPAC nucleotide code provides symbols for different degenerate symbols and is an example of an extended alphabet.

A possible representation of a motif is to store the alignment of all its instances in the input sequences. Likewise, we can also store the starting position and later recover the correspond strings. This way we capture all the diversity of the motif, conserving all the information. A drawback of this representation is that we need to store as many strings as occurrences of the motif.

An alternative representation is to take advantage of an extended alphabet and capture the motif with regular expression syntax. This is called a *consensus* motif. While this provides a more compact and intuitive motif representation, we loose information regarding the number of times each instance of the motif occurs.

Another possibility is to build a profile or matrix of the frequencies of the symbols at each position of the motif, represented as columns of the matrix. This representation only depends on the size of the alphabet and the length of the motif and is independent on the number motif occurrences. The negative aspect comes from the fact that we no longer preserve the order in which the symbols occur. In the next chapter, we will discuss in detail the representation based on matrix of frequencies along the motif positions.

One important measure of interest of a motif is its recurrence, also denoted as *frequency* or *support*. For an input set of sequences, D, this can be measured as the number of matches of M in different sequences of D (each sequence counts once) or in the total number of matches of M in D (more than one count per sequence). This leads to the notion of *frequent motif*, which is a motif that has frequency higher or equal than a user defined threshold; and infrequent otherwise.

For the sake of generality, we can define an abstract measure of interest, *Score*, as measure of M with relation to D. The frequency of a motif M is an example of such score measure. For a profile representation, we can think of a scoring measure that takes into account the frequencies of the most frequent symbol at each position. Overall, these frequencies can be either summed or multiplied. For a profile matrix *count* and motif length L the score function for a motif M can be defined the equation:

$$Score(M) = \sum_{i=1}^{L} max_{k \in \Sigma} count(k, i) \tag{10.1}$$

We are now able to formalize the motif discovery problem: given an input a set of sequences $D = \{S_1, S_2, ..., S_t\}$, defined over an alphabet Σ, a motif length L, an optional minimum score value σ: find all the motifs in D that maximize a given score function or that have a score greater or equal than σ.

To implement the algorithms for deterministic motif finding, we will create a class called **DeterministicMotifFinding**. The class will contain two attributes, the *motif_size* that corresponds to the length of the motif and the vector *seqs* that contains the input sequences where the motif is to be found. The function **read_file** reads the input sequences from a given file.

In Fig. 10.1, we show that from a set of aligned motif matches, we can build a matrix with a number of columns equal to the motif length and number of rows equal to the alphabet size. Each cell of the matrix contains the frequency of the symbol in the given position. The function **create_motif_from_indexes** is defined to build such matrix. It receives as input the indices of the motif along the input sequences. Variable *res* consists of a bi-dimensional matrix. Next, it iterates over all indices to retrieve the respective sequence matches. Note the use of the built-in function **enumerate** that returns a sequential index and the associated value in the vector. Then, with a nested loop the count on respective cell of the matrix is incremented.

These functions provide a suitable motif representation to approach the motif finding problem.

```
class DeterministicMotifFinding:
    """ Class for deterministic motif finding. """

    def __init__(self, size = 8, seqs = None):
        self.motif_size = size
        if (seqs != None):
            self.seqs = seqs
            self.alphabet = seqs[0].alphabet()
        else:
            self.seqs = []

    def __len__ (self):
        return len(self.seqs)

    def __getitem__(self, n):
        return self.seqs[n]

    def seq_size (self, i):
        return len(self.seqs[i])

    def read_file(self, fic, t):
        for s in open(fic, "r"):
            self.seqs.append(MySeq(s.strip().upper(),t))
        self.alphabet = self.seqs[0].alphabet()

    def create_motif_from_indexes(self, indexes):
        pseqs = []
        res = [[0]*self.motif_size for i in range(len(self.alphabet))
   ]
        for i,ind in enumerate(indexes):
            subseq = self.seqs[i][ind:(ind+self.motif_size)]
            for i in range(self.motif_size):
                for k in range(len(self.alphabet)):
                    if subseq[i] == self.alphabet[k]:
                        res[k][i] = res[k][i] + 1
        return res
```

The function **score** defined below implements the scoring function defined in Eq. (10.1). This function iterates through all the positions of the motif and, for each position, determines the maximum value which is added to the score of the motif. The function **score_multiplicative** implements a similar scoring, but instead of summing, the score is obtained by multiplying the maximum value at each motif position.

```
def score(self, s):
    score = 0
    mat = self.create_motif_from_indexes(s)
    for j in range(len(mat[0])):
        maxcol = mat[0][j]
        for i in range(1, len(mat)):
            if mat[i][j] > maxcol:
                maxcol = mat[i][j]
        score += maxcol
    return score

def score_multiplicative(self, s):
    score = 1.0
    mat = self.create_motif_from_indexes(s)
    for j in range(len(mat[0])):
        maxcol = mat[0][j]
        for  i in range(1, len(mat)):
            if mat[i][j] > maxcol:
                maxcol = mat[i][j]
        score *= maxcol
    return score
```

10.2 Brute-Force Algorithms: Exhaustive Search

To tackle the motif discovery problem as described in the previous section, we need to implement a solution that finds the vector s with the best initial positions of the motif in each sequence. The algorithm will be defined as:

- Inputs: t – sequences (length n); L – length of motif; and σ – optional minimum score.
- Outputs: A vector with the initial positions $s = (s_1, s_2, ..., s_t)$ of the motif M in D that maximize the $Score(s, D)$.

We can devise a solution based on an exhaustive search for the best positions. Since the goal is to maximize $Score(s, D)$, we can scan all possible vectors of initial positions and calculate the respective score. The best motif will correspond to the best scoring position vector. From this, we can then derive the profile and the consensus sequence of the motif. The function **exhaustive_search** provides an implementation for this approach. For each possible solution, i.e. vector of initial positions, it keeps track of the solution with the highest score. The **next_solution** function provides a way to iterate through all the $n - L + 1$ possible values of the position in input sequence in D.

```
def next_solution (self, s):
    next_sol= [0]*len(s)
    pos = len(s) - 1
    while pos >=0 and s[pos] == self.seq_size(pos) - self.
motif_size:
        pos -= 1
    if (pos < 0):
        next_sol = None
    else:
        for i in range(pos):
            next_sol[i] = s[i]
        next_sol[pos] = s[pos]+1;
        for i in range(pos+1, len(s)):
            next_sol[i] = 0
    return next_sol

def exhaustive_search(self):
    best_score = -1
    res = []
    s = [0]* len(self.seqs)
    while (s!= None):
        sc = self.score(s)
        if (sc > best_score):
            best_score = sc
            res = s
        s = self.next_solution(s)
    return res
```

10.3 Branch-and-Bound Algorithms

In the exhaustive approach, the solution space is composed of $(n - L + 1)^t$ possible solutions. As the length and the number of the input sequences increases, the number of candidates ex-

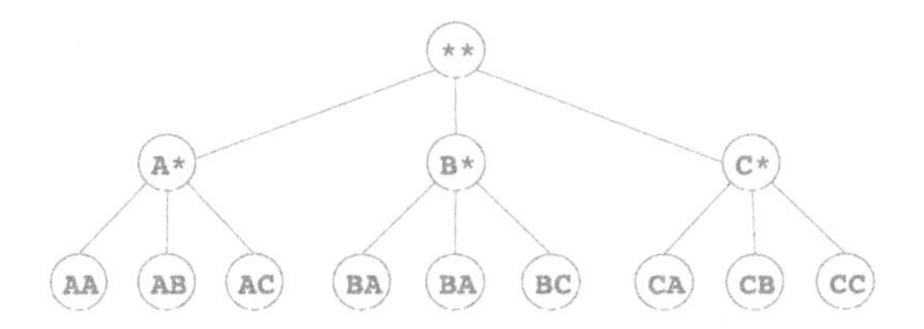

Figure 10.2: Tree structure representation for motif candidates with length 2 and alphabet with three symbols.

plodes and the problem becomes computationally intractable. In practice, this approach is not feasible with real life data. We, therefore, need a more efficient search procedure. This will require a clever way to traverse the solution space and reduce the number of motif candidates.

The branch and bound algorithm belongs to a group of combinatorial optimization algorithms. It enumerates the candidates, but uses an intelligent mechanism of *lookahead* to avoid the explicit enumeration of some of the solutions. It is, therefore, well suited for this task.

The algorithm organizes the search space in a tree structure. This allows a partition of the search space into sub-sets. Each leaf of the tree corresponds to a candidate and the internal nodes are partial sub-spaces. Traversing all the leaves of the tree provides an exhaustive enumeration. More interestingly, the branch and bound algorithm seeks a more efficient enumeration of the candidate solutions by exploring tree branches. Partial solutions are related to the solutions at the nodes of the tree. Thus, each branch also carries information related to the objective function values of the leaves below. Thus, at each internal node or branch, an upper or a lower bound of the objective function score for the respective sub-space can be calculated. The algorithm will expand the search within that branch of the tree, if it has the potential to yield a better solution than the current one. Otherwise, all the candidates within that sub-space are discarded (not enumerated). The elimination of some candidates below a branch is typically called as *pruning* the search space. The algorithm borrows its name from the branching task that allows exploring the sub-tree of a given node and bounding that estimates a bound on the candidates of the sub-tree rooted at that node.

For a motif of length L and an alphabet Σ, there are $|\Sigma|^L$ possible candidates. Consider for instance a hypothetical alphabet of three letters $\Sigma = \{A, B, C\}$ and a motif of length 2. There are $3^2 = 9$ candidates, AA, AB, AC, BA, BB, ... Fig. 10.2 provides a representation of the search space organized as a tree for this specific case.

In order to adapt the branch and bound algorithm for our motif discovery problem, we need to implement functionalities that allow us to traverse the tree and to calculate the bounds of the objective function values of the candidates. In this case, the objective function will correspond to the *Score* function previously introduced.

(A) Vertex enumeration

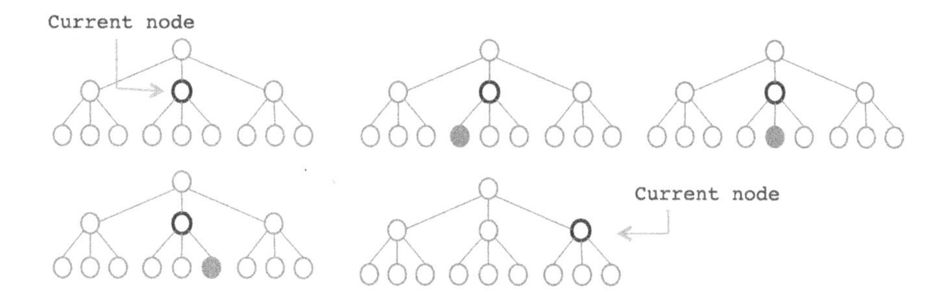

(B) Branch bypass

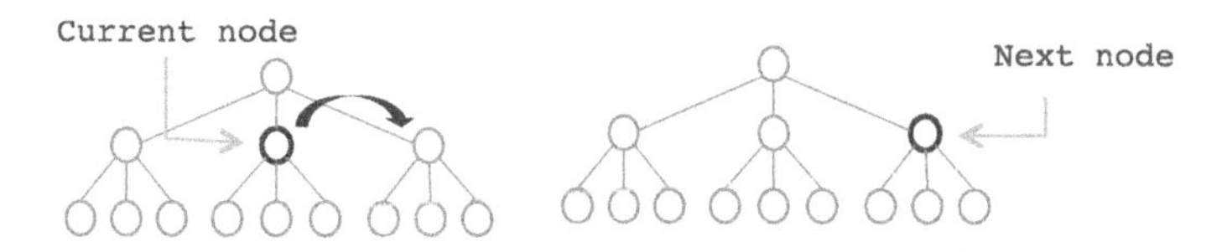

Figure 10.3: Representation of the workflow for the next vertex enumeration and bypass operations on tree structure arrangement of motif search space.

For the traversal of the search space, we need an operation that allows enumerating the candidates by visiting the relevant leaves in the tree. A function that points to next leaf to visit is required. If the current traversal point is an internal node, the function should point to the first leaf of its sub-space. If current point is a leaf, the function should return the subsequent leaf of the sub-space. If this leaf is the last one within the sub-space, then it should point to next internal node that is at the root of the next sub-space to visit. Fig. 10.3A shows a representation of the sub-space enumeration.

As we will see next, in a bounding operation we will ignore the sub-space corresponding to the sub-tree rooted at the given node and jump to the next node found at the same level of the current one. This is called a *bypass* operation. Fig. 10.3B shows an example of branch bypass where the pointer to current node moves to the next equivalent node.

In order to decide if the sub-space should be ignored or not, we need to calculate the upper bound of the sub-space. This value will estimate if the candidates within the sub-space represent or not a solution that may improve over the current one.

For our motif discovery problem, we try to maximize the *Score* function by selecting the best initial positions of the motif $s = (s_1, s_2, ..., s_t)$. Therefore, instead of working on a motif representation search space we will work on a positional representation space. For t input

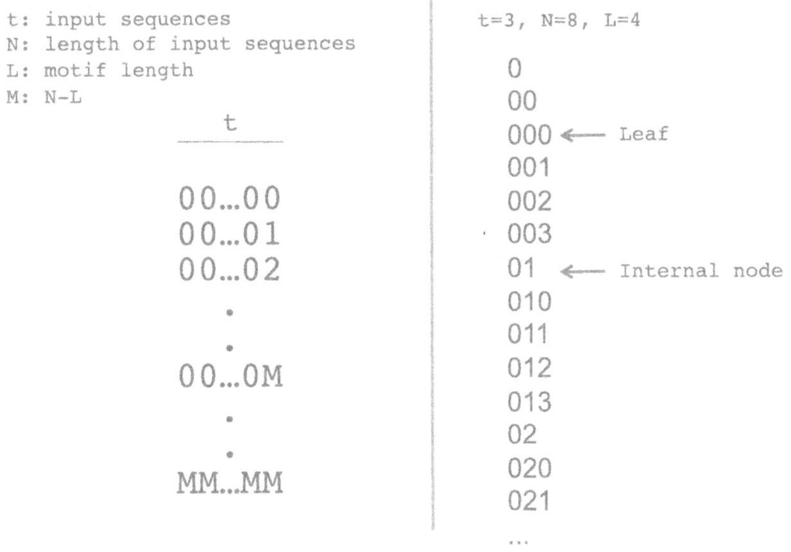

```
t: input sequences
N: length of input sequences
L: motif length
M: N-L
         t
      -------
      00...00
      00...01
      00...02
         .
         .
      00...0M
         .
         .
      MM...MM
```

```
t=3, N=8, L=4

0
00
000 <-- Leaf
001
002
003
01  <-- Internal node
010
011
012
013
02
020
021
...
```

Figure 10.4: Representation of the enumeration space based on the vector of motif starting positions.

sequences of length N and a motif length L, we can search each starting position, from the 0 to $M = N - L$ index in the t input sequences.

Fig. 10.4 shows the representation for all possible solutions represented as leaves of the tree. Each leaf is represented as a vector of length t. An internal node corresponding to the level i is represented by a vector of length i, as also shown in the figure.

The function **next_vertex**, provided below, implements the retrieval of the next leaf according to this representation. It tests if the length of current position vector is shorter than the number of sequences t. If it is, then it represents an internal node. In this case, it copies the current solution and goes down one level by setting the next level as position zero. If the length of current position vector is equal to the number of sequences then it searches at leaf level. Here, it tests if the position of the current sequence is already maximal, i.e. equal to $M = N - L$. If it is not, it copies the current positional solution and increments by one the position of the current sequence. Otherwise, it increments the position for the previous sequence.

```
def next_vertex (self, s):
    res =  []
    if len(s) < len(self.seqs): # internal node -> down one level
        for i in range(len(s)):
            res.append(s[i])
        res.append(0)
```

```
        else: # bypass
            pos = len(s)-1
            while pos >=0 and s[pos] == self.seq_size(pos) - self.
    motif_size:
                pos -= 1
            if pos < 0: res = None # last solution
            else:
                for i in range(pos): res.append(s[i])
                res.append(s[pos]+1)
        return res
```

The code for the **bypass** function is similar to the previous function, but in this case there is no test for internal nodes.

```
    def bypass (self, s):
        res =  []
        pos = len(s)-1
        while pos >=0 and s[pos] == self.seq_size(pos) - self.
    motif_size:
            pos -= 1
        if pos < 0: res = None
        else:
            for i in range(pos): res.append(s[i])
            res.append(s[pos]+1)
        return res
```

We now need to implement a way to estimate the bound for a sub-space, and test if it is worth to explore the solutions within the corresponding branch. Let's suppose that we are traversing the tree from the top to the bottom and we are currently at level i. For our t sequences the position vector can be divided as: $(s_1, s_2, ..., s_i)$ and $(s_{i+1}, ..., s_t)$. Each sequence can contribute at most with a score of L. Now, considering fixed all the initial positions in the first i sequences, for the best-case scenario the subsequent $t - i$ sequences can contribute with $(t - i) * L$ for the current score. The bound for level i is, thus, given by $Score((s_1, s_2, ..., s_i), D) + (t - i) * L$. If this sum is smaller than the best score seen so far, iterating through this sub-space will never reach the best score and, therefore, can be skipped. This bounding operation can be easily performed with the **bypass** function that jumps from the current internal node to the subsequent internal node.

The **branch_and_bound** function iterates through the initial position solution space to find the best motif. The function **next_vertex** provides at each step the next solution. If currently at an internal node, the bound is estimated and a bypass is performed if the conditions explained above are met. If positioned at a leaf, the current score is tested; if it exceeds the best score then the new best motif is updated to the current motif solution.

```
def branch_and_bound (self):
    best_score = -1
    best_motif = None
    size = len(self.seqs)
    s = [0]*size
    while s != None:
        if len(s) < size:
            # estimate the bound for current internal node
            # test if the best score can be reached
            optimum_score = self.score(s) + (size-len(s)) * self.
motif_size
            if optimum_score < best_score: s = self.bypass(s)
            else: s = self.next_vertex(s)
        else:
            # test if current leaf is a better solution
            sc = self.score(s)
            if sc > best_score:
                best_score = sc
                best_motif = s
            s = self.next_vertex(s)
    return best_motif
```

10.4 Heuristic Algorithms

So far, we have considered two algorithms, exhaustive search and the branch and bound enumeration, which provide a way of considering the whole solution space, offering guarantees of achieving optimal solutions for the provided scoring function. However, the algorithmic complexity of these algorithms is proportional to the number t and length N of the input sequences and the length L of the motif. The exhaustive enumeration requires $L * (N - L + 1)^t$ operations. In the branch and bound approach for i bypass operations, the search skips $L * (N - L + 1)^{(t-i)}$ candidates. In both cases, when the number of sequences reaches over 10–12 and their length hundreds or thousands of symbols the problem becomes intractable.

In order to overcome this, heuristics have been introduced in different methods, with a trade-off between the optimality of the solution and computational efficiency. One of the earliest solutions was proposed by Stormo, Hertz, and Hartzell in the CONSENSUS algorithm [75, 145].

The algorithm starts by considering only the first two input sequences to search for the initial positions s_1 and s_2 with the best score. These provide the best partial contribution to the overall score. For the subsequent input sequences $i = 3, ..., t$, the algorithm iterates through each sequence and selects the initial position that maximizes the score. Positions for the previous sequences $(s_1, s_2, ..., s_{i-1})$ are fixed. The function **heuristic_consensus** implements this approach.

```
def heuristic_consensus(self):
    res = [0]* len(self.seqs)
    max_score = -1;
    partial = [0,0]
    for i in range(self.seq_size(0)-self.motif_size):
        for j in range(self.seq_size(1)-self.motif_size):
            partial[0] = i
            partial[1] = j
            sc = self.score(partial);
            if(sc > max_score):
                max_score = sc
                res[0] = i
                res[1] = j
    for k in range(2, len(self.seqs)):
        partial = [0]*(k+1)
        for j in range(k):
            partial[j] = res[j]
        max_score = -1
        for i in range(self.seq_size(k)-self.motif_size):
            partial[k] = i
            sc = self.score(partial)
            if(sc > max_score):
                max_score = sc
                res[k] = i
    return res
```

While this algorithm greatly improves the computational efficiency, the reported results are highly dependent on the first two initial positions, which on its turn depend on the order of the input sequences. A way to overcome this is to present to the algorithm the input sequences in a different order. The CONSENSUS algorithm keeps track of a large number of partial solutions for each iteration and selects the result with the best score at the end.

We finally wrap-up the different functions and algorithms within the class building a simple testing function:

```
def test():
    seq1 = MySeq("ATAGAGCTGA","dna")
    seq2 = MySeq("ACGTAGATGA","dna")
    seq3 = MySeq("AAGATAGGGG","dna")
    mf = DeterministicMotifFinding(3, [seq1,seq2,seq3])

    print ("Exhaustive:")
    sol = mf.exhaustive_search()
    print ("Solution: ", sol)
    print ("Score: ", mf.score(sol))

    print ("\nBranch and Bound:")
    sol2 = mf.branch_and_bound()
    print ("Solution: ", sol2)
    print ("Score:", mf.score(sol2))

    print ("\nHeuristic consensus: ")
    sol3 = mf.heuristic_consensus()
    print ("Solution: ", sol3)
    print ("Score:", mf.score(sol3))
```

We can use a more realistic example using the input sequences in the file "exampleMotifs.txt" provided in the book's website. Note the difference in the running times of the different algorithms.

```
def test2():
    mf = DeterministicMotifFinding()
    mf.read_file("exampleMotifs.txt","dna")
    print ("Branch and Bound:")
    sol = mf.branch_and_bound()
    print ("Solution: ", sol)
    print ("Score:", mf.score(sol))

    print ("\nHeuristic consensus: ")
```

```
sol2 = mf.heuristic_consensus()
print ("Solution: ", sol2)
print ("Score:", mf.score(sol2))
```

Bibliographic Notes and Further Reading

The problem of deterministic motif finding represents an interesting computational challenge that has attracted the attention of many computational biologists and computer scientists over the years. It is out of the scope of this book to make here an exhaustive review of the work in this area, which has already been done many authors, see for instance [136], [96], [41] and [17].

The book of Dan Gusfield [70] is an essential work for all of those with particular interest in these topics, providing core formalisms and discussing strategies and data structures commonly used. The book by Jones and Pevzner [84] has undoubtedly influenced the work here presented. While each topic is motivated by real life examples it is approached from an analytical and algorithmic point of view.

Sagot [134], and Pevzner and Sze [126] have formulated the deterministic motif finding problem as the task of finding in a set of N input sequences, the occurrences of a motif of length L, that should occur in all or a given minimal number of sequences. The motif occurrences should have a distance (number of different nucleotides) from each other less than d symbols. In the former work, this was called as the *common motif problem* and in the later work as the Planted (L, d)-Motif problem. Both studies have proposed efficient algorithms for this task. The Planted (L, d)-Motif problem became a well known challenge, since for large values of L (> 14) and d (> 4) it is a difficult problem. Methods based on enumeration strategies do not perform well on this problem given their exponential complexity with relation to motif length. Several methods have been proposed based on other strategies, including graph-based methods (e.g. Winnover and SP-Star) [126], projection strategies [30], or even statistical distributions and Markov models [109], just to name a few. Other works, tackling related research problems, have followed, including those that have looked to the limits in which a motif can be still considered significant [88] or for more complex motif formulations such as the composite motif discovery problem [59].

Exercises and Programming Projects

Exercises

1. Implement a function to visualize the consensus motif. This function should scan the motif frequency matrix and for each column output the most frequent symbol.

2. Extend the previous function to include the IUPAC nucleotide ambiguity code. For columns with multiple symbols use the corresponding symbol from IUPAC.
3. The result of the **heuristic_consensus** algorithm is highly dependent on the first two input sequences. Create a wrapper function that performs multiple iterations presenting the input sequences in shuffled order and keep track of the resulting IUPAC consensus motif at each iteration.

Programming Projects

1. The motif finding algorithms presented in this chapter assume a *one or more motif occurrences* per sequence. It can happen that the input set of sequences contain sequences that are not target of a motif (present in the rest of the sequences). Try to adapt the above implementations to handle *zero or more motif occurrences*. The initial position vector for a sequence without a motif occurrence will have an offset of -1. One possible criteria to define the absence of a motif hit in a sequence is that all its strings of length L have approximately the same score. Thus, one could define a motif absence when the difference between the best scoring string and the worst scoring string is lower than a given threshold score.
2. Adapt the function implementing the CONSENSUS algorithm to be more robust, considering at each stage several possible alternatives and exploring those in further iterations. Keep in each iteration a maximum of K alternatives, where K could be user-defined.

CHAPTER 11

Probabilistic Motifs and Stochastic Algorithms

In this chapter, we present probabilistic representations of biological motifs, mainly position weighted matrices, allowing to assign probabilities of symbols for specific positions of a pattern. First, we will address how to use these representations to search for motif occurrences over DNA and protein sequences. Then, we will discuss algorithms for probabilistic motif discovery, including stochastic algorithms based on Expectation-Maximization strategies and Gibbs sampling.

11.1 Representing and Searching Probabilistic Motifs

Probabilistic motifs are typically expressed through *Probabilistic Weight Matrices (PWM)* [47,68,144] also called *Templates* or *Profiles*. A PWM is a matrix of the weighted matches of each of the biological symbols (nucleotides or aminoacids) as the rows, and each position of the motif, as columns. Fig. 11.1 shows the scheme of a PWM, which represents a DNA motif of size N. The value of the cell P_{ij} denotes the probability of the nucleotide i be found in position j of the motif P. In this model, independence between symbols of the motif is assumed. For a sequence $S = S_1 S_2 ... S_N$ of length N, the likelihood of being matched by a PWM P is given by formula (11.1).

For sequence comparison purposes the logarithms are handled easier than probability values, therefore the log-odds of the P_{ij} cells are usually used, creating the so-called *Position Specific*

	1	2	...	N
A	P_{A1}	P_{A2}	...	P_{An}
T	P_{T1}	P_{T2}	...	P_{Tn}
C	P_{C1}	P_{C2}	...	P_{Cn}
G	P_{G1}	P_{G2}	...	P_{Gn}

Figure 11.1: Representation of a *Position Weight Matrix*. Rows represent symbols and columns the positions along the motif. Each cell contains the probability of finding the respective symbol at a certain position.

Bioinformatics Algorithms. DOI: 10.1016/B978-0-12-812520-5.00011-0

```
GATCAT
GATGAT
GAAGAA
CAAGAC
AAACTT
GGACCT
GCAAAG
CTGCAT
```

⟹

	1	2	3	4	5	6
A	1/8	5/8	5/8	1/8	6/8	1/8
T	0	1/8	3/8	0	1/8	5/8
C	1/8	1/8	0	4/8	1/8	1/8
G	5/8	1/8	1/8	3/8	0	1/8

S = GCGGATCATCAA

Scanned Sequence	Probability calculation	p(a|P)
GCGGATCATCAA	5/8 x 1/8 x 1/8 x 3/8 x 6/8 x 1/8	3.4×10^{-4}
G**CGGATC**ATCAA	1/8 x 1/8 x 1/8 x 5/8 x 1/8 x 1/8	3.8×10^{-6}
GC**GGATCA**TCAA	5/8 x 1/8 x 5/8 x 0 x 1/8 x 1/8	0
GCG**GATCAT**CAA	5/8 x 5/8 x 3/8 x 4/8 x 6/8 x 5/8	**0.03433**
GCGG**ATCATC**AA	1/8 x 1/8 x 0 x 1/8 x 1/8 x 1/8	0
GCGGA**TCATCA**A	0 x 1/8 x 5/8 x 0 x 1/8 x 1/8	0
GCGGAT**CATCAA**	1/8 x 5/8 x 3/8 x 4/8 x 6/8 x 1/8	1.4×10^{-3}

Figure 11.2: Position weight matrix derived from a set of eight sequences and scanning of input sequence for finding the most probable motif match.

Scoring Matrices. The likelihood of S being matched by P can be written as a score given by formula (11.2).

$$p(S, P) = \prod_{i=1}^{N} P(S_i, i) \tag{11.1}$$

$$score(S, P) = \sum_{i=1}^{N} logP(S_i, i) \tag{11.2}$$

Fig. 11.2 presents the PWM generated from a set of 8 sequences. The probability of a sequence $a = GATCAT$ matching this motif P is given by the product of all its position probabilities as: $p(GATCAT|P) = \frac{5}{8} \times \frac{7}{8} \times \frac{3}{8} \times \frac{4}{8} \times \frac{6}{8} \times \frac{5}{8} = 0.03433$.

Given a sequence S and PWM P, we can calculate the sub-sequence of S with length N with the highest likelihood of being matched by P. This can be done by scanning S with a sliding window of length N and applying the calculation as in formula (11.1). Fig. 11.2 describes the process and the respective probability calculation when scanning an input sequence.

We will now define the class **MyMotifs** that handles position weight matrices and implements the calculations described above. Before that, we implement two functions: the first one is required to create a matrix of given dimensions and the second for visualization of a matrix.

```
def create_matrix_zeros (nrows, ncols):
    res = [ ]
    for i in range(0, nrows):
        res.append([0]*ncols)
    return res

def print_matrix(mat):
    for i in range(0, len(mat)): print(mat[i])
```

The class **MyMotifs** implements the data structure and methods required to create a PWM, to derive different deterministic representations of the motif, and to determine the probability of occurrence of the motif along a sequence.

The class contains five attributes: i) sequences used to build the PWM model; ii) total number of sequences; iii) alphabet; iv) a matrix containing the absolute counts, and v) a matrix with the frequency of the symbols at each position of the motif. These matrices are created calling the functions **do_counts** and **create_pwm**.

```
class MyMotifs:
    """Class to handle Probabilistic Weighted Matrix"""
    def __init__(self, seqs = [], pwm = [], alphabet = None):
        if seqs:
            self.size = len(seqs[0])
            self.seqs = seqs # objet from class MySeq
            self.alphabet = seqs[0].alphabet()
            self.do_counts()
            self.create_pwm()
        else:
            self.pwm = pwm
            self.size = len(pwm[0])
            self.alphabet = alphabet

    def __len__ (self):
        return self.size
```

```
    def do_counts(self):
        self.counts = create_matrix_zeros(len(self.alphabet), self.
size)
        for s in self.seqs:
            for i in range(self.size):
                lin = self.alphabet.index(s[i])
                self.counts[lin][i] += 1

    def create_pwm(self):
        if self.counts == None: self.do_counts()
        self.pwm = create_matrix_zeros(len(self.alphabet), self.size)
        for i in range(len(self.alphabet)):
            for j in range(self.size):
                self.pwm[i][j] = float(self.counts[i][j]) / len(self.
seqs)
```

From a PWM, we can generate a consensus sequence that captures the most conserved symbols at each position of the motif. The **consensus** function scans every position of the motif and selects the most frequent symbol at each position. The **masked_consensus** function works in a similar way, but for each position it outputs a symbol from the alphabet in the case its frequency is at least 50% of the sequences, otherwise it outputs the symbol "-".

```
    def consensus(self):
        """ returns the sequence motif obtained with the most
frequent symbol at each position of the motif"""
        res = ""
        for j in range(self.size):
            maxcol = self.counts[0][j]
            maxcoli = 0
            for i in range(1, len(self.alphabet) ):
                if self.counts[i][j] > maxcol:
                    maxcol = self.counts[i][j]
                    maxcoli = i
            res += self.alphabet[maxcoli]
        return res

    def masked_consensus(self):
        """ returns the sequence motif obtained with the symbol that
occurs in at least 50% of the input sequences"""
```

```
    res = ""
    for j in range(self.size):
        maxcol = self.counts[0][j]
        maxcoli = 0
        for i in range(1, len(self.alphabet) ):
            if self.counts[i][j] > maxcol:
                maxcol = self.counts[i][j]
                maxcoli = i
        if maxcol > len(self.seqs) / 2:
            res += self.alphabet[maxcoli]
        else:
            res += "-"
    return res
```

We then implement in the class the functions to calculate the probability of a sequence matching a given motif. This is implemented in function **probability_sequence** where for the input sequence we calculate the product of the probability of occurrence in that position of all its symbols.

We can then generalize this procedure for a longer sequence and apply the probability of occurrence to every sub-sequence of motif length N (function **probability_all_positions**). Note that the scanning is done from the first index to the one corresponding to position $|S| - N + 1$, i.e. to the length of the input sequence minus the length of the motif plus one. The probabilities of every sub-sequence are then stored in a list. This function can be easily adapted to find the sub-sequence with highest probability of matching the motif. In the function **most_probable_sequence**, the index of such sub-sequence is returned.

Knowing the sub-sequences where the motif has higher likelihood of occurrence will allow to update and refine our motif. In function **create_motif** we scan the input sequences and select the most probable sub-sequences to build a new motif with an object of the class **MyMotifs** being returned.

```
def probability_sequence(self, seq):
    res = 1.0
    for i in range(self.size):
        lin = self.alphabet.index(seq[i])
        res *= self.pwm[lin][i]
    return res

def probability_all_positions(self, seq):
```

```
        res = []
        for k in range(len(seq)-self.size+1):
            res.append(self.probability_sequence(seq))
        return res

    def most_probable_sequence(self, seq):
        """ Returns the index of the most probable sub-sequence of
    the input sequence"""
        maximum = -1.0
        maxind = -1
        for k in range(len(seq)-self.size):
            p = self.probability_sequence(seq[k:k+ self.size])
            if(p > maximum):
                maximum = p
                maxind = k
        return maxind

    def create_motif(self, seqs):
        from MySeq import MySeq
        l = []
        for s in seqs:
            ind = self.most_probable_sequence(s.seq)
            subseq = MySeq ( s[ind:(ind+self.size)], s.
    get_seq_biotype() )
            l.append(subseq)

        return MyMotifs(l)
```

Finally, code is provided to test our class. We start by providing eighth sequences to build the motif and visualize the matrix of absolute counts and frequencies. The consensus and masked consensus motifs are also found. Probability calculation is also performed for several input sequences.

```
def test():
    from MySeq import MySeq
    seq1 = MySeq("AAAGTT")
    seq2 = MySeq("CACGTG")
    seq3 = MySeq("TTGGGT")
    seq4 = MySeq("GACCGT")
```

```
seq5 = MySeq("AACCAT")
seq6 = MySeq("AACCCT")
seq7 = MySeq("AAACCT")
seq8 = MySeq("GAACCT")
lseqs = [seq1, seq2, seq3, seq4, seq5, seq6, seq7, seq8]
motifs = MyMotifs(lseqs)
print ("Counts matrix")
print_matrix (motifs.counts)
print ("PWM")
print_matrix (motifs.pwm)
print ("Sequence alphabet")
print(motifs.alphabet)

[print(s) for s in lseqs]
print ("Consensus sequence")
print(motifs.consensus())
print ("Masked Consensus sequence")
print(motifs.masked_consensus())

print(motifs.probability_sequence("AAACCT"))
print(motifs.probability_sequence("ATACAG"))
print(motifs.most_probable_sequence("CTATAAACCTTACATC"))
```

In the first part of this chapter, we discussed how a PWM provides a probabilistic representation of a motif by capturing the frequency of each symbol along its sequence positions. PWM can then be used to search for novel motifs matches within the input sequences and can be refined by incorporating other high-scoring matches.

When the number of sub-sequences from which the PWM model is derived is relatively small, some symbols may not be observed at some positions, i.e. have an observed frequency of zero. This will lead to situations where the resulting motif probability will be zero. To avoid this and to increase numerical stability a pseudo-count is typically added to the PWM values.

Visualization is another important aspect of motif representation. A common way to visualize a PWM is through a sequence logo, introduced by Schneider and Stephens in 1990 [47,137]. This type of graphics represents a stack of letters with a height proportional to the conservation level of the motif positions. Tools such as Weblogo [40] allow to easily create appealing representations of PWMs. In sequence logos, the information content of position i for every symbol b, is given by formula (11.3).

$$I_i = 2 + \sum_b P_{b,i} \times log_2 P_{b,i} \tag{11.3}$$

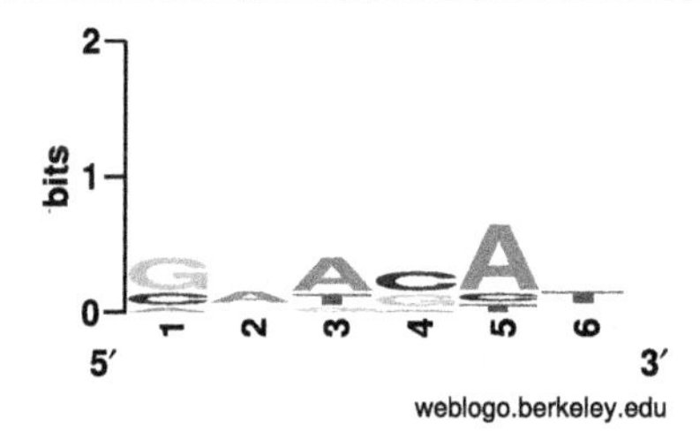

Figure 11.3: Weblogo for the eight sub-sequences.

For a DNA motif, positions perfectly conserved contain 2 bits of information, while 1 bit corresponds to positions where two of the four bases occur in less than 50% of the positions. Fig. 11.3 shows the sequence logo created by Weblogo for the eight sub-sequences in Fig. 11.2.

11.2 Stochastic Algorithms: Expectation-Maximization

So far, we were given a set of sub-sequences that represent at least a good initial guess for the best motif location along the input sequences. From these sub-sequences we can derive the PWM that captures the characteristics of the motif. But, as we already discussed in the previous chapter, this information is typically not known in advance. Therefore, methods for the discovery of motifs need to be applied.

Recall from the definition of the motif finding problem in Chapter 10 that the input consists of a set D with t input sequences (DNA or protein) of length L (possibly variable among sequences) and a motif length N. We expect the method to output a vector s with length t containing the initial positions of the sub-sequences in t that maximize the scoring function $score(s, D)$ (objective function that maximizes the quality of the motif).

While the problem setting is the same for deterministic and probabilistic motifs, the algorithmic approach for these two types of motifs will be different since they require different ways to scan the search space. In the previous chapter, we have analyzed how we can use enumeration methods for exhaustive search, and how certain heuristics can improve the efficiency of this search. While these approaches tend to find the best possible motif solutions, they require certain model abstractions to achieve an enumerable search space and, therefore, may overlook less evident (possibly more degenerate) motifs.

Algorithms based on the *Expectation Maximization* (EM) can overcome this problem by providing a way to traverse the search space and to optimize the motifs. This strategy relies on an iterative approach that can simultaneously determine the most representative sub-sequences, i.e. the best hits of the motif along the input sequences, while updating and refining the motif model.

For a motif of length L, the EM algorithm scans every sub-sequence of length L in the input sequences and initializes the PWM based on the symbols of that single sub-sequence adding a small fraction of the background frequencies. Next, for each sub-sequence of length L in the input sequences it calculates the probability of being generated by the motif rather than a random model of the input sequences. The best hits are then used to improve the motif by averaging the frequencies of all the selected sub-sequences. The procedure stops when the improvement on the score is zero or minimal.

This approach typically converges to a near-optimal solution. Additional motifs can be found by masking (replacing the symbols) the sequences for the best hits of the motif and re-running the method. *MEME* [18] is one of the most representative implementations of the EM algorithm for motif finding.

Fig. 11.4 depicts the EM process of initializing the motif with a fraction of the background frequencies, in this case assuming a fraction of 20% of a uniform distribution of the nucleotides. The motif model is iteratively updated by scanning the input sequences and refining the motif with its new best hits.

By scanning all the sub-sequences of length L in the input sequences, the EM algorithm is performing a deterministic optimization. While a convergence to an optimal solution typically occurs, it requires an exhaustive enumeration of all initial solutions. Frequently, the search process can be optimized by introducing a stochastic component on the search. For instance, the selection of the initial sub-sequences used to derive the motif can be randomly selected.

We now describe the basic steps that form the basis of the algorithms based on stochastic heuristic search:

1. Start by randomly selecting the initial positions $s = (s_1, \ldots, s_t)$ along the input sequences D.
2. A PWM profile P is created from the sequences generated in step 1.
3. The motif P is used to search for the most probable sub-sequences n in D. The vector s is then updated for these new initial positions according to n.
4. A new PWM profile P is created using the positions calculated in step 3. Steps 3 and 4 are repeated the score $score(s, D)$ is no longer improved.

We now extend the class **MotifFinding** that we started developing in the previous chapter. Recall that this class will contain methods for deterministic and probabilistic motif search. Additive and multiplicate scores are provided by the functions **score** and **score_multiplicative**.

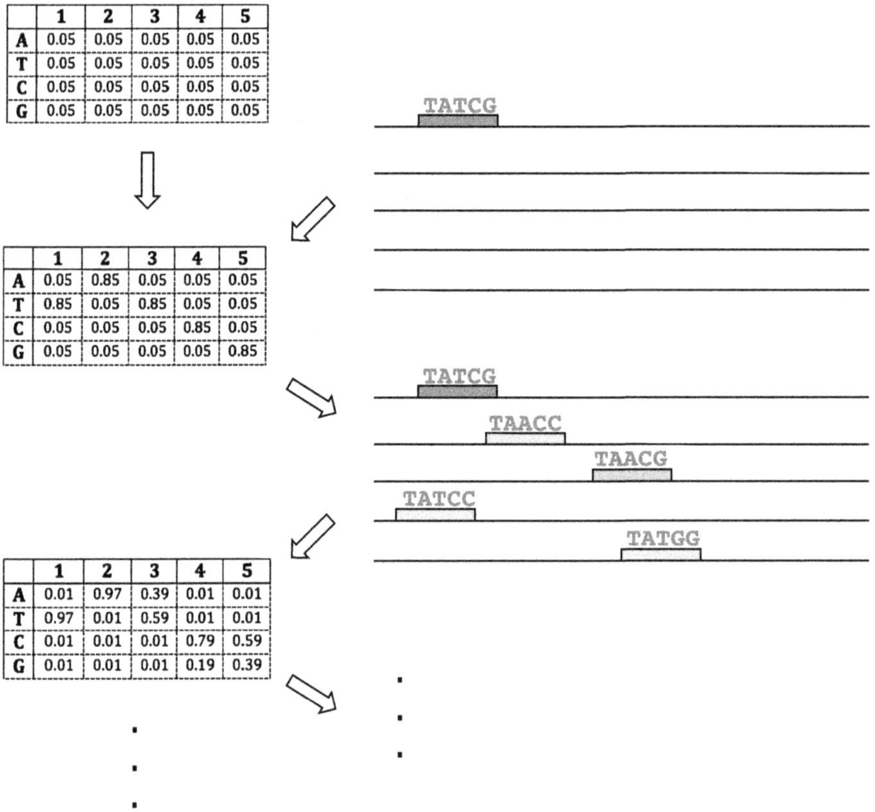

	1	2	3	4	5
A	0.05	0.05	0.05	0.05	0.05
T	0.05	0.05	0.05	0.05	0.05
C	0.05	0.05	0.05	0.05	0.05
G	0.05	0.05	0.05	0.05	0.05

TATCG

	1	2	3	4	5
A	0.05	0.85	0.05	0.05	0.05
T	0.85	0.05	0.85	0.05	0.05
C	0.05	0.05	0.05	0.85	0.05
G	0.05	0.05	0.05	0.05	0.85

TATCG
TAACC
TAACG
TATCC
TATGG

	1	2	3	4	5
A	0.01	0.97	0.39	0.01	0.01
T	0.97	0.01	0.59	0.01	0.01
C	0.01	0.01	0.01	0.79	0.59
G	0.01	0.01	0.01	0.19	0.39

Figure 11.4: Expectation-Maximization strategy for motif finding. Representation of how a motif model is created from an initial sub-sequence and then iteratively updated by scanning the input sequences. The PWM is initialized with a fraction (20%) of uniform background frequencies. A sub-sequence is used to update the PWM. Input sequences are then scanned for motif matches, which are then used to update the PWM. The process is repeated until no improvements on the motif score are obtained.

In the following code, we introduce the function **create_motif_from_indexes** that creates a probabilistic motif of type **MyMotifs**, given a list of indexes containing the initial positions of the sub-sequences used to build the motif model. Several objects of the class **MySeq** are created based on the sub-sequence and biological type and then passed to class **MyMotifs**.

The previously described algorithm is now implemented in function **heuristic_stochastic**. It creates a motif object from randomly selected initial positions on the input sequences, the corresponding PWM and its multiplicative score. It proceeds with the iterative process of refining the motif with the most probable sub-sequence from each of the input sequences, until the score of the resulting new motif can no longer be improved, i.e. does not increase with relation to the previous score.

```
def create_motif_from_indexes(self, indexes):
    pseqs = []
    for i,ind in enumerate(indexes):
        pseqs.append( MySeq(self.seqs[i][ind:(ind+self.motif_size)],
   self.seqs[i].get_seq_biotype()) )
    return MyMotifs(pseqs)

def heuristic_stochastic (self):
        from random import randint
        s = [0]* len(self.seqs)
        for k in range(len(s)):
            s[k] = randint(0, self.seq_size(k)- self.motif_size)
        motif = self.create_motif_from_indexes(s)
        motif.create_pwm()
        sc = self.score_multiplicative(s)
        bestsol = s
        improve = True
        while(improve):
            for k in range(len(s)):
                s[k] = motif.most_probable_sequence(self.seqs[k])
            if self.score_multiplicative(s) > sc:
                sc = self.score_multiplicative(s)
                bestsol = s
                motif = self.create_motif_from_indexes(s)
                motif.create_pwm()
            else: improve = False
        return bestsol
```

In this approach, the random choice of the initial positions will certainly have an impact on the final solution. If these initial positions are close to the best possible ones, the search will converge faster to an optimal solution, while if they are far away, the algorithm will take longer and may not always converge to the desired solution. Thus, it is a frequent procedure to run the method multiple times to increase the likelihood of finding the best possible solution.

11.3 Gibbs Sampling for Motif Discovery

The previous algorithm can be improved by introducing a strategy called "Gibbs sampling". This idea, proposed by Lawrence et al. [94], is used by several motif finding algorithms and as a heuristic method it does not guarantee the best solution, but in practice when run multiple times it works well.

As in the previous approach, the motif model is initialized with randomly selected sub-sequences, which are then scored against the initial model. At each iteration, the algorithm performs a local search by probabilistically deciding if one of the motif hits should be updated. For a given input sequence, it removes the corresponding sub-sequence used to model the motif. Then, it tries to replace with another sub-sequence, computes the score of the motif, and decides if the update is kept. More formally, the steps of the Gibbs sampling approach can be enumerated as:

1. Start by randomly selecting the initial positions $s = (s_1, \ldots, s_t)$ along the input sequences D.
2. Randomly select a sequence i from D.
3. Create a PWM P excluding in s the sub-sequence from the sequence selected in step 2.
4. For each position p in sequence i, calculate the probability of sub-sequence started in p with length L being generated by P.
5. Select p in a stochastic way according to the probabilities calculated in step 4.
6. Repeat steps 2 to 5 until the score of s in motif P keeps improving.

Fig. 11.5 provides a graphical overview of the Gibbs sampling approach. Note that in this representation for the selected sequence (step 2), all its sub-sequences are tested and the corresponding scores calculated. The sub-sequence with the highest score is the one selected to incorporate the new motif version.

The following **gibbs** function implements the above algorithm. It receives as an input an optional parameter with number of iterations.

```
def gibbs (self, iterations = 100):
    from random import randint
    s = []
    for i in range(len(self.seqs)):
        s.append(randint(0, len(self.seqs[0]) - self.motif_size -
1))
    best_s = list(s)
    best_score = self.score_multiplicative(s)
    for it in range(iterations):
        # randomly pick a sequence
        seq_idx = randint(0, len(self.seqs)-1)
        seq_sel = self.seqs[seq_idx]
        s.pop(seq_idx)
```

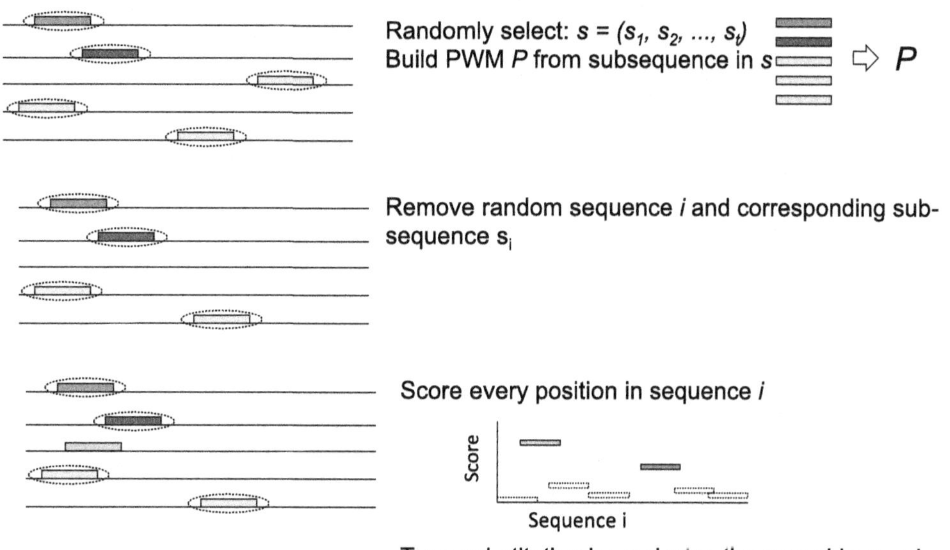

Figure 11.5: Schematic representation of Gibbs sampling approach. A motif model is created by randomly sampling one sub-sequence from each of the input sequences. The motif is iteratively updated by randomly selecting one of the input sequences and including in the model the corresponding sub-sequence with the highest potential to be a motif hit.

```
            removed = self.seqs.pop(seq_idx)
            motif = self.create_motif_from_indexes(s)
            motif.create_pwm()
            self.seqs.insert(seq_idx, removed)
            r = motif.probability_all_positions(seq_sel)
            pos = self.roulette(r)
            s.insert(seq_idx, pos)
            score = self.score_multiplicative(s)
            if score > best_score:
                best_score = score
                best_s = list(s)
        return best_s

    def roulette(self, f):
        from random import random
        tot = 0.0
```

```
    for x in f: tot += (0.01 + x)
    val = random() * tot
    acum = 0.0
    idx = 0
    while acum < val:
        acum += (f[idx] + 0.01)
        idx += 1
    return idx-1
```

After determining the probabilities of all the sub-sequences along the removed sequence, one could determine the new best position as the one with the highest score with relation to the motif. Instead, a different strategy is used. With the **roulette** function a sample space for the experiment of spinning once a roulette wheel is created. All positions have a chance of being picked, but this chance is proportional to their score.

We can now compare the results obtained with a general stochastic algorithm and those from Gibbs sampling.

```
def test_heuristic_algorithms():
    mf = MotifFinding()
    mf.read_file("exampleMotifs.txt","dna")

    print("Heuristic stochastic")
    sol = mf.heuristic_stochastic()
    print ("Solution: ", sol)
    print ("Score:", mf.score(sol))
    print ("Score mult:", mf.score_multiplicative(sol))
    print("Consensus:", mf.create_motif_from_indexes(sol).consensus()
)

    sol2 = mf.gibbs(1000)
    print ("Score:", mf.score(sol2))
    print ("Score mult:", mf.score_multiplicative(sol2))
    print("Consensus:", mf.create_motif_from_indexes(sol2).consensus
())
```

11.4 Probabilistic Motifs in BioPython

The package *Bio.Motifs* of *BioPython* provides a set of functionalities to work with probabilistic motifs, their representation and search over target sequences. We will show here a few examples, while a more complete coverage can be found in Chapter 14 of the tutorial.

The following code example shows how to create an object of the class **Motif**, which represents a motif in *BioPython*, from a set of instances (sequences represented as a list of objects of class **Seq**). The content of the motif *m* is then printed, including its consensus, the raw counts of each symbol for each position, the PWM, and the created PSSM. The last line will save a logo of the motif to your current folder.

```
from Bio.Seq import Seq
from Bio import motifs
from Bio.Alphabet import IUPAC

instances = []
instances.append(Seq("TATAA",IUPAC.unambiguous_dna))
instances.append(Seq("TATTA",IUPAC.unambiguous_dna))
instances.append(Seq("TTTAT",IUPAC.unambiguous_dna))
instances.append(Seq("TATAC",IUPAC.unambiguous_dna))

m = motifs.create(instances, IUPAC.unambiguous_dna)

print(m)
print(len(m))
print(m.consensus)
print(m.counts)
print(m.pwm)
print(m.pssm)

m.weblogo("mymotif.png")
```

Notice in the results of the previous script that the PSSM has a few positions filled with "-inf", representing minus infinity. This happens due to the presence of zeros in the PWM. One way to prevent this situation is the use of pseudo-counts, as discussed above. The following code shows how to add pseudo-counts in the example, re-generating the respective PWM and PSSM.

```
pwm = m.counts.normalize(pseudocounts=0.5)
pssm = pwm.log_odds()
print(pwm)
print(pssm)
```

It is important to mention that there are also functions that allow to read motifs in different formats, with the functions **motifs.read** or **motifs.parse** (for multiple instances from the same file). These allow to load motifs from databases and tools as JASPAR, MEME, or TRANSFAC. Also, the function **motifs.write** allows to write motifs in a selected format.

Another important functionality of this module is the ability to search for motif occurrences in target sequences. One possible type of search is to find exact occurrences of the instances used to create the motif. This is done in the next example for a toy sequence to exemplify its use.

```
test_seq=Seq("TTTTATACACTGCATATAACAACCCAAGCATTATAA", IUPAC.
    unambiguous_dna)

for pos, seq in m.instances.search(test_seq):
     print (pos, " ", seq)
```

Probably a more useful feature is the ability to look for matches of the motif using the PSSM to score those matches and defining proper thresholds for acceptance. This is done in the example below defining as the minimum threshold a score of 4 to accept the match.

```
for position, score in pssm.search(test_seq, threshold=4.0):
     print ("Position %d: score = %5.3f" % (position, score) )
```

Notice that in the result some positions are returned as negative numbers. These are matches of the motif in the reverse complement of the original sequence, which is also scanned.

Finally, we will show how we can provide the scores for all possible positions of the motif in the sequence in a list by using the function **calculate**.

```
print(pssm.calculate(test_seq))
```

The support of *BioPython* for motif discovery is quite limited, mainly consisting on the ability to parse the results of the MEME tool, using the input/output functions mentioned above.

Bibliographic Notes and Further Reading

There is an extensive research on motif finding with many algorithms proposed over the last decades. This prompts the question of which of the methods and algorithms are better for the task. The answer to this question depends on many aspects including the characteristics of the input data and the defined constraints on the search process. In order to provide a more generic view on the performance of different methods several small and large-scale comparisons have been done. We refer to the work of Tompa et al. [148] that provides a comparison of thirteen methods for the identification of transcription factor binding sites.

Beyond sequence conservation, there are other features that may indicate that a motif represents a true biological signal. This includes among other aspects location specificity that captures the bias on motif localization. Example of this includes motifs that occur preferentially on the promoter region close to the transcription start site. Motif co-occurrence is another interesting aspect, where motifs may occur in modules in combination with other motifs. Several algorithms take advantage of one or more of these features to determine the relevance of the discovered motifs and discard spurious discoveries [153].

Here, we quantified the interest of a motif based on its probability or respective score. Nevertheless, other measures were developed to quantify the interest of a motif and can be used as complementary measures. For instance, measures like information gain, surprise, mutual information capture how unexpected is the motif occurrence given the background composition of the input sequences. We refer to the work of Li and Tompa [101] and Ferreira and Azevedo [65] which have focused on the evaluation of different measures of motif interest.

Exercises and Programming Projects

Exercises

1. Within the class **MyMotifs** create a function that receives as input a constant and adds this constant as a pseudo-count of the PWM.
2. Write a function for the class **MyMotifs** that given an input sequence and a score threshold, reports all the motif matches with a score higher than a given threshold.
3. In the Expectation-Maximization algorithm, the PWM is initialized with a fraction of the background frequencies. Write a function that given a fraction ($(0, 1]$) and the set of input sequences, initializes the PWM accordingly.

Programming Projects

1. Add to the class **MyMotifs** the possibility of calculating PSSM and using those to score hits of the motif in sequences. Consider using pseudo-counts to avoid problems with the logarithm of zero. Implement search of motifs over sequences ranking the results and filtering through the use of a minimum score threshold. Consider the option to search for motifs not only in the sequence, but also in the reverse complement (assuming DNA sequences).
2. Implement a function to generate a random solution for a motif finding problem. Based on this function, implement a random search algorithm, that generates N solutions using the previous function, and selects the one with highest score (N is a parameter). Compare the results of this algorithm with the ones developed in this chapter, providing the same maximum number of solutions to each.

CHAPTER 12

Hidden Markov Models

In this chapter we present Hidden Markov Models (HMMs), as stochastic models that capture statistical regularities from the input sequences and allow to devise algorithms for motif finding and database search. We will discuss the elements that compose an HMM and how input sequences can be evaluated in terms of their likelihood. How to improve the HMMs parameters from additional training sequences will also be discussed. Implementations for simplified versions of the main algorithms in Python are provided. Finally, examples of some of the many applications of HMM in biological sequence analysis are provided.

12.1 Introduction: What Are Hidden Markov Models?

In Chapter 5, we have seen that the string matching problem consists of finding in a sequence S, defined over an alphabet Σ, all the occurrences of pattern P (defined in the same alphabet). In Section 5.4, we introduced the Deterministic Finite Automaton (DFA), an abstract mathematical concept, that can be used to represent a string language, i.e. the set of strings that match a given pattern. The basic idea is to build an automaton from the pattern. Then, once the DFA is built multiple strings can be scanned to look for the match of the pattern. The DFA is represented as a graph with a finite number of states. Each symbol in the pattern corresponds to a state and every matched symbol sends the automaton to a new state. The DFA has a start and an end state. If the end state is reached, it means that the scanned string is accepted. Using a DFA, only one pass over the input string is required, making it a very efficient method for string matching. DFAs provide a deterministic pattern search, i.e. either match or not the scanned sequence, as it happened also with regular expressions.

However, in biological sequence analysis, the events associated to a sequence are often probabilistic. Examples of this include for instance, the classification of the subcellular localization (e.g. cytosol, nucleus or the membrane) of a protein given its sequence; the presence of protein domains within a sequence or the binding probability of a transcription factor on the sequence of a gene promoter region.

Hidden Markov Models (HMMs) are probabilistic models that capture statistical regularities within a linear sequence. As we will see, HMMs are very well suited for the mentioned tasks. Indeed, HMMs have been largely used in many different problems in molecular biology, including gene finding, profile searches, multiple sequence alignment, identification of regulatory sites, or the prediction of secondary structure in proteins.

Bioinformatics Algorithms. DOI: 10.1016/B978-0-12-812520-5.00012-2

Let us now examine the structure of an HMM by looking at a specific example. Consider the elements of a mature mRNA. It contains a tripartite structure that starts with a 5' untranslated region (5' UTR), a region that will be translated and code for a protein (CDS) and ends with a 3' untranslated region (3'UTR) [112]. Moreover, we also know that the CDS region should start with a start codon (ATG) and end with one of the three stop codons. Here, for sake of simplicity we will omit the codon details.

So, in this case, we have a grammatical structure that defines a mature mRNA. We can view the untranslated and CDS regions as words. The sentences of our grammar will have certain rules, like the fact that the 5'UTR should always precede the CDS and the 3'UTR should always come after the CDS. We can also assume that the three elements will always be present in a mature mRNA. Our goal will be to scan a sequence and label each of its nucleotide symbols with the corresponding region. A probability will be associated at every labeling point. The representation of each region or label will be called state of the HMM.

When scanning a symbol at a given state, there is a certain probability that the next symbol will belong to the current state or not. The probabilities associated with a change of state are called *transition probabilities* and depend on several factors including, for instance, the typical sequence length of the corresponding region. In this case, let's consider that, when scanning a symbol on the 5'UTR, we have 0.8 probability that the next symbol will also be part of the 5'UTR and 0.2 probability that it will change to the CDS region. We can ascertain probabilities to the other regions in a similar manner. To model our HMM we will need three states, plus two additional states that define the start and the end of the sequence. The start state represents the probability of starting in any of the defined states.

It is well known that the composition of mRNA regions differ depending on the species [159]. Considering, for instance, that the $G + C$ content in the 5'UTR is of 60%, in the CDS of 50% and in the 3'UTR of 30%, these compositional frequencies will determine the *emission probabilities* that provide the probability of symbol being emitted by a state.

To determine the probability of a sequence, we just need to multiply the probability of the symbol in the respective state and the respective transition probability. Consider for instance the string TAGTTA with 3 nucleotides within the CDS and 3 nucleotides in the 3'UTR. The probability of this sequence will be given by:

$$\begin{aligned}
&P_{CDS}(T) \times P(CDS|CDS) \times P_{CDS}(A) \times P(CDS|CDS) \times P_{CDS}(G) \\
&\times P(CDS|3U) \times P_{3U}(T) \times P(3U|3U) \times P_{3U}(T) \times P(3U|3U) \times P_{3U}(A) \\
&= 0.25 \times 0.9 \times 0.25 \times 0.9 \times 0.25 \times 0.1 \times 0.35 \times 0.8 \times 0.35 \times 0.8 \times 0.35 \\
&= 3.47287 \times 10^{-5}
\end{aligned}$$

where $P_S(N)$ is the emission probability of symbol N in state S and $P(S_A|S_B)$ the transition probability from state S_A to S_B, which can also be denoted as P_{S_A,S_B}.

Since the multiplication of the probabilities can lead to very small numbers, it is often more convenient to use a score based on the logarithm of the probability, multiplied by -1. If transition and emission probabilities are converted to logarithmic values the score can be calculated as a sum of these components. In this case, the score will be 10.26. The scanning of the input sequence to calculate its probability or score can also be viewed a sequence generation process by the HMM.

At a given state, a probability will be emitted based on the current symbol of the observed sequence. The transition to the next state will depend on the distribution of transition probabilities of the current state. Thus, we can consider that the HMM generates two strings of information. A visible sequence of symbols emitted at each state, called the *observed sequence* and the *state path*, the hidden sequence of states corresponding to the labels of each symbol. The above probability calculation follows a process known as *Markov chain*, which defines a sequence of events based on a finite set of states with a serial dependence, i.e. the next state depends only on the current state.

Similarly to a deterministic finite automaton, an HMM is also defined by five different elements:

- Alphabet of symbols, $A = \{a_1, a_2, ..., a_k\}$.
- Set of states, $S = \{S_1, S_2, ..., S_n\}$.
- Initial state probability, as the probability of being at state S_i at instant $t = 0$, $I(S_i)$.
- Emission probabilities, with $e_{i,a}$ as the probability of emitting symbol a in state i, with $\sum_{a \in A} e_i(a) = 1$.
- Transition probabilities, with $T_{i,j}$ as the transition probability of changing from state i to state j including itself, with $\sum_{i \in S, j \in S} T_i(j) = 1$.

While the first three parameters are represented as uni-dimensional vectors or lists, the emission and transition probabilities are bi-dimensional matrices. The parameters of an HMM are frequently denoted as θ. In certain situations one may refer only to a subset of parameters, e.g. $\theta = (e_{i,a}; T_{i,j})$ if only the emission and transition probabilities are relevant for the context. Additionally, an observed sequence of length T will be denoted as $O = o_1, o_2, .., o_T$. Each o_t denotes the symbol observed at position t. The sequence of states or state path will be denoted as π, with π_t being the state corresponding to position t. We will use the big P notation with square brackets to refer to a conditional probability. For instance $P[O, \pi|M]$ refers to the probability of observing the sequence O and state path π given the model M.

Fig. 12.1 depicts the HMM structure for a simplified version of mature mRNA region labeling. Consider an initial state probabilities of $I(5) = 1$, $I(C) = 0$ and $I(3) = 0$. For a possible

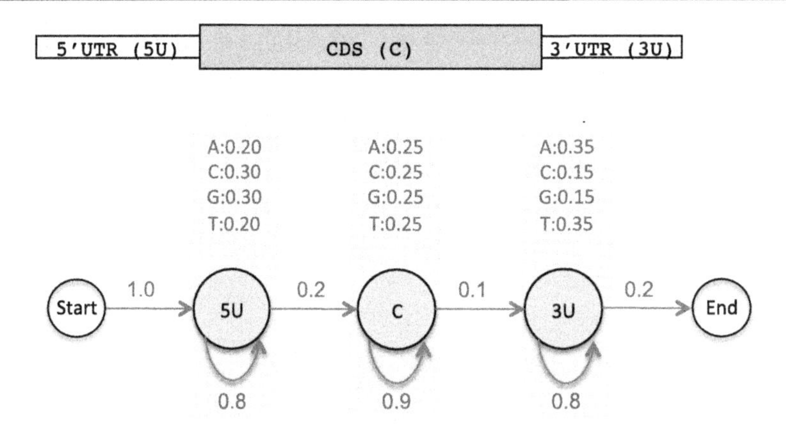

Figure 12.1: HMM scheme to detect three gene regions. (Top) Division of a gene in three regions containing a 5' and 3' untranslated regions flanking the coding regions (CDS). (Middle) Schematic representation for an HMM with 5 states to detect the 3 gene regions. Includes transition and emission probabilities. (Bottom) Example of a segment of DNA sequence and corresponding state of each symbol.

state transition sequence $55CCCC33$ and an emitted sequence $ATCGCGAA$, we can calculate the corresponding probabilities as follows:

$$\begin{aligned} P_T\,[55CCCC33] &= I(5) \times T_{5,5} \times T_{5,C} \times T_{C,C} \times T_{C,C} \times T_{C,C} \times T_{C,C} \times T_{C,3} \times T_{3,3} \\ &= 1 \times 0.8 \times 0.2 \times 0.9 \times 0.9 \times 0.9 \times 0.1 \times 0.8 \approx 0.0093312 \end{aligned}$$

and

$$P_e\,[ATCGCGAA] = 0.2 \times 0.2 \times 0.25 \times 0.25 \times 0.25 \times 0.35 \times 0.35 \approx 1.9141e-05$$

Thus:

$$P\,[e(ATCGCGC) \wedge T(55CCCC33)] = 0.0093312 \times 1.9141e-05 \approx 1.78e-07$$

The tasks that can be solved by an HMM can be divided into three main groups: *scoring* or probability calculation, *decoding*, and *learning*. These functions can be applied to a specific state path or to the universe of all possible state paths.

Scoring functions provide the calculation of the probabilities of an observed sequence. For an observed sequence O and a state path π, the joint probability of S and π under the HMM M

is $P[O, \pi|M]$. This can be calculated as in the example above where we obtain the joint product of emission and transition probabilities. If no state path is provided, the goal is to calculate the total probability of the sequence, $P(O|M)$. This probability can be yield by summing the probabilities of all possible paths, which is an exponentially large number of paths, but can be efficiently calculated with the *Forward algorithm*. This algorithm provides the probability that the HMM M will output a sequence $o_1...o_t$ and reach the state S_i at position t. The Forward probability is denoted as $\alpha_t(S_i)$ and equivalently as $P[O = o_1...o_t, \pi_t = S_i|M]$.

Another important calculation is to determine the backward probability. This provides the probability of starting at state S_i and position t and generating the remainder of the observed sequence $o_t...o_T$. The Backward probability is denoted as $\beta_t(S_i)$ and equivalently as $P[O = o_t...o_T|\pi_t = S_i|M]$.

Decoding refers to the process of determining the most likely underlying sequence of states given an observed sequence. In our previous example this refers to the task of determining the sequence of states 5U, CDS, or 3U for each of the observed sequence symbol. Given an HMM M and an observed sequence O, the task of the decoder is to find the most probable state path π. Here, we will apply a dynamic programming algorithm, called *Viterbi* algorithm, to find the path with the maximum score over all paths. A trace-back function can be used to keep track of all visited paths and to determine the most probable path.

Another relevant question is to determine the total probability of the symbol o_t to be emitted by the state k, when considering all state paths. This is called posterior probability and denoted as $P[\pi_t = k|O, M]$. This probability is given by the product of the forward probability $\alpha_t(k)$ and the backward probability $\beta_t(k)$ divided by the probability of O, $P(O)$.

In our mRNA HMM example we have defined *a priori* the parameters of the model, which we could have determined either using previous knowledge or estimating from observed data. If during the analysis process we acquire more data we may want to adjust these parameters and optimize our model. By analyzing a given dataset we can also learn these parameters. Our goal is to find the parameters $\theta = (e_{i,a}; T_{i,j})$ that maximize the probability of the observed sequence under the model M, $P(O|M)$. If the observed sequence has been labeled, i.e. the state path is known, then a supervised learning is performed. In this case, the emission and transition probabilities are directly inferred from the training data. If the training sequences are unlabeled, then unsupervised learning is performed.

Here, an Expectation-Maximization algorithm called *Baum-Welch* will be used for learning. Briefly, the algorithm starts with best guess for the parameters θ of the model M. It then estimates the emission and transition probabilities from the data. It updates the model according to these new values. It iterates on this procedure until the parameters θ of the model M converge.

Initial

5	0.8
M	0.15
3	0.05

Transition

	5	M	3
5	0.8	0.2	0.0
M	0.0	0.9	0.1
3	0.0	0.0	1.0

Emission

	A	C	G	T
5	0.20	0.30	0.30	0.20
M	0.25	0.25	0.25	0.25
3	0.35	0.15	0.15	0.35

Figure 12.2: Parameters for an example HMM with three states, including initial, transition and emission probabilities matrices. The 5'UTR is represented by the letter 5, the CDS by M and the 3'UTR by 3.

12.2 Algorithms and Python Implementation

In this section, we will look in more detail how the previous tasks are solved with HMMs. The detailed mathematical formulation for the different analysis is out of scope of this book and can be found in the references provided at the end of this chapter. Nevertheless, we provide the sufficient formulation that allows the reader to follow the rational of the implementation which is presented and discussed together.

We will consider a simplified example model based on the example analyzed in the last section. For simplicity, only three states are represented and each state is denoted with one single letter. Fig. 12.2 provides parameters that define our example HMM.

We start by defining the class **HiddenMarkovModel** and its constructor containing five attributes. The initial, emission, and transition probability matrices are represented as dictionaries of dictionaries. The attributes states and symbols are represented as lists and can be inferred from the emission matrix. We define three methods to set and get (not shown here) values from the initial matrices. For this class, some of the methods adapt code from: [82, 157].

```
class HiddenMarkovModel:

    def __init__(self, init_probs, emission_probs, trans_prob):
        """Constructor based on five different attributes: initial,
    emission probabilities and transition probabilities matrices.
    States and symbols lists."""
        self.initstate_prob = init_probs
        self.emission_probs = emission_probs
        self.transition_probs = trans_prob
        self.states = self.emission_probs.keys()
        self.symbols = self.emission_probs[self.emission_probs.keys()
    [0]].keys()
```

```
    def get_init_prob(self, state):
        '''Initial probability of a given state'''
        if state in self.states:
            return (self.initstate_prob[state])
        else:
            return 0

    def get_emission_prob(self, state, symbol):
        '''Probability of a given state emit a symbol'''
        if state in self.states and symbol in self.symbols:
            return (self.emission_probs[state][symbol])
        else:
            return 0

    def get_transition_prob(self, state_orig, state_dest):
        '''Probability of transition from an origin state to
destination state'''
        if state_orig in self.states and state_dest in self.states:
            return (self.transition_probs[state_orig][state_dest])
        else:
            return 0
```

The example in Fig. 12.2 can be represented in our code as follows:

```
initial_probs = {"5": 0.8,"M": 0.15,"3": 0.05}

emission_probs = {"5" : {"A": 0.20, "C": 0.30, "G":0.30, "T"
:0.20}, "M" : {"A":0.25, "C":0.25, "G":0.25, "T":0.25}, "3": {"A"
:0.35, "C":0.15, "G":0.15, "T":0.35}}

transition_probs = {"5":{"5": 0.8,"M": 0.2,"3": 0.0},"M":{"5":
0.0,"M": 0.9,"3": 0.1}, "3":{"5": 0.0,"M": 0.0,"3": 1.0}}
```

12.2.1 Joint Probability of an Observed Sequence and State Path

If we are given an observed sequence O labeled by the state path π and the HMM M, we can calculate the joint probability of O and π under M as $P[O, \pi|M]$. As we have already analyzed in the previous section, this is simply the chained multiplication of the respective transition and emission probabilities. Note that the first element of the multiplication is the initial probability of the first state.

The function **joint_probability** receives an observed sequence and a state path and obtains the chained multiplication of the transition and emission probabilities.

```
def joint_probability(self, sequence, path):
    '''Given an observed sequence and a corresponding state path
calculate the probability of the sequence given the path under
the model'''
    seq_len = len(sequence)
    if seq_len == 0:
        return None

    path_len = len(path)
    if seq_len != path_len:
        print ("Observed sequence and state path of different
lengths!")
        return None

    prob = self.get_init_prob(path[0]) * self.get_emission_prob(
path[0], sequence[0])
    for i in range(1, len(sequence)):
        prob = prob * self.get_transition_prob(path[i-1], path[i
]) * self.get_emission_prob(path[i], sequence[i])

    return prob
```

12.2.2 Probability of an Observed Sequence Over All State Paths

In this section, we approach the task of determining the probability of an observed sequence $O = o_1...o_T$ under an HMM M, $P[O|M]$. In this case the state path is not known. To obtain this probability, we must calculate: $P[O|M] = \sum(P[O|\pi, M] \times P[\pi|M])$. It represents the total probability of the observed sequence over the universe of all possible state paths. Since there is an exponential number of possible state paths, we use dynamic programming approach with the Forward Algorithm, for this calculation. The forward probability provides the probability of observing a sequence $o_1...o_t$, reaching the state S_i and position t for an HMM model M:

$$\alpha_t(k) = P[O = o_1...o_t, \pi_t = S_i|M] \tag{12.1}$$

The algorithm consists of three parts:

- Initialization step:

$$\alpha_1(i) = I(i) \times e_i(o_1), \forall i \in S \tag{12.2}$$

- Iteration step:

$$\alpha_{t+1}(i) = \sum_{j=1}^{n} \alpha_t(i) \times T_{i,j} \times e_i(o_{t+1}), \forall i \in S \tag{12.3}$$

  with n as number of states in S.
- Finalization step:

$$P[O|M] = \sum_{i=1}^{n} \alpha_T(i), \forall i \in S \tag{12.4}$$

The **forward** method returns a matrix with the state probabilities at each position of the sequence. The total probability can be obtained by summing all the probabilities at the last position of the matrix.

```
def forward(self, sequence):
    '''Given an observed sequence calculate the list of forward
probabilities of the sequence
    using the chain rules'''
    seq_len = len(sequence)
    if seq_len == 0:
        return []

    # calculate the product of the initial probability of each
state and the first symbol of the sequence
    prob_list = [{}]
    for state in self.states:
        prob_list[0][state] = self.get_init_prob(state) * self.
get_emission_prob(state, sequence[0])
    # iterate through the sequence and for each state multiply by
 the transition probability with any other of the possibles
states; this corresponds to a jump to a new state
    # once in this new state multiply by the corresponding
emission probability of the sequence symbol in that state
    for i in range(1, seq_len):
        prob_list.append({})
```

```
            for state_dest in self.states:
                prob = 0
                for state_orig in self.states:
                    prob += prob_list[i-1][state_orig] * self.
    get_transition_prob(state_orig, state_dest)
                prob_list[i][state_dest] = prob * self.
    get_emission_prob(state_dest, sequence[i])
        return prob_list
```

12.2.3 Probability of the Remainder of an Observed Sequence

The backwards probability $\beta_t(i) = P[o_{t+1}...o_T | \pi_t = S_i]$ provides the probability of generating the $o_{t+1}...o_T$ when starting at state S_i at position t. It can be derived as a variant of the Forward algorithm running in a backwards manner. The algorithm starts at the last observed symbol and goes backwards in the network until it reaches the position $t + 1$. By definition, the probability of each state i at position T is 1. The algorithm steps are:

- Initialization step:

$$\beta_T(i) = 1, \forall i \in S \tag{12.5}$$

- Iteration step:

$$\beta_t(i) = \sum_{j=1}^{n} T_{i,j} \times e_i(o_{t+1}) \times \beta_{t+1}(j), \forall j \in S \tag{12.6}$$

The **backward** function calculates the probability for an observed sequence to start in a state S_i at position t of the sequence and generate the remainder of the sequence from t to the end. It starts the calculation from the end of the sequence and computes it as a variant of the forward algorithm following the above iteration formula. The probability of the states corresponding to the last symbol of the sequence are by definition equal to 1.

```
    def backward(self, sequence):
        '''Given an observed sequence calculate the list of backward
    probabilities of the sequence'''
        seq_len = len(sequence)
        if seq_len == 0:
            return []

        beta = [{}]
        for state in self.states:
```

```
        beta[0][state] = 1

    for i in range(seq_len - 1, 0, -1):
        beta.insert(0, {})
        for state_orig in self.states:
            prob = 0
            for state_dest in self.states:
                prob += beta[1][state_dest] * self.
get_transition_prob(state_orig, state_dest) * self.
get_emission_prob(state_dest, sequence[i])
            beta[0][state_orig] = prob
    return beta
```

12.2.4 Finding the Optimal State Path

We now tackle the problem of finding the most likely state at each position of the observed sequence. The number of possible paths is exponential with relation to the length of the sequence and the alphabet size. In practice, we are only interested in the state path $s_1...s_n$ that maximizes the probability of our sequence, $P[\pi|O, M]$. Since the enumeration of all possible state paths is unfeasible, an algorithm based on dynamic programming similar to the ones presented in Chapter 6 is applied to solve this problem. The *Viterbi* algorithm is a dynamic programming algorithm that, given an HMM and an input sequence, finds the most probable or optimal state path.

At each position t of the observed sequence, the algorithm tries to find the highest probability along every single state path ending in state S_i:

$$\delta_t(i) = max_{\pi_1...\pi_{t-1}} P[\pi_1...(\pi_{t-1} = S_i), o_1...o_t|M] \tag{12.7}$$

By using a backtracking vector $\phi_t(i)$, it keeps track of the most likely state at each position. The algorithm steps are the following:

- Initialization step:

$$\delta_1(i) = I(i) \times e_i(o_1), \forall i \in S \tag{12.8}$$

$$\phi_1(i) = 0 \tag{12.9}$$

- Iteration step:

$$\delta_t(i) = max_{i=1}^{n} \delta_{t-1}(i) \times T_{i,j} \times e_j(o_t), \forall j \in S, 1 < t \leq T \tag{12.10}$$

$$\phi_t(i) = argmax_{t=1}^{n} [\delta_{t-1}(i) \times T_{i,j}] \tag{12.11}$$

- Finalization step: find the optimal state path $\pi_1^*, ..., \pi_T^*$,

$$\pi_T^* = argmax_{i=1}^n \delta_T(i) \tag{12.12}$$

$$\pi_t^* = \phi_{t+1}(\pi_{t+1}^*) \tag{12.13}$$

In the finalization step, the algorithm starts by selecting at the last position of the observed sequence the state with the highest probability. It then backtracks following the optimal path of hidden states at each position t. The function **viterbi** returns the optimal probability and the best state path for the observed sequence.

```
def viterbi(self, sequence):
        '''Viterbi algorithm calculates the most probable state path
    for an observed sequence.'''
        seq_len = len(sequence)
        if seq_len == 0:
            return []

        viterbi = {}
        state_path = {}
        # Initialize the probabilities for the first symbol
        for state in self.states:
            viterbi[state] = self.get_init_prob(state) * self.
    get_emission_prob(state, sequence[0])
            state_path[state] = [state]

        # compute recursively until the last element
        for t in range(1, seq_len):
            new_state_path = {}
            new_path = {}
            viterbi_tmp = {}
            for state_dest in self.states:
                intermediate_probs = []
                for state_orig in self.states:
                    prob = viterbi[state_orig] * self.
    get_transition_prob(state_orig, state_dest)
                    intermediate_probs.append((prob, state_orig))

                (max_prob, max_state) = max(intermediate_probs)
                prob = self.get_emission_prob(state_dest, sequence[t
    ]) * max_prob
```

```
                viterbi_tmp[state_dest] = prob
                new_state_path[state_dest] = max_state
                new_path[state_dest] = state_path[max_state] + [
state_dest]

            viterbi = viterbi_tmp
            state_path = new_path # just keep the optimal path

        max_state = None
        max_prob = 0
        # among the last states find the best prob and the best path
        for state in self.states:
            if viterbi[state] > max_prob:
                max_prob = viterbi[state]
                max_state = state

        return (max_prob, state_path[max_state])
```

12.2.5 Learning the Parameters of an HMM Model

So far we have considered models where the respective parameters are known in advance. We can try to infer or learn these parameters from the observed data. Given the topology of an HMM, the goal is to learn θ from a set of observed sequences, W, such that we derive $maxP[W|\theta]$. If the training sequences are labeled with the corresponding state path, which represents a case of supervised learning, the best estimate is given by the frequencies of the transitions and emissions that occur in the training sequences. The formal demonstration of how the parameters θ are calculated is beyond the scope of this book, but we provide here an intuition. To estimate the maximum likelihood of the transition probabilities $T_{i,j}$, we count the number of times a transition between state i and state j occurs and then normalize by the total count of transitions from state i:

$$\hat{T}_{i,j} = \frac{\textit{expected number of transitions } S_i \rightarrow S_j}{\textit{expected number of transitions from } S_i} \tag{12.14}$$

Similarly for the emission probabilities:

$$\hat{e}_i(o_t) = \frac{\textit{expected number of times in } S_i \textit{ observing symbol } o_t}{\textit{expected number of times in } S_i} \tag{12.15}$$

For small training sets, where few observations are available, some frequencies may be zero. In order to avoid probabilities of zero, pseudo-counts should be added. These may either be informed pseudo-counts if some *a priori* knowledge on the emissions and transitions frequencies is available. Otherwise, fixed pseudo-counts ϵ, can be added, typically with $\epsilon < 1$. This is called *smoothing* of probabilities.

If the state paths are not known for the training sequences, an unsupervised learning strategy needs to be applied to estimate θ. The *Baum-Welch* algorithm is based on an Expectation-Maximization approach, that provides an estimate and update for the emission and transition probabilities. The expected emission and transition probabilities, often denoted as γ and ξ respectively, calculated in the Expectation step, can be defined in terms of forward and backward probabilities:

$$\gamma_i(w) = \frac{\sum_{t=1,o_t=w}^{T} \alpha_t(i) \times \beta_t(i)}{\sum_t \alpha_t(i) \times \beta_t(i)}, \forall i \in S \tag{12.16}$$

$$\xi_{i,j} = \frac{\sum_t \alpha_t(i) \times T_{i,j} \times e_j(o_{t+1}) \times \beta_{t+1}(j)}{\sum_t \alpha_t(i) \times \beta_t(i)}, 1 < t \leq T, \forall i, j \in S \tag{12.17}$$

In the Maximization step, the γ and the ξ are used to recompute and update the $\hat{e}_i(w)$ and $\hat{T}_{i,j}$. The new values are calculated as follows:

$$\hat{e}_i(w) = \frac{\sum_{t=1,o_t=w}^{T} \gamma_t(i)}{\sum_{t=1}^{T} \gamma_t(i)}, \forall i \in S \tag{12.18}$$

$$\hat{T}_{i,j} = \frac{\sum_{t=1}^{T-1} \xi_t(i,j)}{\sum_{k=1}^{N} \sum_{t=1}^{T-1} \gamma_t(i)}, \forall i, j \in S \tag{12.19}$$

The learning algorithm that will optimize the parameters of the model, receives an observed sequence O of length T and updates the transition and emission matrices. It calculates the forward and backward probabilities. Next, it updates the emission and transition probabilities. It has the following steps:

- Initialization:
  - Randomly initialize parameters $\theta = (e_{i,a}; T_{i,j})$ of model M or retrieve existing ones.
- Iteration until convergence:
  - Expectation: calculate $\gamma_i(w)$ and $\xi_{i,j}$.
  - Maximization: calculate $\hat{e}_i(w)$ and $\hat{T}_{i,j}$.

As one might expect, the improvements on the estimated probabilities will depend largely on the amount of available training data and the initial conditions. Therefore, the random initialization may not be optimal and providing additional information may be desirable. For instance, the transition probabilities can be initially inferred from an independent labeled training sequences, even if a smaller sample size is available.

The function **baum_welch** learns and updates the emission and transition probabilities based on observed unlabeled sequence data. It has an Expectation phase, where the expected emission and transition probabilities are calculated based on the γ and ξ formulas, which is followed by a Maximization phase where updates to the emission and transition probabilities are made based on $\hat{e}$ and $\hat{T}$ formulas.

```
def baum_welch(self, sequence):
    '''Computes an update of the emission and transition
  probabilities based on observed sequence.
    '''

    seq_len = len(sequence)
    if seq_len == 0:
        return []

    alpha = self.forward(sequence)
    beta = self.backward(sequence)

    # Expectation phase

    # gamma: probs. for finding a symbol w at position t in state
   i; product of the forward (alpha) and backward (beta) probs for
   all t and all i
    gamma = [{} for t in range(seq_len)]
    # xi: probs for transition between state i at position t and
  state j at position t+1, for all i,j and t
    xi = [{} for t in range(seq_len - 1)]
    for t in range(seq_len):
        # compute gamma
        sum_alpha_beta = 0
        for i in self.states:
            gamma[t][i] = alpha[t][i] * beta[t][i]
            sum_alpha_beta += gamma[t][i]
```

```
            for i in self.states:
                gamma[t][i] = gamma[t][i] / sum_alpha_beta

            # set initial probs
            if t == 0:
                self.set_init_prob(i, gamma[t][i])

            # compute xi values up to T - 1
            if t == seq_len - 1:
                continue

            sum_probs = 0
            for state_orig in self.states:
                xi[t][state_orig] = {}
                for state_dest in self.states:
                    p = alpha[t][state_orig] * self.
get_transition_prob(state_orig, state_dest) * self.
get_emission_prob(state_dest, sequence[t + 1]) * beta[t + 1][
state_dest]
                    xi[t][state_orig][state_dest] = p
                    sum_probs += p

            for state_orig in self.states:
                for state_dest in self.states:
                    xi[t][state_orig][state_dest] /= sum_probs
        # Maximization step: with gamma and xi calculated re-estimate
 emissions and transitions

        # re-estimate emissions
        for i in self.states:
            denominator = 0
            for t in range(seq_len):
                denominator += gamma[t][i]

            for w in self.symbols:
                numerator = 0.0
                for t in range(seq_len):
                    if sequence[t] == w:
                        numerator += gamma[t][i]
```

```
                if denominator > 0:
                    self.set_emission_prob(i, w, numerator /
denominator)
                else:
                    self.set_emission_prob(i, w, 0.0)

        # re-estimate transitions
        # now that we have gamma and zi let us re-estimate
        for i in self.states:
            for j in self.states:
                denominator = 0.0
                for t in range(seq_len -1):
                    denominator += gamma[t][i]
                numerator = 0.0
                for t in range(seq_len -1):
                    numerator += xi[t][i][j]
                self.set_transition_prob(i, j, numerator /
denominator)
                if denominator > 0:
                    self.set_transition_prob(i, j, numerator /
denominator)
                else:
                    self.set_transition_prob(i, j, 0.0)
```

12.3 HMMs for Database Search

Proteins with related sequences and structures can be arranged into protein families. They often share similar function and clear evolutionary relationship. In Chapter 11 we have seen that by aligning conserved regions across multiple sequences of the same family, we can derive a position weight matrix (PWM) that captures the patterns of conservation along the different positions of the alignment.

One of the most popular applications of HMMs in biological sequence analysis are the HMM-profiles. They provide a probabilistic profile of a protein family and resemble PWMs but provide a more flexible way to deal with insertions and deletions. A topology of an HMM-profile contains, apart from the start and end states, three different groups of states: main, insertion, and deletion states. The main states model the columns and these are as many as positions in the alignment. The insert states model highly variable regions of the alignment.

The deletion states, also called silent states, do not match any residue and make possible the jump between one or more columns of the alignments (corresponding to highly variable regions).

A typical bioinformatics application is to build an HMM-profile from a family of proteins and to search a database of sequences for other unseen family members. For an HMM-profile M representing a protein family, all sequences W in database D are scanned. Sequences with a probability or a score higher than a threshold are potential members of the family. The probability $P[W|M]$ can be calculated with the Forward algorithm. For further information on theory and applications of HMM-profiles we refer the reader to the textbooks [50] and Chapter 4 by Anders Krogh in [6]. The Pfam database [143] contains a large collection of multiple alignments and HMM-profiles for a large number for manually verified protein families.

Bibliographic Notes and Further Reading

A general introduction to hidden Markov models can be found in [130] and Chapter 9 of [85] provides a very detailed algorithmic introduction. Materials for more oriented biological sequence analysis with HMMs can be found in the reference textbooks [50], [6], [84] and [19]. A quick introduction to the topic can be found in [54].

HMMs are very well suited for many tasks in biological sequence analysis problems. This includes the tasks of gene finding, profile search, multiple sequence alignment, regulatory site identification and secondary structure prediction or copy-number detection. Here, we refer the reader to a few examples of applications on each of these topics. This is certainly not exhaustive and many other interesting works can be found elsewhere. In gene finding we emphasize the HMMgene [90] and GENSCAN [31] methods. For HMM-profiles we refer to the seminal work of David Haussler, Anders Krogh and colleagues [73,91], an extensive review by Sean Eddy [53] and the theory behind HMM-profiles can be found in [19,50] and Chapter 4 of [6]. For multiple sequence alignment with HMM the work of Sean Eddy [52] and for prediction of secondary structure from sequence [14,86]. HMMs have also been applied to detect copy-number alterations from genome sequencing data [152].

The software suite *hmmer*, developed by Sean Eddy, is possibly the most complete tool implementing different methods using hidden Markov models for multiple biological sequence analysis tasks, including searching for homologous sequences in databases and multiple sequence alignments. The algorithmic details behind the methods can be found in [50] and [55–57] provide description of the different functionalities implemented in the package.

Exercises and Programming Projects

1. Adapt the code to convert probabilities into scores by taking the log_2 of each probability and summing instead of multiplying.
2. The probability of a sequence depends directly on its length. The log-odds measure is generally a more convenient way to represent the score of a sequence. For an observed sequence O of length T, the measure can be defined as: $logodds(O) = log\frac{P[O]}{0.25^T} = log P[O] - T \times log 0.25$. The 0.25 is derived from a background model where all the nucleotides are expected to occur with equal frequency.
3. Adapt the code of the Viterbi algorithm to keep track of all possible paths, returning them in addition to the optimal path.

CHAPTER 13

Graphs: Concepts and Algorithms

In this chapter, we present the mathematical concept of graph and its computational representation. We address some of the main algorithms over graphs and develop a set of Python classes to implement different types of graphs and underlying algorithms. Graphs will be central in the development of algorithms for handling biological networks and genome assembly, tasks addressed in the next chapters.

13.1 Graphs: Definitions and Representations

A graph can be defined, in Mathematics, as a set of objects in which some of the pairs of the objects in this set are related. While they can be easily defined and have a simple structure, they are powerful and flexible data structures, with a huge set of applications in computer science and many fields of science and engineering.

Formally, a graph G can be defined by two sets: (V, E), where V is the set of objects, named as *vertices* or *nodes* of the graph, and E is a set of pairs $(u.v)$ of vertices from V, named *edges* or *arcs*, indicating the existence of a relationship between u and v.

The edges in E may have an orientation, i.e. the pairs are ordered, in which case the graph is classified as *directed*, or *digraph*. Otherwise, the pairs are unordered and the graph is termed *undirected*. Another common variant is to associate numerical weights to the edges of a graph, in which case the non-existence of an edge is typically associated with the value 0 for that weight.

It is typical to represent graphs in figures by drawing nodes as circles (or other similar shapes), and edges as lines connecting the two circles representing the vertices belonging to the corresponding node pair. These lines may have an arrow, when representing directed graphs, indicating the direction of the edge. For weighted graphs, the weight of the edge is typically represented over the respective line. Fig. 13.1A shows a simple example, where the three main types of graphs aforementioned are illustrated.

Graphs can be represented computationally in different ways. The choice depends on several factors, as the type of application of the graph and the required algorithms to be developed over them or the density of possible edges among all possible pairs of vertices. The most common representations are incidence matrices and adjacency lists, while in many practical cases a combination of both might be the best solution.

Bioinformatics Algorithms. DOI: 10.1016/B978-0-12-812520-5.00013-4

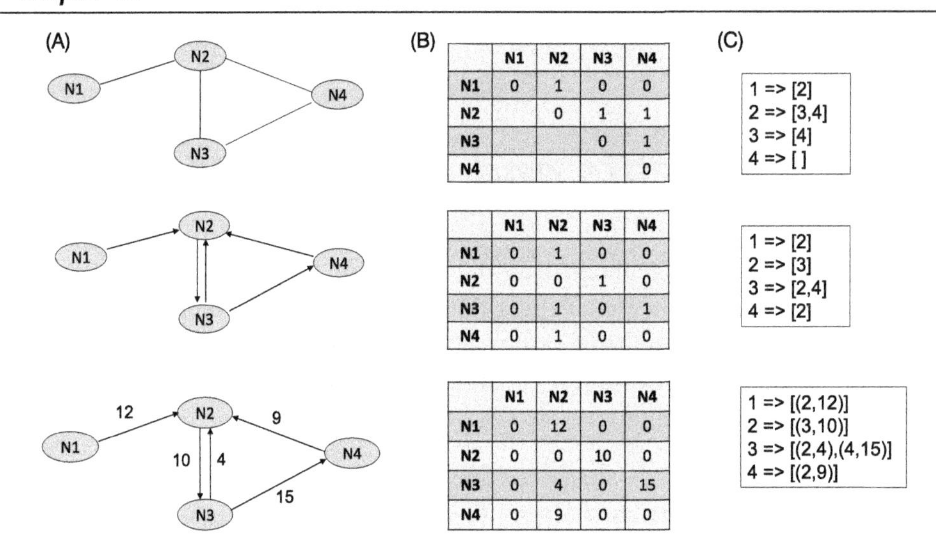

Figure 13.1: Different types of graphs (A) and their representation using matrices (B) and adjacency lists (C). The upper row shows an undirected graph, the middle row shows a directed graph, while the bottom row shows a weighted graph.

In the case of matrices, all possible combinations of vertices are represented, with both rows and columns representing the vertices of the graph. Each cell of the matrix in row i and column j will represent the existence of an edge connecting vertex i to vertex j. In case of directed graphs, the matrix is complete and the rows represent origin vertices, while columns represent destination vertices. When the graph is undirected, the matrix is triangular. In the case of weighted graphs, the cell in the matrix keeps the value of the weight, while cells with 0 still indicate non-existent edges.

Fig. 13.1B shows examples of the three types of graphs represented as matrices. This type of representation brings advantages in dense graphs, where vertices are highly connected, but can be a waste of memory if the resulting matrix is sparse, i.e. mostly filled with zeros, representing a graph with few edges.

On the other hand, adjacency lists only represent the existing edges. In this case, each vertex has an associated list with the neighbor nodes. If the graph is directed, the list for a vertex v includes all the destination nodes for the edges where v is the origin. In the case of undirected graphs, the edge will only exist for one of the directions; if nodes can be ordered, the edge will be represented in the list of the lower rank node. For weighted graphs, the edges can be represented as pairs with destination node and weight values. Fig. 13.1C shows examples of adjacency lists for the previous examples.

13.2 A Python Class for Graphs

Here, we will implement graphs using Python object-oriented features, implementing a base class with the behavior of general purpose graphs, which will be extended in subsequent chapters to address specific tasks. The option will be to implement directed graphs without weights, since these are the most suitable to the Bioinformatics tasks tackled in this book, as will become clearer in the next chapters. Also, given that these graphs are typically sparse, we will use an implementation based on adjacency lists.

Thus, in the proposed class, **MyGraph**, the graph will be represented as a Python dictionary, which will be the only attribute of the objects in the class. Here, keys will be the identifiers of the nodes, and the values will be the lists of neighbors, i.e. the vertices that are destinations in edges with origin in the key vertex.

The next code chunk shows the class definition, the constructor, and some methods that allow to get information from the graph and print it (in the form of an adjacency list). Note that the constructor receives a dictionary as input; if no dictionary is passed, the graph is created empty.

```
class MyGraph:

    def __init__(self, g = {}):
        ''' Constructor - takes dictionary to fill the graph as input
    ; default is empty dictionary '''
        self.graph = g

    ## get basic information about the graph

    def get_nodes(self):
        return list(self.graph.keys())

    def get_edges(self):
        edges = []
        for v in self.graph.keys():
            for d in self.graph[v]:
                edges.append((v,d))
        return edges

    def size(self):
        ''' Returns number of nodes, number of edges '''
```

```
        return len(self.get_nodes()), len(self.get_edges())

    def print_graph(self):
        for v in self.graph.keys():
            print (v, " -> ", self.graph[v])
```

If the graph is created empty it needs to be filled by adding vertexes and edges. The next code block defines methods to achieve these tasks, showing an example of a program where these methods are used to create the directed graph in Fig. 13.1.

```
    def add_vertex(self, v):
        ''' Add a vertex to the graph
        Tests if vertex exists not adding if it does. '''
        if v not in self.graph.keys():
            self.graph[v] = []

    def add_edge(self, o, d):
        ''' Add edge to the graph.
        If vertices do not exist, they are added to the graph '''
        if o not in self.graph.keys():
            self.add_vertex(o)
        if d not in self.graph.keys():
            self.add_vertex(d)
        if d not in self.graph[o]:
            self.graph[o].append(d)

if __name__ == "__main__":
    gr = MyGraph()
    gr.add_vertex(1)
    gr.add_vertex(2)
    gr.add_vertex(3)
    gr.add_vertex(4)
    gr.add_edge(1,2)
    gr.add_edge(2,3)
    gr.add_edge(3,2)
    gr.add_edge(3,4)
    gr.add_edge(4,2)
    gr.print_graph()
    print(gr.size())
```

13.3 Adjacent Nodes and Degrees

Since graphs are data structures that represent connectivity between the entities represented by their vertices or nodes, different types of analysis are supported by analyzing which vertices are connected and in what ways. In this section, we will provide a set of definitions and their Python implementations regarding graph connectivity.

In a directed graph $G = (V, E)$, a vertex s is a *successor* of a vertex v if the ordered pair (s, v) exists in the set E containing the edges of the graph. In this case, s is a *predecessor* of v. Also, the two vertices s and v are named *adjacent*, i.e. two vertices are adjacent if one is the successor of the other.

Note that in an undirected graph, only the last definition makes sense and, in this case, two vertices s and v are adjacent if the unordered pair (s, v) (i.e. (s, v) or (v, s)) exists in E.

Given these definitions, we will address their implementation on the context of the class **MyGraph** we started to develop in the previous section. In this case, we will add methods to collect the set of successors, predecessors and adjacent vertices of a node v, given the graph represented by the object in the class (*self*). An example of the use of those methods is provided below, considering the same example of the previous code chunk.

```
    def get_successors(self, v):
        return list(self.graph[v])      # avoids list being
  overwritten

    def get_predecessors(self, v):
        res = []
        for k in self.graph.keys():
            if v in self.graph[k]:
                res.append(k)
        return res

    def get_adjacents(self, v):
        suc = self.get_successors(v)
        pred = self.get_predecessors(v)
        res = pred
        for p in suc:
            if p not in res: res.append(p)
        return res
```

```
if __name__ == "__main__":
    gr = MyGraph()
    gr.add_vertex(1)
    (...)
    print (gr.get_successors(2))
    print (gr.get_predecessors(2))
    print (gr.get_adjacents(2))
```

A related concept is that of vertex or node *degree*, which is defined as the number of adjacent vertices of a given vertex. In the case of directed graphs, we can also defined the *in-degree*, which counts the number of predecessors of a node, and the *out-degree*, which counts the number of its successors. Note that, in this case, the degree is not always the sum of these two values as a vertex can be at the same successor and predecessor of the same node, if the edges in both directions exist.

```
    def out_degree(self, v):
        return len(self.graph[v])

    def in_degree(self, v):
        return len(self.get_predecessors(v))

    def degree(self, v):
        return len(self.get_adjacents(v))

if __name__ == "__main__":
    gr = MyGraph()
    gr.add_vertex(1)
    (...)

    print (gr.in_degree(2))
    print (gr.out_degree(2))
    print (gr.degree(2))
```

If there is the need to compute the degrees of a certain type for the whole set of nodes in the graph, then the next function might be used, which saves computation time, compared to calling the previous methods for all nodes of the graph. It receives the degree type as a string ("inout" is used for the node degree), and returns a dictionary where the keys are the node identifiers and the values are their degrees.

```
    def all_degrees(self, deg_type = "inout"):
        ''' Computes the degree for all nodes.
        deg_type can be "in", "out", or "inout"
        Returns a dictionary: node -> degree.'''
        degs = {}
        for v in self.graph.keys():
            if deg_type == "out" or deg_type == "inout":
                degs[v] = len(self.graph[v])
            else: degs[v] = 0
        if deg_type == "in" or deg_type == "inout":
            for v in self.graph.keys():
                for d in self.graph[v]:
                    if deg_type == "in" or v not in self.graph[d]:
                        degs[d] = degs[d] + 1
        return degs

if __name__ == "__main__":
    gr = MyGraph()
    (...)

    print(gr.all_degrees("inout"))
    print(gr.all_degrees("in"))
    print(gr.all_degrees("out"))
```

13.4 Paths, Searches, and Distances

In a directed graph, a *path* can be defined as an ordered list of nodes, where consecutive nodes in the list need to be connected by an edge. Mathematically, in a directed graph $G = (V, E)$, a path P between nodes x and y is a list $P = p_1, p_2, ..., p_n$, where $p_1 = x$, $p_n = y$, and all pairs of consecutive nodes on P, $(p_i, p_{i+1}), \forall i \in 1, \ldots, n-1$ belong to E. Note that if the graph is undirected, either $(p_i, p_{i+1}) \in E$ or $(p_{i+1}, p_i) \in E$.

In a graph, two nodes s and v are considered to be *connected* if there is a valid path between both. In this case, node v is said to be reachable from s. In many situations, it is useful to compute all the nodes that can be reached from a source node s, i.e. all nodes for which there are paths starting in s and reaching the target node.

It is important to notice that, in graphs, given nodes s and v, there might be different valid paths connecting the two nodes, i.e. starting in s and reaching v. In many cases, it is important to consider just one, that is typically the *shortest path*. This is defined as the path connecting the two nodes with the shorter length, i.e. the minimum number of intermediate nodes. The *distance* between the two nodes s and v is given by the number of edges contained on the shortest path between these two nodes.

To be able to address the definitions put forward in this section, we need to develop algorithms that are able to traverse the graph, starting in a given source node, gathering the visited nodes. There are mainly two alternative strategies that can be used to address this task:

- *Breadth-first search* (BFS): starts by the source node, then visits all its successors, followed by their successors, until all possible nodes are visited;
- *Depth-first search* (DFS): starts by the source node, explores its first successor, then its first successor, until no further exploration is possible, and then backtracks to explore further alternatives.

The code block below shows two methods to address each of these strategies, both returning the set of reachable nodes from a given source vertex. Note that the functions are quite similar, using a list of visited nodes to return the result (*res*) and a list of nodes to be handled (*l*). The way this working list is handled makes the difference of the two versions, since it implements two different data structures: in the BFS it implements a queue (last in, last out), while in the DFS it implements a stack (first in, first out), thus changing the order of the results.

The two functions are tested with a simple example, that is graphically illustrated in Fig. 13.2, where the two strategies are clearly distinguished.

```
def reachable_bfs(self, v):
    l = [v]
    res = []
    while len(l) > 0:
        node = l.pop(0)
        if node != v: res.append(node)
        for elem in self.graph[node]:
            if elem not in res and elem not in l:
                l.append(elem)
    return res

def reachable_dfs(self, v):
    l = [v]
    res = []
    while len(l) > 0:
        node = l.pop(0)
```

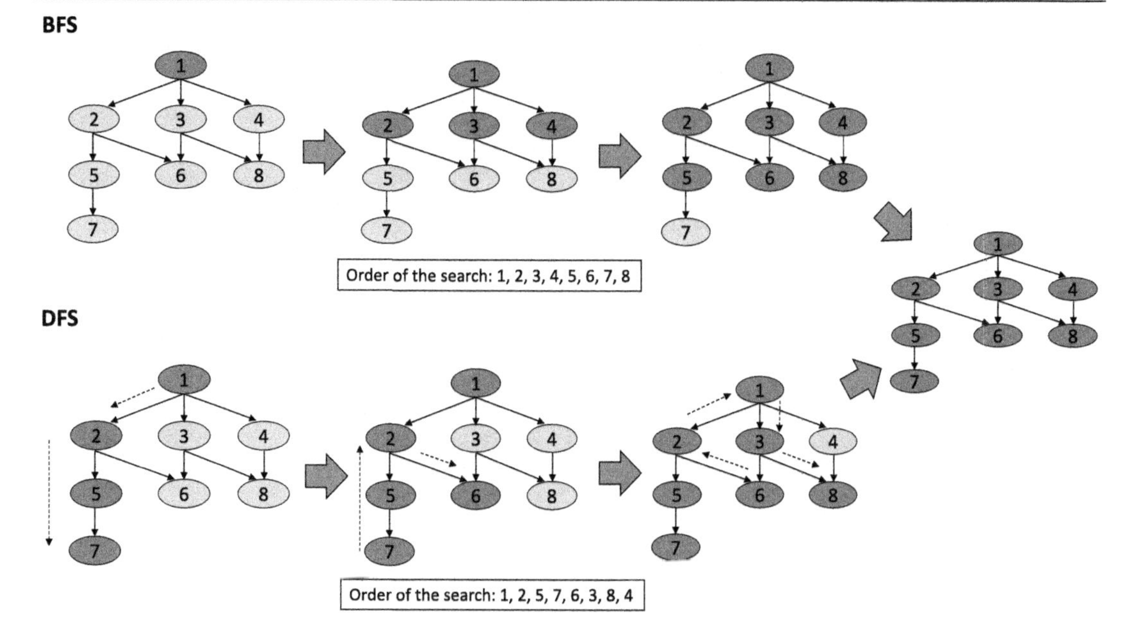

Figure 13.2: Illustration of the two strategies for traversing a graph with an example. In the upper part, the breadth-first strategy is shown, and in the bottom the depth-first search is illustrated.

```
                if node != v: res.append(node)
                s = 0
                for elem in self.graph[node]:
                    if elem not in res and elem not in l:
                        l.insert(s, elem)
                        s += 1
            return res

if __name__ == "__main__":
    gr2 = MyGraph({1:[2,3,4],
   2:[5,6],3:[6,8],4:[8],5:[7],6:[],7:[],8:[]})
    print(gr2.reachable_bfs(1))
    print(gr2.reachable_dfs(1))
```

Based on these strategies, we can devise a strategy to calculate the distance between two nodes, or similarly the shortest path between them. Looking at the two previous strategies, it becomes clear that the one suitable in this case is the BFS, since in this case all nodes at a distance of n from the source are always covered before the nodes at a distance $n + 1$, as it is clear looking at Fig. 13.2.

Thus, the two following functions, in the next code chunk, show how to implement the distance and shortest path algorithms, in the class we have been building in this chapter. In both cases, the strategy is similar to the **reachable_bfs** function above. In the case of the distance, the working list keeps already visited nodes and their distances to the source, while in the case of the shortest path this list keeps the visited nodes and the path to the source node (in this case in the form of a list). The main difference, in this case, is that we will stop the search process when the destination node is reached. In both these functions return *None*, if the destination node is unreachable.

```
    def distance(self, s, d):
        if s == d: return 0
        l = [(s,0)]
        visited = [s]
        while len(l) > 0:
            node, dist = l.pop(0)
            for elem in self.graph[node]:
                if elem == d: return dist + 1
                elif elem not in visited:
                    l.append((elem,dist+1))
                    visited.append(elem)
        return None

    def shortest_path(self, s, d):
        if s == d: return 0
        l = [(s,[])]
        visited = [s]
        while len(l) > 0:
            node, preds = l.pop(0)
            for elem in self.graph[node]:
                if elem == d: return preds+[node,elem]
                elif elem not in visited:
                    l.append((elem,preds+[node]))
                    visited.append(elem)
        return None

if __name__ == "__main__":
    gr2 = MyGraph({1:[2,3,4],
   2:[5,6],3:[6,8],4:[8],5:[7],6:[],7:[],8:[]})
```

```
    print(gr2.distance(1,7))
    print(gr2.shortest_path(1,7))
    print(gr2.distance(1,8))
    print(gr2.shortest_path(1,8))
    print(gr2.distance(6,1))
    print(gr2.shortest_path(6,1))
```

We can also define a function that combines the full BFS search with the distance, gathering all reachable nodes and their distance from the source.

```
    def reachable_with_dist(self, s):
        res = []
        l = [(s,0)]
        while len(l) > 0:
            node, dist = l.pop(0)
            if node != s: res.append((node,dist))
            for elem in self.graph[node]:
                if not is_in_tuple_list(l,elem) and not
   is_in_tuple_list(res,elem):
                    l.append((elem,dist+1))
        return res

def is_in_tuple_list(tl, val):
    res = False
    for (x,y) in tl:
        if val == x: return True
    return res

if __name__ == "__main__":
    gr2 = MyGraph({1:[2,3,4],
   2:[5,6],3:[6,8],4:[8],5:[7],6:[],7:[],8:[]})
    print(gr2.reachable_with_dist(1))
```

Note that, in some cases, there might be no path between the nodes, a situation where the distance is considered to be infinite. If all pairs of nodes have a finite distance, the graph is said to be *strongly connected*. This definition is implemented by the following function.

```
    def is_connected(self):
        total = len(self.graph.keys()) - 1
        for v in self.graph.keys():
```

```
            reachable_v = self.reachable_bfs(v)
            if (len(reachable_v) < total): return False
        return True
```

13.5 Cycles

In a graph, a path is defined as being *closed* if it starts and ends in the same vertex. If within a closed path, there are no repeated nodes or edges, the path is called a *cycle*. The existence or not of cycles in graphs is an important property that divides the graphs between *cyclic* and *acyclic*. Directed acyclic graphs are an important class of graphs in many fields, including Bioinformatics, including as a special case the trees, already covered in this book.

To close this section, we will define functions that allow to check of a graph is *cyclic*. These are put forward in the next code block.

```
    def node_has_cycle (self, v):
        l = [v]
        res = False
        visited = [v]
        while len(l) > 0:
            node = l.pop(0)
            for elem in self.graph[node]:
                if elem == v: return True
                elif elem not in visited:
                    l.append(elem)
                    visited.append(elem)
        return res

    def has_cycle(self):
        res = False
        for v in self.graph.keys():
            if self.node_has_cycle(v): return True
        return res

if __name__ == "__main__":
    gr = MyGraph()
    gr.add_vertex(1)
    (...)
```

```
gr2 = MyGraph({1:[2,3,4],
2:[5,6],3:[6,8],4:[8],5:[7],6:[],7:[],8:[]})
print(gr.has_cycle())
print(gr2.has_cycle())
```

The function **node_has_cycle** checks if there is a cycle starting (and ending) on a given node, while the **has_cycle** function tests if there is a cycle starting on any node.

Bibliographic Notes and Further Reading

An introduction to graph theory may be found in numerous textbooks; we cite here the classical one by Harary [71] and the more recent by Diestel [48] and Even [61]. Most of the algorithms described in this chapter are covered in many textbooks on introductory algorithms, including for instance the book by Cormen et al. [38]. The breadth-first search described here has been invented independently by several authors, including Konrad Zuse [161] and [116], while depth first search has been used since the 19th century as a strategy to solve maizes [61].

Exercises and Programming Projects

Exercises

1. In a graph, an *isolated node* is a node that does not have any successors or predecessors.
   a. Write a function **is_isolated**(*self, node*), to add to the **MyGraph** class, the given a node identifier verifies if it is an isolated node, returning a Boolean value (`True` or `False`).
   b. Based on the previous, write another function **isolated_nodes**(*self*) which returns a list of all isolated nodes in the graph.
2. A *clique* in a graph corresponds to a sub-graph (i.e. set of nodes and edges from the graph) where all pairs of nodes are adjacent. Write a function **is_clique**(*self, list_nodes*) to add to the **MyGraph** class, that given a list of nodes as input, checks if the sub-graph containing those nodes is a clique or not. The result should be Boolean (`True` or `False`).
3. Write a method to add to the class **MyGraph** that identifies the components of a graph, returning as a result a list of lists, each with the set of nodes of each component. Notice that in a directed graph, a component is a sub-graph that includes a set of nodes that are strongly connected and not connected to other nodes in the graph (i.e. there are paths connecting each pair of nodes). Notice that the union of the lists of nodes in the result may not be the set of nodes in the graph.

Programming Projects

1. Write a class to represent undirected graphs. Check which methods from the class **MyGraph** still work over undirected graphs and re-implement those methods that do not apply.
2. Write a class to represent weighted directed graphs as a sub-class to the class **MyGraph**. Check which methods need to be redefined. Add methods to calculate shortest paths and distances between a pair of nodes, taking into account the weights of the edges (search for the details of the Dijkstra algorithm if you are not familiar with this problem).

CHAPTER 14

Graphs and Biological Networks

In this chapter, we briefly review the importance of networks in biological discovery and discuss how biologically relevant networks can be represented using graphs, similar to the ones presented in the last chapter. We present and implement different techniques for network topological analysis, including degree distributions, mean distances, clustering coefficients and hub analysis, and discuss their relevance in biological research.

14.1 Introduction

Over thc last years, different types of representations, based in the concept of graph presented in the previous chapter, have been used to characterize and simulate different types of biological systems, mainly within cells. These graphs, globally denominated as *biological networks*, typically represent in their nodes biological entities, such as genes, proteins, or chemical compounds, while edges represent different types of interactions or relationships between these entities, with a well-defined biological meaning (e.g. genes encoding proteins, metabolites participating in reactions, proteins that regulate gene expression or bind to compounds).

These networks have provided powerful tools, within the field of Systems Biology, to support for the global analysis of cell organization and phenotypical behavior, and their main subsystems with an emphasis in metabolism, gene expression regulation and signal transduction.

Metabolic networks typically encompass reactions (or their encoding enzymes) and metabolites (compounds involved in metabolic reactions), representing the set of (enzyme-catalyzed) chemical transformations that sustain life. These networks may include only one type of entity (either reactions/enzymes or metabolites) or include both. Fig. 14.1 represents a very simple metabolic system and its representation in different types of networks.

On the other hand, regulatory networks encompass genes and proteins, representing the processes that allow cells to control the levels of gene expression in different situations (e.g. different cell types or distinct stimuli), while signaling networks represent the protein cascades typically involved in the transduction of signals from the outside of the cell to the internal regulatory and metabolic systems.

Even without having the capability to support the simulation of cellular behavior directly, the structural analysis of these networks has provided, in the last few years, some insights into

Bioinformatics Algorithms. DOI: 10.1016/B978-0-12-812520-5.00014-6

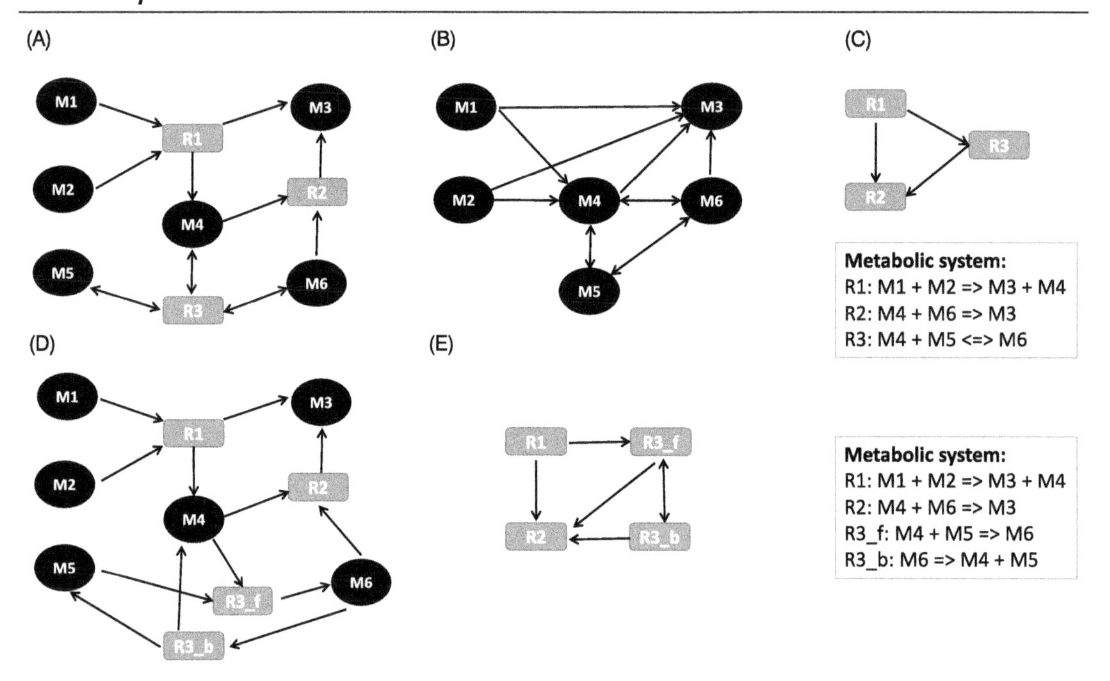

Figure 14.1: Different metabolic networks for a simple metabolic system with three reactions: (A) metabolite-reaction network; (B) metabolite-metabolite network; (C) reaction-reaction network; (D) and (E) show the same networks as (A) and (C) considering the split of reversible reactions into the two directions. In the metabolic system on the right, => represents an irreversible reaction, while <=> represents a reversible one.

important biological properties. Therefore, in the remaining of this chapter, we will discuss some of the possible algorithms that can be implemented over these graphs to provide meaningful properties of the underlying networks and biological systems.

Some attempts have also been made to represent different sub-systems simultaneously, combining two or several sub-systems into integrated networks. Although powerful, these can become more difficult to analyze, and thus in this text we will focus on simpler networks, mostly using metabolic networks as the predominant illustrative example.

14.2 Representing Networks With Graphs

14.2.1 A Python Class for Metabolic Networks

As stated above, we will represent networks using graphs assuming the representation put forward in the previous chapter. As before, we will provide a Python implementation of the mentioned concepts, in the form of a class to represent metabolic networks, which will extend the core class to represent directed graphs presented before.

It is important to notice that metabolic networks encompassing both types of nodes (reactions and metabolites), as shown in Fig. 14.1A, bring an additional requirement, since the graph, in this case, has two different types of nodes. This is a particular case of a class of graphs named *bipartite graphs*.

Formally, a graph $G = (V, E)$ is a *bipartite graph*, if V can be split into two disjoint sets V_1 and V_2 ($V_1 \cup V_2 = V$ and $V_1 \cap V_2 = \emptyset$), and E only contains pairs where the first element belongs to V_1 and the second belongs to V_2, i.e. there are no edges connecting nodes of the same node set. It is easy to check that metabolic networks (metabolite-reaction) are a special case of a bipartite graph where V_1 represents the set of metabolites and V_2 the set of reactions.

To represent metabolic networks, considering all the types of networks illustrated by Fig. 14.1, we will define the class **MetabolicNetwork**, which is a sub-class of the class **MyGraph**. As shown below, this will add to the base class a variable specifying the type of network (the options will be "metabolite-reaction", "metabolite-metabolite" and "reaction-reaction", which match the networks depicted in Fig. 14.1A, B and C, respectively), an attribute (a dictionary) which will keep the nodes of each type (in the case it is a bipartite "metabolite-reaction" graph), and a Boolean variable specifying if the reversible reactions are represented split in their two directions (as it is the case of the networks in Fig. 14.1D and E) or as a single reaction.

```
from MyGraph import MyGraph

class MetabolicNetwork (MyGraph):

    def __init__(self, network_type = "metabolite-reaction",
   split_rev = False):
        MyGraph.__init__(self, {})
        self.net_type = network_type
        self.node_types = {}
        if network_type == "metabolite-reaction":
            self.node_types["metabolite"] = []
            self.node_types["reaction"] = []
        self.split_rev =  split_rev
```

To add some content to these networks, we can insert nodes and edges "by hand", i.e. using the methods defined in the parent class. In this case, we will implement a new method to allow to add a node to a bipartite graph, specifying its type. This is illustrated in the code chunk below where the network in Fig. 14.1A is created and printed.

```
class MetabolicNetwork (MyGraph):

    (...)

    def add_vertex_type(self, v, nodetype):
        self.add_vertex(v)
        self.node_types[nodetype].append(v)

    def get_nodes_type(self, node_type):
        if node_type in self.node_types:
            return self.node_types[node_type]
        else: return None

def test1():
    m = MetabolicNetwork("metabolite-reaction")
    m.add_vertex_type("R1","reaction")
    m.add_vertex_type("R2","reaction")
    m.add_vertex_type("R3","reaction")
    m.add_vertex_type("M1","metabolite")
    m.add_vertex_type("M2","metabolite")
    m.add_vertex_type("M3","metabolite")
    m.add_vertex_type("M4","metabolite")
    m.add_vertex_type("M5","metabolite")
    m.add_vertex_type("M6","metabolite")
    m.add_edge("M1","R1")
    m.add_edge("M2","R1")
    m.add_edge("R1","M3")
    m.add_edge("R1","M4")
    m.add_edge("M4","R2")
    m.add_edge("M6","R2")
    m.add_edge("R2","M3")
    m.add_edge("M4","R3")
    m.add_edge("M5","R3")
    m.add_edge("R3","M6")
    m.add_edge("R3","M4")
    m.add_edge("R3","M5")
    m.add_edge("M6","R3")
    m.print_graph()
    print("Reactions: ", m.get_nodes_type("reaction") )
```

```
    print("Metabolites: ", m.get_nodes_type("metabolite") )

test1()
```

A more useful alternative that allows to work with large-scale networks is to load information from files and build the network based on the file's content. Here, we will assume information is given in a text file with a reaction per row, encoded in the form shown in the boxes in the right-hand side of Fig. 14.1.

A function **load_from_file** is provided in the **MetabolicNetwork** class to read the file and build the desired network (shown in the following code). This function loads information from the file and creates the network according to the network type provided by the attribute in the class, and also taking into account the flag that specifies how to handle reversible reactions.

Note the use of the **split** function, which applied over a string *str* and providing a separator string (actually a regular expression), splits *str* in several strings cutting in every occurrence of the separator, being the result a list of the resulting sub-strings (typically named as *tokens*).

```
    def load_from_file(self, filename):
        rf = open(filename)
        gmr = MetabolicNetwork("metabolite-reaction")
        for line in rf:
            if ":" in line:
                tokens = line.split(":")
                reac_id = tokens[0].strip()
                gmr.add_vertex_type(reac_id, "reaction")
                rline = tokens[1]
            else: raise Exception("Invalid line:")
            if "<=>" in rline:
                left, right = rline.split("<=>")
                mets_left = left.split("+")
                for met in mets_left:
                    met_id = met.strip()
                    if met_id not in gmr.graph:
                        gmr.add_vertex_type(met_id, "metabolite")
                    if self.split_rev:
                        gmr.add_vertex_type(reac_id+"_b", "reaction")
                        gmr.add_edge(met_id, reac_id)
                        gmr.add_edge(reac_id+"_b", met_id)
```

```
                else:
                    gmr.add_edge(met_id, reac_id)
                    gmr.add_edge(reac_id, met_id)
            mets_right = right.split("+")
            for met in mets_right:
                met_id = met.strip()
                if met_id not in gmr.graph:
                    gmr.add_vertex_type(met_id, "metabolite")
                if self.split_rev:
                    gmr.add_edge(met_id, reac_id+"_b")
                    gmr.add_edge(reac_id, met_id)
                else:
                    gmr.add_edge(met_id, reac_id)
                    gmr.add_edge(reac_id, met_id)
        elif "=>" in line:
            left, right = rline.split("=>")
            mets_left = left.split("+")
            for met in mets_left:
                met_id = met.strip()
                if met_id not in gmr.graph:
                    gmr.add_vertex_type(met_id, "metabolite")
                gmr.add_edge(met_id, reac_id)
            mets_right = right.split("+")
            for met in mets_right:
                met_id = met.strip()
                if met_id not in gmr.graph:
                    gmr.add_vertex_type(met_id, "metabolite")
                gmr.add_edge(reac_id, met_id)
        else: raise Exception("Invalid line:")

    if self.net_type == "metabolite-reaction":
        self.graph = gmr.graph
        self.node_types = gmr.node_types
    elif self.net_type == "metabolite-metabolite":
        self.convert_metabolite_net(gmr)
    elif self.net_type == "reaction-reaction":
        self.convert_reaction_graph(gmr)
    else: self.graph = {}
```

```
    def convert_metabolite_net(self, gmr):
        for m in gmr.node_types["metabolite"]:
            self.add_vertex(m)
            sucs = gmr.get_successors(m)
            for s in sucs:
                sucs_r = gmr.get_successors(s)
                for s2 in sucs_r:
                    if m != s2:
                        self.add_edge(m, s2)

    def convert_reaction_graph(self, gmr):
        for r in gmr.node_types["reaction"]:
            self.add_vertex(r)
            sucs = gmr.get_successors(r)
            for s in sucs:
                sucs_r = gmr.get_successors(s)
                for s2 in sucs_r:
                    if r != s2: self.add_edge(r, s2)
```

Note that the information is loaded and a "metabolite-reaction" is created; if another network type is required, it is then converted using the two provided auxiliary methods. This is a reasonable approach, since the best way to create networks with reactions or metabolites only is to create the bipartite graph firstly. Indeed, we will connect two metabolites M_1 and M_2 (in the "metabolite-metabolite" network) if they share a reaction R as neighbor (R is a successor of M_1 and a predecessor of M_2). Also, when building a "reaction-reaction" network, two reactions R_1 and R_2 will be connected if a metabolite M is produced by R_1 (successor) and consumed by R_2 (predecessor).

To allow checking the behavior of these functions, a simple example is provided in the file "example-net.txt". This file has the following content:

```
R1: M1 + M2 => M3 + M4
R2: M4 + M6 => M3
R3: M4 + M5 <=> M6
```

Thus, the metabolic system in the file is the same as provided by Fig. 14.1. In the example below, we use this file to create all the different types of networks represented in this figure.

```
def test2():
```

```
    print("metabolite-reaction network:")
    mrn = MetabolicNetwork("metabolite-reaction")
    mrn.load_from_file("example-net.txt")
    mrn.print_graph()
    print("Reactions: ", mrn.get_nodes_type("reaction") )
    print("Metabolites: ", mrn.get_nodes_type("metabolite") )
    print()

    print("metabolite-metabolite network:")
    mmn = MetabolicNetwork("metabolite-metabolite")
    mmn.load_from_file("example-net.txt")
    mmn.print_graph()
    print()

    print("reaction-reaction network:")
    rrn = MetabolicNetwork("reaction-reaction")
    rrn.load_from_file("example-net.txt")
    rrn.print_graph()
    print()

    print("metabolite-reaction network (splitting reversible):")
    mrsn = MetabolicNetwork("metabolite-reaction", True)
    mrsn.load_from_file("example-net.txt")
    mrsn.print_graph()
    print()

    print("reaction-reaction network (splitting reversible):")
    rrsn = MetabolicNetwork("reaction-reaction", True)
    rrsn.load_from_file("example-net.txt")
    rrsn.print_graph()
    print()

test2()
```

14.2.2 An Example Metabolic Network for a Real Organism

To illustrate some of the concepts put forward in this chapter with a real world scenario, we will create a metabolic network for a known model organism, the bacterium *Escherichia coli*.

As a basis for this network, we will consider one of the most popular metabolic networks reconstructed by the research community, the *iJR904* metabolic model by Reed and co-workers [133].

The reactions included in this model are provided, in the same format discussed in the previous section, in the file "ecoli.txt" (available in the book's website). The following code allows to load this file and create each of the three types of networks used in this work, allowing to confirm that the network has 931 reactions and 761 metabolites, resulting in over 5000 edges in the "metabolite-reaction" network and over 130,000 in the "reaction-reaction" network. This larger and more realistic network will be used in the next sections to illustrate some results of the topological analysis functions.

```
def test3():
    print("metabolite-reaction network:")
    ec_mrn = MetabolicNetwork("metabolite-reaction")
    ec_mrn.load_from_file("ecoli.txt")
    print(ec_mrn.size())

    print("metabolite-metabolite network:")
    ec_mmn = MetabolicNetwork("metabolite-metabolite")
    ec_mmn.load_from_file("ecoli.txt")
    print(ec_mmn.size())

    print("reaction-reaction network:")
    ec_rrn = MetabolicNetwork("reaction-reaction")
    ec_rrn.load_from_file("ecoli.txt")
    print(ec_rrn.size())

test3()
```

14.3 Network Topological Analysis

In this section, we will overview a number of different forms of network topology analysis, i.e. metrics computed taking into account the graph structure. These metrics can be used to check properties of the network, which may lead to biological insights. As before, we will focus on the specific case of metabolic networks, while other types of networks may be analyzed in similar ways.

14.3.1 Degree Distribution

In Section 13.3, we have already defined the concepts related to node degrees and provided functions for their calculation. Given the set of the degrees for all nodes in a graph, some global metrics of interest may be calculated.

The *mean degree* (denoted by $< k >$) is simply defined as the mean value of the degrees over all nodes in the graph. In directed graphs, we can calculate this value for in-degrees, out-degrees, or both. The mean degree can be used to assess the overall level of connectivity of a network.

Another important metric is the distribution of the node degrees, allowing to estimate $P(k)$, the probability that a node has a degree of k, by the frequencies in the graph. The node degree distribution is an important parameter that allows to check if a network has a structure similar to a randomly generated network (where $P(k)$ would approximate a Normal distribution with mean $< k >$), or follows a different structure.

A popular type of networks, target of many studies in the last years, have been *scale-free* networks, where $P(k)$ typically approximates a power law, i.e. $P(k) \simeq k^{-\gamma}$, where $2 < \gamma < 3$. In these networks, most nodes have a small degree, while there are a few nodes with large degrees. The representation of $P(k)$ in a logarithmic scale is a straight line.

Both these concepts are implemented, within the **MyGraph** class presented in the previous chapter, since these are general concepts that can be calculated over any graph. Note that, by defining these methods in the parent class, they will also be available in the **Metabolic-Network** class.

```
class MyGraph:

    (...)

    def mean_degree(self, deg_type = "inout"):
        degs = self.all_degrees(deg_type)
        return sum(degs.values()) / float(len(degs))

    def prob_degree(self, deg_type = "inout"):
        degs = self.all_degrees(deg_type)
        res = {}
        for k in degs.keys():
            if degs[k] in res.keys():
                res[degs[k]] += 1
```

```
            else:
                res[degs[k]] = 1
        for k in res.keys():
            res[k] /= float(len(degs))
        return res

    (...)
```

A simple example of usage over the toy networks created in previous code can be seen in the next block, where some lines are added to the **test2** function provided in a previous example.

```
def test2():
    (...)

    print(mmn.mean_degree("out"))
    print(mmn.prob_degree("out"))

test2()
```

These functions may also be applied to the *Escherichia coli* network built above, by adding similar lines to the definition of the respective test function, as shown below. Note that we only show the example for one of the network types, while similar code may be applied for the other network types (and also to obtain the same values for in- or out-degrees). We have put some code to print the results of the dictionary returned by the **prob_degree** function sorting by the k value (note that the keys of a dictionary do not have a specific order).

```
def test3():
    ec_mrn = MetabolicNetwork("metabolite-reaction")
    ec_mrn.load_from_file("ecoli.txt")

    print("Mean degree: ", ec_mrn.mean_degree("inout"))
    pd = ec_mrn.prob_degree("inout")
    pdo = sorted(list(pd.items()), key=lambda x : x[0])
    print("Histogram of degree probabilities")
    for (x,y) in pdo: print(x, "->", y)

test3()
```

An analysis of the results shows that most of the nodes have a small value of the degree, but there are a few nodes with high degrees. This is consistent with the hypothesis of this metabolic network being a scale-free network.

14.3.2 Shortest Path Analysis

In Section 13.4, we have covered the concepts of paths and distances in graphs, putting forward ways to search and traverse a graph. Based on the definition of distance between two nodes, the number of edges traversed by the shortest path between both nodes, we can calculate global metrics of the network. Again, the simplest one is the *mean distance*, denoted by $< L >$, which is calculated as the mean of the length of the shortest paths between all pairs of nodes in the graph. Note that, if the graph is not strongly connected (as defined in Section 13.4), some distances are infinite, since the nodes are not connected. In this case, we normally choose to ignore these pairs in the calculation of $< L >$, but it is important to know in how many cases this occurs.

This metric may help in assessing some properties of the underlying networks. One important class are *small-world* networks, which are defined since they have a value of $< L >$ much smaller than expected, i.e. when compared to a randomly generated network of the same size.

This concept is implemented by function **mean_distances**, defined in the class **MyGraph** as shown below. The function returns a tuple with $< L >$ calculated over all node pairs where there is a path, and also the percentage of these (connected) pairs over all pairs in the network (if the graph is strongly connected this second value will be 1).

```
class MyGraph:

    (...)

    def mean_distances(self):
        tot = 0
        num_reachable = 0
        for k in self.graph.keys():
            distsk = self.reachable_with_dist(k)
            for _, dist in distsk:
                tot += dist
            num_reachable += len(distsk)
        meandist = float(tot) / num_reachable
        n = len(self.get_nodes())
        return meandist, float(num_reachable)/((n-1)*n)
```

As above, this can be tested by adding a couple of lines to the **test2** function for the toy example.

```
def test2():
    (...)

    print(mmn.mean_distances())
    print(mrn.mean_distances())

test2()
```

It is also interesting to investigate the results of applying these functions to the *E. coli* networks. This is done in the code chunk below. Note that, given the size of these networks, it may take some time to run.

```
def test3():
    (...)

    print("Mean distance (M-R): ", ec_mrn.mean_distances())
    print("Mean distance (M-M): ", ec_mmn.mean_distances())

test3()
```

The results show values of $< L >$ that are quite low (between 3 and 5) thus corroborating the hypothesis that metabolic networks are indeed within the class of small world networks.

14.3.3 Clustering Coefficients

In many applications, it is important to measure the tendency of the nodes of a graph to create local groups of highly connected nodes, or clusters. To measure this, the *clustering coefficient* of a node can be defined as the ratio between the number of edges existing between its neighbors and the total number of possible edges that might exist between the same nodes. Indeed, a node with a high clustering coefficient has neighbors that are highly connected between themselves.

A function to calculate the clustering coefficient of a node was added to the class **MyGraph**, as shown below. Note that in this function, we assume that a node is connected to another if they are neighbors, regardless of the directionality of the connecting edge(s).

```
class MyGraph:

    (...)
```

```
    def clustering_coef(self, v):
        adjs = self.get_adjacents(v)
        if len(adjs) <=1: return 0.0
        ligs = 0
        for i in adjs:
            for j in adjs:
                if i != j:
                    if j in self.graph[i] or i in self.graph[j]:
                        ligs = ligs + 1
        return float(ligs)/(len(adjs)*(len(adjs)-1))
```

Based on this concept, a global metric can be defined for the network, the *mean clustering coefficient*, denoted as $< C >$, calculated as the mean of the clustering coefficient for all nodes in the network. Another metric of interest relates the clustering coefficient with the node degrees. Thus, $C(k)$ is defined as the mean of the clustering coefficients for all nodes with degree k.

These concepts are also implemented in the class **MyGraph** through three distinct functions: **all_clustering_coefs** that calculates the clustering coefficients of all nodes, **mean_clustering_coef** that calculates $C(k)$, and **mean_clustering_perdegree** that computes $C(k)$ for all values of k.

```
class MyGraph:

    (...)

    def all_clustering_coefs(self):
        ccs = {}
        for k in self.graph.keys():
            ccs[k] = self.clustering_coef(k)
        return ccs

    def mean_clustering_coef(self):
        ccs = self.all_clustering_coefs()
        return sum(ccs.values()) / float(len(ccs))

    def mean_clustering_perdegree(self, deg_type = "inout"):
        degs = self.all_degrees(deg_type)
        ccs = self.all_clustering_coefs()
        degs_k = {}
```

```
    for k in degs.keys():
        if degs[k] in degs_k.keys(): degs_k[degs[k]].append(k)
        else: degs_k[degs[k]] = [k]
    ck = {}
    for k in degs_k.keys():
        tot = 0
        for v in degs_k[k]: tot += ccs[v]
        ck[k] = float(tot) / len(degs_k[k])
    return ck
```

As before, these functions may be tested adding a few rows to the test function previously defined for the toy example. Note that these functions should not be applied over bipartite graphs (as the "metabolite-reaction" networks), since by definition, in this case, all coefficients will be zero (remember that the vertices of the same type cannot be connected). This can be easily checked by the reader.

```
def test2():
    (...)
    print(mmn.all_clustering_coefs())
    print(mmn.mean_clustering_coef())
    print(mmn.mean_clustering_perdegree())

test2()
```

Applying these functions over the metabolites network of *E. coli* is shown in the code below. Given the extensive output, the **all_clustering_coefs** is not applied in this case.

```
def test3():
    (...)
    print(ec_mmn.mean_clustering_coef())
    cc = ec_mmn.mean_clustering_perdegree()
    cco = sorted(list(cc.items()), key=lambda x : x[0])
    print("Clustering coefficients per degree")
    for (x,y) in cco: print(x, "->", y)

test3()
```

An analysis of the output of the clustering coefficient by degree shows an interesting pattern, where the $C(k)$ value decreases with the increase of k. This is a typical pattern of a hierarchical modular network, composed of modules with mostly nodes with low degree, that are highly connected among themselves. These modules are typically connected among themselves by a few nodes of higher degree. This behavior is expected in the case of a metabolic network.

14.3.4 Hubs and Centrality Measures

One important task in network analysis is to identify nodes that may be considered as the most important. In biological networks, these nodes may be of increased importance in understanding the behavior of the system, and related to important properties, for instance network robustness and resistance to failure.

There are many different criteria to evaluate the importance of nodes, which are named as *centrality* measures or indexes. We will here only cover a few of the most important centrality measures, leaving the remaining for the reader to explore. Notice that the applicability of these centrality indexes are related to the application field, which may define how to identify important nodes in the network.

Probably the simplest way to rank nodes by importance is to consider as most important the nodes that have more neighbors, i.e. use the node degree as the centrality measure, thus known as *degree centrality*. Applying this measure to a metabolic network (namely to a "metabolite-metabolite" network) will allow to identify highly-connected metabolites, typically co-factors in many reactions, which are also known as currency metabolites. In some applications of these networks, it is common to remove these metabolites to make easier some types of analyses.

In this case, the functions to calculate the highest ranked nodes are based on the ones for the calculation of node degrees, put forward in Section 14.3.1 above. The function **highest_degrees** shown below allows to select the nodes with the highest rank in the degree centrality measure.

```
class MyGraph:

    (...)

    def highest_degrees(self, all_deg= None, deg_type = "inout", top=
    10):
        if all_deg is None:
            all_deg = self.all_degrees(deg_type)
        ord_deg = sorted(list(all_deg.items()), key=lambda x : x[1],
   reverse = True)
        return list(map(lambda x:x[0], ord_deg[:top]))
```

Notice a few details about this function's code. The `lambda` notation allows to easily define functions, typically used within other functions and therefore with no need for an explicit name. In the cases of the previous code, these functions are used to extract the second and the first component of tuples, respectively. Another peculiarity is the use of the **map** function, which allows to apply the same function the all elements of a list, returning a new list with the results (i.e. a list with the same size as the original).

Applying the **highest_degrees** function over the same network for *E. coli*, as shown in the code below, returns the set of the most connected metabolites. Here, we may find metabolites that are present in many reactions, mainly as co-factors, such as hydrogen or water, but also ADP, ATP, NAD+, NADH, or Acetyl-CoA. Notice that the metabolites are given as identifiers of the model. To know which metabolite (or reaction) is associated with a given identifier, the reader show consult the BiGG database in the URL: `http://bigg.ucsd.edu/`.

```
def test3():

    (...)

    print(ec_mmn.highest_degrees(top = 20))

test3()
```

An important alternative to degree centrality measures are measures that identify the most important nodes in a graph as the ones which are closest to the other nodes. The normalized *closeness centrality* for directed graphs may be defined as:

$$CC(v) = \frac{N-1}{\sum_{x \in V} d(v,x)} \tag{14.1}$$

where $d(x, y)$ represents the distance between nodes x and y, and N is the number of nodes in the graph ($|V|$).

Therefore, according to this definition, the highest ranked nodes will be the ones with the smallest mean distances between themselves and the remaining nodes in the graph. In this case, we considered the distance from outgoing links of the target node, while in directed graphs the same measure may be applied for incoming links, where the previous definition would consider $d(x, v)$.

Notice that this metric applies well to connected graphs (or nearly connected graphs). In the case there are unconnected pairs of nodes, the usual is to ignore those in the above computation, while if these pairs are in a large proportion in the network, this will affect the results heavily.

The implementation of the closeness centrality for a single node (considering outgoing edges) and for the whole network is provided by the functions given in the code below.

```
class MyGraph:

    (...)

    def closeness_centrality(self, node):
        dist = self.reachable_with_dist(node)
        if len(dist)==0: return 0.0
        s = 0.0
        for d in dist: s += d[1]
        return len(dist) / s

    def highest_closeness(self, top = 10):
        cc = {}
        for k in self.graph.keys():
            cc[k] = self.closeness_centrality(k)
        ord_cl = sorted(list(cc.items()), key=lambda x : x[1],
  reverse = True)
        return list(map(lambda x:x[0], ord_cl[:top]))
```

The **highest_closeness** function can be applied to the *E. coli* network as in the previous case. The results will overlap significantly with the ones from the degree centrality. It is left as an exercise to the reader to implement the closeness centrality for incoming edges and check the results on the same network.

The last example of a centrality measure that we will cover here is the *betweenness centrality*, which is based on the shortest paths between pairs of nodes in the network. This metric is defined as the proportion of the shortest paths between all pairs of nodes in the network that pass through the target node. Thus, it takes values between 0 and 1.

The following function calculates the betweenness centrality of a node in the graph.

```
class MyGraph:

    (...)

    def betweenness_centrality(self, node):
```

```
    total_sp = 0
    sps_with_node = 0
    for s in self.graph.keys():
        for t in self.graph.keys():
            if s != t and s != node and t != node:
                sp = self.shortest_path(s, t)
                if sp is not None:
                    total_sp += 1
                    if node in sp: sps_with_node += 1
    return sps_with_node / total_sp
```

The generalization of this function to the calculation of the values of the betweenness centrality for all nodes in the network, as well as the application to the provided examples are left as exercises to the reader.

14.4 Assessing the Metabolic Potential

The representation of metabolic networks as bipartite graphs, apart from enabling the structural analysis provided by the metrics described in the previous section, allows the study of the metabolic potential of a metabolic system, in terms of the metabolites that the system can produce.

This analysis requires to look at the network in a different way, by considering that each reaction node may be active or inactive, depending on the availability of the metabolites that act as its inputs. Indeed, a reaction will only occur within a cell if all required substrates (including co-factors) are present.

To provide for such analysis, from the networks considered above, we need to select a "metabolite-reaction" network, where reversible reactions are split into two, one for each direction of the reaction (see Fig. 14.1D). This is necessary to be able to clearly mark substrates (inputs) and products (outputs) of each reaction.

Given this representation, we can build a function to identify all reactions that are active, providing as input a list of metabolites assumed to be available within the cell (for the sake of simplicity, we are ignoring here the existence of possible cell compartments). Also, symmetrically, we can also implement a function that identifies all metabolites that can be produced given a set of active reactions.

These functions are given in the next code block, added to the **MetabolicNetwork** class, given its specificity for this type of networks.

```
class MetabolicNetwork:

    (...)

    def active_reactions(self, active_metabolites):
        if self.net_type != "metabolite-reaction" or not self.
    split_rev:
            return None
        res = []
        for v in self.node_types['reaction']:
            preds = set(self.get_predecessors(v))
            if len(preds)>0 and preds.issubset(set(active_metabolites
    )):
                res.append(v)
        return res

    def produced_metabolites(self, active_reactions):
        res = []
        for r in active_reactions:
            sucs = self.get_successors(r)
            for s in sucs:
                if s not in res: res.append(s)
        return res
```

Having these functions implemented, we can implement a function that can determine the complete set of metabolites that can be produced given an initial list of available metabolites. This implies to iterate calls over the previous functions checking if the metabolites that can be produced lead to the activation of other reactions, which in turn can lead to new metabolites being produced, and so on. The process ends when no reactions can be added to the list of active reactions. This is implemented by the function **all_produced_metabolites** provided below.

```
    def all_produced_metabolites(self, initial_metabolites):
        mets = initial_metabolites
        cont = True
        while cont:
            cont = False
            reacs = self.active_reactions(mets)
            new_mets = self.produced_metabolites(reacs)
```

```
            for nm in new_mets:
                if nm not in mets:
                    mets.append(nm)
                    cont = True
        return mets
```

This function can be applied to the toy network given in the last section for testing.

```
def test4():
    mrsn = MetabolicNetwork("metabolite-reaction", True)
    mrsn.load_from_file("example-net.txt")
    mrsn.print_graph()

    print(mrsn.produced_metabolites(["R1"]))
    print(mrsn.active_reactions(["M1","M2"]))
    print(mrsn.all_produced_metabolites(["M1","M2"]))
    print(mrsn.all_produced_metabolites(["M1","M2","M6"]))

test4()
```

An exploration of these functions using the *E. coli* network is left as an exercise for the interested reader.

Bibliographic Notes and Further Reading

A paper by Jeong et al. in 2000 [83] analyzed the metabolic network of 43 organisms from different domains of life, concluding that, despite the differences, they shared similar topological properties, including those related to their scale-free nature. The concept of scale-free networks and their characterization were firstly put forward in a 1999 paper by Barabási et al. [21]. In 2002, Ravasz et al. [132] introduced the concept of hierarchical networks, in the context of metabolism.

Interesting examples of the analysis of other types of network might be found, for instance, in [81], which studies the transcriptional network, and [16] that analyzes the signaling network, both of the baker's yeast (*Saccharomyces cerevisiae*). A recent discussion of integrated networks, spanning different sub-systems, may be found in [67].

An important concept in network analysis, not covered here, is the topic of network motifs, which are patterns that are over-represented in different types of biological networks, is addressed in a paper by Milo et al. in 2002 [114].

Barabási and Olivai wrote, in 2004 [23], a very comprehensive review on the field of *network biology*, where they review these results over metabolic networks, but also consider other types of biological networks, considering their properties and discussing their evolutionary origin. In a more recent review, the same author and colleagues have published a very interesting review on some applications of network biology related methods to medical research [22].

Finally, many other representation formalisms, many based in graphs, have been used to model cells and their sub-systems in the wider field of Systems Biology, many of those allowing for methods to perform several types of simulation leading to phenotype prediction. An interesting review of such paradigms has been done in [106].

Exercises and Programming Projects

Exercises

1. Considering the networks of the type "metabolite-reaction" presented in this chapter, which correspond to bipartite graphs, write a method to add to the class **MetabolicNetwork** that can detect the set of "final" metabolites, those that are produced by at least one reaction, but are not consumed by any reaction in the network.
2. Write a method to add to the class **MetabolicNetwork** that, given a set of initial metabolites (assumed to be available) and a target metabolite, returns the shortest path to produce the target metabolite, i.e. the shortest list of reactions that activated in that order allow to produce the target metabolite from the initial set of metabolites. The reactions in the list are only valid if they can be active in the given order (i.e. all their substrates exist). The method returns `None` if the target metabolite cannot be produced from the list of initial metabolites.

Programming Projects

1. Considering the class to implement undirected graphs proposed in an exercise in the previous chapter, add methods to implement the concepts related to network topological analysis described in this chapter.
2. Consider the database *Kyoyo Encyclopedia of Genes and Genoems* (KEGG) (`http://www.genome.jp/kegg/`). Explore the Python interfaces for this database. Implement code to get the reactions and metabolites for a given organism and build its metabolic network. Explore networks of different organisms.
3. Explore one of the Python libraries implementing graphs, such as *NetworkX* (`http://networkx.github.io`) or *igraph* (`http://igraph.org/python`). Implement

metabolic networks using those libraries, with similar functionality to the ones implemented in this chapter.

4. Build an implementation of a regulatory network, containing regulatory genes (e.g. those encoding transcription factors) and regulated genes. Design a network able to represent activation and inhibition events.

CHAPTER 15

Assembling Reads Into Genomes: Graph-Based Algorithms

In this chapter, we will address some of the challenges posed by genome assembly, i.e. rebuilding a sequence from the overlapping fragments (reads) returned by the DNA sequencing equipment. We will show that the most effective algorithms for these challenging problems are based on graphs, most precisely on the concepts of Hamiltonian and Eulerian paths, and related algorithms. We will implement, in Python, algorithms for simplified versions of genome assembly problems and discuss their efficiency. As well as discuss some of the challenges involved in real-word genome assembly tasks and the their solutions from state-of-the-art programs.

15.1 Introduction to Genome Assembly and Related Challenges

Sequencing a genome has been many times compared with the reading of a book, the book of life. However, sequencing a genome is technically a quite different process from reading. Indeed, there are no current technologies that allow to sequence full genomes, being these techniques constrained to sequence small fragments of the DNA molecules of at most a few hundreds of base pairs. These overlapping fragments, called *reads*, need then to be put together to reconstruct the original sequence (or genome), in pretty much the same way one puts together the pieces of a puzzle (Fig. 15.1).

The complexity of this problem can be huge, which can be easily understood if one considers that the size of the reads is typically of a few hundred nucleotides, while full genomes typically contain from a few to thousands of millions of nucleotides. Until the 1990's, this was indeed thought of as an impossible mission, and even today some larger genomes are still out of reach for the current software tools.

Apart from the size of the genomes, other factors make this a very complex problem, in practice. Some are enumerated below:

- DNA molecules have two complementary strands and there is no way to tell from which strand a given fragment was taken;
- all types of genome sequencing equipment have relatively high error rates, and thus in many cases the fragments can be different from the original sequence, which makes the overlap with other adjacent fragments more difficult to infer;

Bioinformatics Algorithms. DOI: 10.1016/B978-0-12-812520-5.00015-8

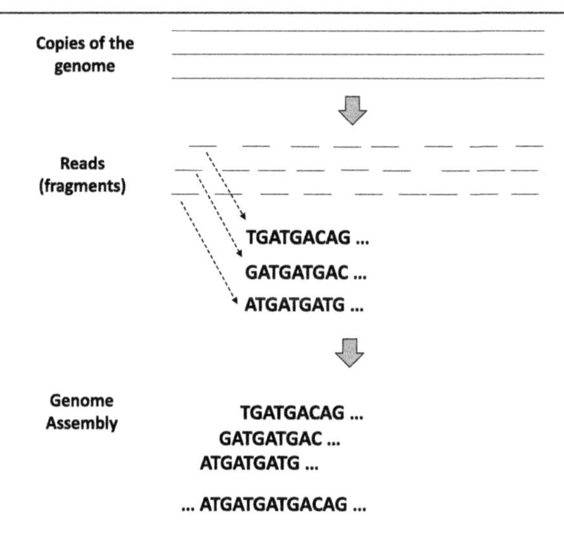

Figure 15.1: Overall representation of the process of genome assembly. Multiple copies of the genome (DNA molecule) are broken to generate fragments of short size (reads), which are sequenced and assembled to reach the original genome sequence.

- there are regions of the genome that are very difficult, or even impossible, to sequence, and thus are not covered by any reads.

In this book, we will not address the intricacies of genome sequencing technologies, nor seek to address all available solutions for these problems. Still, in the end of this chapter, we will discuss some of these problems and how they can be handled.

Before that, we will simplify the problem to an idealized version to discuss possible algorithms for genome assembly. These form the core of current software applications, although in practice these tools need to take into account the previously mentioned problems. In our simplified problem definition, we will consider the whole genome to be a unique sequence with a single strand, that the genome is fully covered by reads, and that these reads are perfect copies of the original genome (there are no sequencing errors). So, in the next section we will start looking at the algorithms available to address this task.

15.2 Overlap Graphs and Hamiltonian Cycles

15.2.1 Problem Definition and Exhaustive Search

Before starting to address the algorithms, let us put forward a useful definition: the *composition* of a sequence (string) s in segments of size k, denoted $comp_k(s)$, is defined as the collection of all sub-sequences of s, of size k, possibly including repeated sub-sequences. The resulting fragments are normally ordered lexicographically.

It is trivial to implement this definition as a Python function as provided in the next code chunk.

```
def composition(k, seq):
    res = []
    for i in range(len(seq)-k+1):
        res.append(seq[i:i+k])
    res.sort()
    return res

def test1():
    seq = "CAATCATGATG"
    k = 3
    print (composition(k, seq))

test1()
```

When performing genome sequencing, we have the complementary problem, i.e. we do not know the original sequence, but we know its composition in fragments. Thus, a possible problem definition is to try to recover the original sequence, given as an input its composition and the value of k.

A natural solution for this problem would be to start with one of the fragments and try to find another fragment that has an overlap of $k - 1$ positions with this fragment. This would allow adding one nucleotide to the reconstructed sequence (the last of the second sequence). Iterating over this strategy could lead to cover all fragments. Notice that a fragment of size k has an overlap of $k - 1$ with another of the same size when the $suffix_{k-1}$ (last $k - 1$ characters) of the first is equal to the $prefix_{k-1}$ (first $k - 1$ characters) of the latter.

The following code implements these concepts as Python functions:

```
def suffix (seq):
    return seq[1:]

def prefix(seq):
    return seq[:-1]
```

Consider an example with $k = 3$ and the following set of fragments: ACC, ATA, CAT, CCA, TAA. Starting with ACC, we can find CCA with an overlap of 2 and this allows to augment

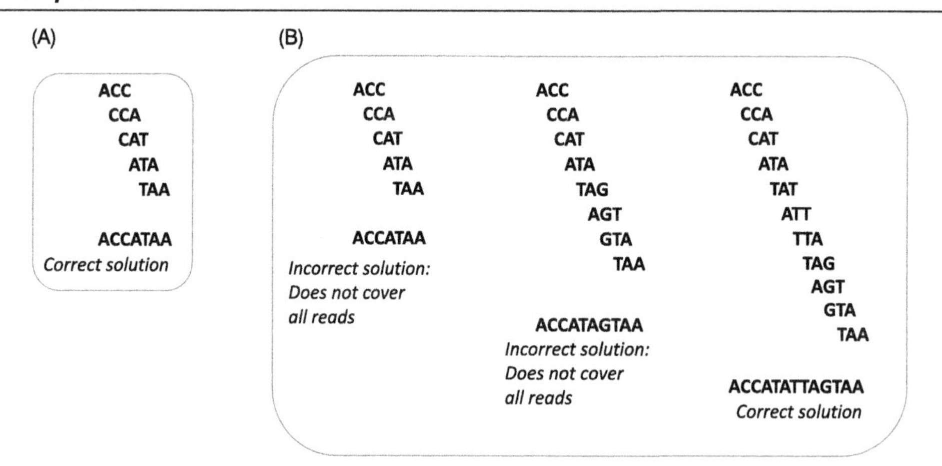

Figure 15.2: Example of the reconstruction of a sequence from overlapping fragments of size $k = 3$. (A) An example where a simple algorithm selecting fragments with overlap of 2 nucleotides returns the correct solution. (B) An example where there are repetitions of sub-sequences of size $k - 1$ in the fragments, making more difficult the selection of the next fragment: three solutions are tried, but only the last one returns a correct solution covering all fragments.

the sequence to ACCA; following the same strategy we add CAT, ATA, and, finally TAA, to reach the final sequence of ACCATAA (Fig. 15.2A).

So, we seem to have solved our problem! However, a more careful analysis will show otherwise. Notice that in this case, we never had a doubt of which sequence to select next, since there were no repeated $k-1$-mers. Also, try to run the same algorithm starting with a different fragment!

Let us try a different example adding a few fragments to the previous: ACC, AGT, ATA, ATT, CAT, CCA, GTA, TAA, TAG, TAT, TTA. If we follow the same algorithm we will add fragments ACC, CCA, and CAT. Here, we have the first decision to make as there are two candidates, ATA and ATT. Let us choose ATA and proceed. Now, we have an even harder decision to make, since we can select TAA, TAG, or TAT. Selecting the first (TAA) as before, we will terminate the algorithm, as there are no sequences starting with AA (Fig. 15.2B). However, we have not covered all fragments and, therefore, this solution is incorrect. We can select TAG, and this would lead to select AGT, GTA, and TAA next; again, this leads to an incomplete solution, as the fragments TAT and ATT are not considered. So, the right choice would be to select TAT, which would lead to a solution covering all fragments and returning the sequence ACCATATTAGTAA, as shown in Fig. 15.2B, last column.

Although, we have managed to find a solution in this case, it is clear that the algorithm is not adequate. Indeed, this strategy would make us backtrack in our previous solution in any situation where more than one option exists. In real scenarios, this is very inefficient, as there are

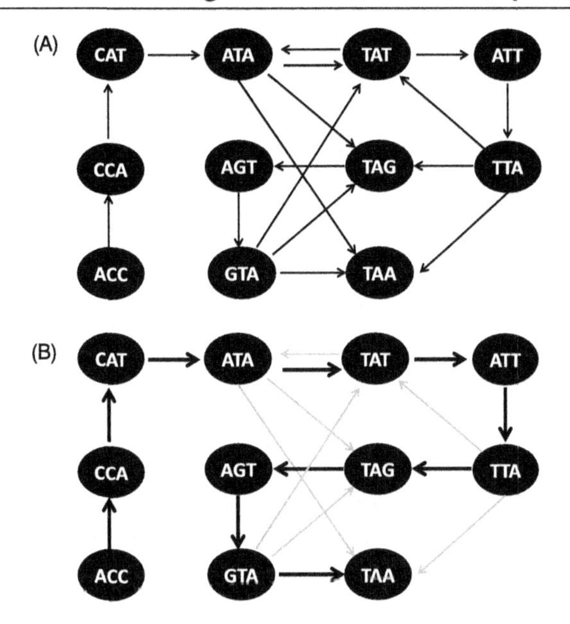

Figure 15.3: Example of the overlap graph for the example in the previous section. (A) The overlap graph. (B) The solution for the reconstruction problem shown as a path over the graph, passing through all nodes; the edges included in this path are shown with wider arrows, while the ones not included are shown in light gray.

many repetitions of $k-1$-mers. As in regular puzzles, where pieces with similar patterns are the most troublesome, also here repeats make our life more difficult.

15.2.2 Overlap Graphs

One alternative to address these problems comes from graph theory, which we have covered in the last two chapters. Indeed, the fragments for this problem can be organized in a special type of graph, called *overlap graph*, which can lead to a different problem formulation and alternative algorithms.

An *overlap graph* is defined as follows:

- Nodes (vertices) correspond to the input fragments, i.e. there will be one vertex per given fragment (read);
- Edges are created between two nodes v and w, if $suffix_{k-1}(v) = prefix_{k-1}(w)$, according to the definition above, i.e. the sequences of nodes v and w have an overlap of $k-1$ nucleotides (and thus the name of the graph).

The graph in Fig. 15.3A represents the overlap graph for the example put forward in the previous section. In this case, there are 11 nodes, one for each fragment and edges exist connecting all pairs of nodes where the last two characters of the first are equal to the first two characters of the second.

Note that a solution to the string reconstruction problem can be represented as a path over this graph. To be a valid solution, this path should pass exactly once over every node of the graph. These paths have a special denomination and interesting properties, as we will explore in the next section.

Fig. 15.3B shows a path for the previous example, which represents the solution found in the previous section (see Fig. 15.2B). Note that, in some cases, there are nodes with multiple edges going out, which make more difficult to select the next node, since there are many choices, as we saw above.

Before proceeding to explore the properties of these graphs and paths, and related algorithms, let us define some Python code that allows to create overlap graphs. To achieve this aim, we will define a Python class, **OverlapGraph**, which will be a sub-class of the class **MyGraph**, defined in Chapter 13.

Following the definition above, the next code chunk shows the definition of the class, and its constructor, which uses an auxiliary function (**create_overlap_graph**) to build the graph, given the set of fragments. Notice that the code uses the functions **suffix** and **prefix** defined before, which need to be defined outside of the class. The function **test2** provides an example, with the set of fragments from the first case discussed in the last section (Fig. 15.2A).

```
from MyGraph import MyGraph

class OverlapGraph(MyGraph):

    def __init__(self, frags):
        MyGraph.__init__(self, {})
        self.create_overlap_graph(frags)

    def create_overlap_graph(self, frags):
        for seq in frags:
            self.add_vertex(seq)
        for seq in frags:
            suf = suffix(seq)
            for seq2 in frags:
                if prefix(seq2) == suf:
                    self.add_edge(seq, seq2)
```

```
def test2():
    frags = ["ACC", "ATA", "CAT", "CCA", "TAA"]
    ovgr = OverlapGraph(frags, False)
    ovgr.print_graph()

test2()
```

Although for the selected example, this code works and builds the desired graph, problems arise when there are repeated fragments in the list. In this case, these repetitions would be ignored and a single node would be created for each unique sequence. However, this does not address the requirements of our problem, since repeated fragments should all be covered, and, therefore, different nodes should be assigned to each fragment.

A possible solution for this problem, while keeping our representation of graphs, that requires nodes to have unique identifiers (those are keys of a dictionary), is to add a numerical identifier to the fragment sequence. Thus, all nodes will have a unique identifier, and still preserve the information from the sequence.

This solution is implemented in the next code block, where the constructor now accounts for both options, to have replicates or not, and a new function is defined for this case of replicates (**create_overlap_graph_with_reps**). The auxiliary function **get_instances** is used to search all nodes containing a given sequence. The code is tested (function **test3**) with an example of a set of fragments with replicates.

```
class OverlapGraph(MyGraph):

    def __init__(self, frags, reps = True):
        MyGraph.__init__(self, {})
        if reps: self.create_overlap_graph_with_reps(frags)
        else: self.create_overlap_graph(frags)
        self.reps = reps

    def create_overlap_graph_with_reps(self, frags):
        idnum = 1
        for seq in frags:
            self.add_vertex(seq+ "-" + str(idnum))
            idnum = idnum + 1
        idnum = 1
        for seq in frags:
```

```
                suf = suffix(seq)
                for seq2 in frags:
                    if prefix(seq2) == suf:
                        for x in self.get_instances(seq2):
                            self.add_edge(seq+ "-" + str(idnum), x)
                idnum = idnum + 1

    def get_instances(self, seq):
        res = []
        for k in self.graph.keys():
            if seq in k: res.append(k)
        return res

def test3():
    frags = [ "ATA", "ACC", "ATG", "ATT", "CAT", "CAT", "CAT", "CCA"
  , "GCA", "GGC", "TAA", "TCA", "TGG", "TTC", "TTT"]
    ovgr = OverlapGraph(frags, True)
    ovgr.print_graph()

test3()
```

15.2.3 Hamiltonian Circuits

In a directed graph, a *Hamiltonian circuit* is a path that goes through all nodes of the graph exactly once. As we hinted above, the problem of reconstructing a string, given its composition, may be defined as the problem of finding a Hamiltonian circuit in the overlap graph. One example of such a circuit is shown in Fig. 15.3B.

Indeed, from the path set by the Hamiltonian circuit, we can easily get the "original" sequence. This is done by simply taking the sequence from the first node in the path, and concatenating the last character of the sequences in the remaining nodes of the path, following its order.

This can be shown in the code below, where we define two new functions to add to the **OverlapGraph** class above, the first to get the sequence represented by a node, and the second to get the full sequence spelled by a path in the graph. The example provided to test complements the code of function **test3**, defining a path over the defined graph, which in this case is a Hamiltonian path, and returns the original sequence that originated the fragments.

```
    def get_seq(self, node):
        if node not in self.graph.keys(): return None
        if self.reps: return node.split("-")[0]
        else: return node

    def seq_from_path(self, path):
        if not self.check_if_hamiltonian_path(path): return None
        seq = self.get_seq(path[0])
        for i in range(1,len(path)):
            nxt = self.get_seq(path[i])
            seq += nxt[-1]
        return seq

def test3():
    frags = [ "ACC", "ATA",  "ATG", "ATT", "CAT", "CAT", "CAT", "CCA
    ", "GCA", "GGC", "TAA", "TCA", "TGG", "TTC", "TTT"]
    ovgr = OverlapGraph(frags, True)
    ovgr.print_graph()
    path = ['ACC-2', 'CCA-8', 'CAT-5', 'ATG-3', 'TGG-13', 'GGC-10',
    'GCA-9', 'CAT-6', 'ATT-4', 'TTT-15', 'TTC-14', 'TCA-12', 'CAT-7',
     'ATA-1', 'TAA-11']
    print (ovgr.seq_from_path(path))

test3()
```

In the previous example, we have shown a path that is Hamiltonian, but have not tested for this condition. It is easy to verify, given a path, if it is a Hamiltonian path or not, checking if it is a valid path, if it contains all nodes and if there are no repeated nodes. Notice that, for a graph $G = (V, E)$ a Hamiltonian path has exactly $|V|$ nodes and $|V| - 1$ edges.

The following code block shows two functions that were added to the class **MyGraph**, defined in Chapter 13. It makes sense to add this code in this class, since these functions will work for any directed graph, and are not restricted to overlap graphs. Also, since the class **OverlapGraph** is a sub-class of **MyGraph**, it will inherit all methods defined there.

The two methods presented can be used, respectively, to check if a given path is valid, and to check if it is a Hamiltonian path. Both return a Boolean result (`True` or `False`).

```
class MyGraph:

## see code in previous chapters
(...)

    def check_if_valid_path(self, p):
        if p[0] not in self.graph.keys(): return False
        for i in range(1,len(p)):
            if p[i] not in self.graph.keys() or p[i] not in self.
    graph[p[i-1]]:
                return False
        return True

    def check_if_hamiltonian_path(self, p):
        if not self.check_if_valid_path(p): return False
        to_visit = list(self.get_nodes())
        if len(p) != len(to_visit): return False
        for i in range(len(p)):
            if p[i] in to_visit: to_visit.remove(p[i])
            else: return False
        if not to_visit: return True
        else: return False
```

These functions may be tested in the **OverlapGraph** class, adding a couple of rows to the previous example.

```
def test3():
    frags = [ "ACC", "ATA", "ATG", "ATT", "CAT", "CAT", "CAT", "CCA"
  , "GCA", "GGC", "TAA", "TCA", "TGG", "TTC", "TTT"]
    ovgr = OverlapGraph(frags, True)
    ovgr.print_graph()
    path = ['ACC-2', 'CCA-8', 'CAT-5', 'ATG-3', 'TGG-13', 'GGC-10',
  'GCA-9', 'CAT-6', 'ATT-4', 'TTT-15', 'TTC-14', 'TCA-12', 'CAT-7',
   'ATA-1', 'TAA-11']
    print (ovgr.seq_from_path(path))
    print(ovgr.check_if_valid_path(path))
    print (ovgr.check_if_hamiltonian_path(path))

test3()
```

So, we checked that it is trivial to verify if a path is Hamiltonian or not. However, unfortunately, it is not easy to search for Hamiltonian circuits in a graph. An alternative is to perform an exhaustive search, which would lead to a variant of a depth-based search over the graph (this was briefly discussed in Chapter 13, Section 13.4). This algorithm tries to build a path moving from each node to one of its successors not previously visited. It will backtrack when a node is reached that has no successors not yet visited.

A function implementing this process is provided below, added to the class **MyGraph**, where the Hamiltonian circuit is sought, taking each possible node as the starting point. The function **search_hamiltonian_path_from_node** tries to build a path starting at that node, that can cover all other nodes. This function implements a search tree by maintaining the current node being processed (variable *current*), the current path being built (variable *path*), as well as the state of the nodes (variable *visited*). This last variable is a dictionary where the keys are nodes, and the values are the indexes of the next successor to explore for that node, thus seeking to minimize the memory used to keep this valuable information.

```
class MyGraph:

    (...)

    def search_hamiltonian_path(self):
        for ke in self.graph.keys():
            p = self.search_hamiltonian_path_from_node(ke)
            if p != None:
                return p
        return None

    def search_hamiltonian_path_from_node(self, start):
        current = start
        visited = {start:0}
        path = [start]
        while len(path) < len(self.get_nodes()):
            nxt_index = visited[current]
            if len(self.graph[current]) > nxt_index:
                nxtnode = self.graph[current][nxt_index]
                visited[current] += 1
                if nxtnode not in path: ## node added to path
                    path.append(nxtnode)
                    visited[nxtnode] = 0
                    current = nxtnode
```

```
            else: ## backtrack
                if len(path) > 1:
                    rmvnode = path.pop()
                    del visited[rmvnode]
                    current = path[-1]
                else: return None
        return path
```

The previous function can be tested in the context of the **OverlapGraph** class, in the previous examples, as shown below.

```
def test4():
    frags = [ "ACC", "ATA", "ATG", "ATT", "CAT", "CAT", "CAT", "CCA",
     "GCA", "GGC", "TAA", "TCA", "TGG", "TTC", "TTT"]
    ovgr = OverlapGraph(frags, True)
    ovgr.print_graph()
    path = ovgr.search_hamiltonian_path()
    print(path)
    print (ovgr.check_if_hamiltonian_path(path))
    print (ovgr.seq_from_path(path))

test4()
```

We can now close the circle by defining a sequence, calculating its composition and trying to recover it by finding a Hamiltonian circuit. This is done in the next example.

```
def test5():
    orig_sequence = "CAATCATGATGATGATC"
    frags = composition(3, orig_sequence)
    print (frags)
    ovgr = OverlapGraph(frags, True)
    ovgr.print_graph()
    path = ovgr.search_hamiltonian_path()
    print (path)
    print (ovgr.seq_from_path(path))

test5()
```

However, as the reader can check by running the example, the recovered sequence is not the same. This happens since there might be many different sequences with the same composition. This becomes increasingly rarer as the value of k becomes larger.

Also, we invite the reader to try to increase the size of the sequence and check that the algorithm starts to behave increasingly worse in terms of its run time. In fact, this algorithm can not be used for large sequences, and respective overlap graphs. Indeed, the problem of finding Hamiltonian graphs is quite complex, belonging to the class of NP-hard optimization problems, which implies that no efficient algorithms exist when the problem scales, in this case, when the graph becomes larger.

This makes it very difficult, in practice, to use this approach for large-scale genome assembly. Still, until a few years ago, these were the strategies used by the existing programs, which used heuristic methods to attain the best possible solutions. These were used, for instance, throughout the *Human Genome Project*.

15.3 DeBruijn Graphs and Eulerian Paths

15.3.1 DeBruijn Graphs for Genome Assembly

Given that the previous attempt to represent the problem data, using overlap graphs, was not successful in finding efficient algorithms, other solutions were sought by the research community. Indeed, approaching similar problems, while at the time not related to genome assembly, the mathematician DeBruijn proposed a distinct representation, which also makes use of directed graphs.

The idea is to represent the fragments of the problem as edges, and not as vertices. Here, the solutions to the problem would, therefore, be paths over the graph containing all edges exactly once. We will explore this in further detail in the next section.

In the case of *DeBruijn graphs*, the nodes contain sequences that correspond to either a $prefix_{k-1}$ or a $suffix_{k-1}$ of one of the fragments. So, each edge, corresponding to a fragment, connects a node representing its $prefix_{k-1}$ to a node representing its $suffix_{k-1}$. In this case, unlike in the previous overlap graphs, prefixes and suffixes of length $k-1$, which share the same sequence, are represented in a single node. So, repeated fragments are represented as multiple edges connecting the same pair of nodes. Fig. 15.4 shows the representation of the example from the previous section, the one implemented in the functions **test3** and **test4** above, as a DeBruijn graph. The set of 15 fragments is ACC, ATA, ATG, ATT, CAT, CAT, CAT, CCA, GCA, GGC, TAA, TCA, TGG, TTC, and TTT. Thus, the graph has 15 edges and 11 unique $k-1$-mers that are either prefixes or suffixes (or both) of these fragments.

As before, we will here implement DeBruijn graphs as a Python class, **DeBruijnGraph**, that also extends the class **MyGraph**. In this case, this class needs to redefine the function **add_adge** to allow repeated edges between the same pair of nodes. The function **create_**

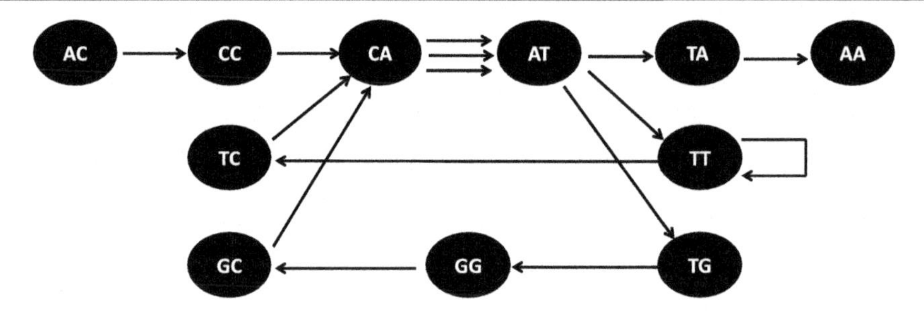

Figure 15.4: Example of the DeBruijn graph for the example in the previous section.

deBruijn_graph implements the creation of the graph according to the explanation provided above. These functions are applied to the example represented in Fig. 15.4.

```
from MyGraph import MyGraph

class DeBruijnGraph (MyGraph):

    def __init__(self, frags):
        MyGraph.__init__(self, {})
        self.create_deBruijn_graph(frags)

    def add_edge(self, o, d):
        if o not in self.graph.keys():
            self.add_vertex(o)
        if d not in self.graph.keys():
            self.add_vertex(d)
        self.graph[o].append(d)

    def create_deBruijn_graph(self, frags):
        for seq in frags:
            suf = suffix(seq)
            self.add_vertex(suf)
            pref = prefix(seq)
            self.add_vertex(pref)
            self.add_edge(pref, suf

def test6():
    frags = [ "ACC", "ATA", "ATG", "ATT", "CAT", "CAT", "CAT", "CCA",
    "GCA", "GGC", "TAA", "TCA", "TGG", "TTC", "TTT"]
```

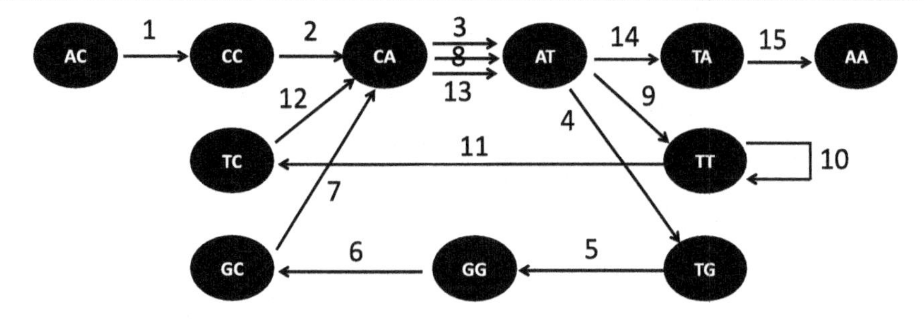

Figure 15.5: Example of an Eulerian cycle over the DeBruijn graph for the previous example. Edges are numbered in the order they appear in the path.

```
    dbgr = DeBruijnGraph(frags)
    dbgr.print_graph()

test6()
```

15.3.2 Eulerian Paths

In a directed graph, an *Eulerian circuit* is a path that goes through all edges in the graph exactly once. On the other hand, an *Eulerian cycle* is a cycle that goes through all edges, in other words it is an Eulerian circuit that starts and terminates in the same node.

The problem of reconstructing a string from its composition in k-mers can be formulated as the search for an Eulerian circuit, in the previously defined DeBruijn graph. An example of such a circuit, over the example in Fig. 15.4, is shown in Fig. 15.5. Note that in the example, the corresponding recovered sequence would be: ACCATGGCATTTCATAA.

We, thus, gain an alternative representation of the problem and its solutions. Does this mean that we have any advantage in terms of the efficiency of the algorithms? To answer this question, we need to understand how to search for Eulerian circuits in a graph.

This problem is one of the oldest problems in graph theory and has been studied by Euler back in the 18th century. Before looking at the algorithms to address this problem, we will see how Euler figured out if a graph has Eulerian cycles or not. Let us start with an important definition: a directed graph is *balanced* if for every vertex the number of edges that reach it (in-degree) is equal to the number of edges that leave it (out-degree).

This leads us to *Euler's theorem*, which states that: *a directed and strongly connected graph has at least one Eulerian cycle, if it is balanced.* Recall the definition of a strongly connected graph from Chapter 13, stating that this implies all pairs of nodes are connected by at least one path. This theorem provides an easy way to check if the graph has an Eulerian cycle or not.

We have seen previously that a solution for our problem is represented by an Eulerian circuit, but the previous theorem states the conditions for the existence of an Eulerian cycle. The difference is subtle but important. Indeed, an extension of the previous theorem states that a directed strongly connected graph has an Eulerian circuit if it is *nearly balanced*, which means it contains, at most, two semi-balanced vertices (i.e. whose difference between the in and the out-degree is −1 and 1, respectively), while all other vertices are balanced.

These definitions can be implemented in Python, by creating functions within the context of the class **MyGraph**, that check if graphs are balanced or semi-balanced. Note that a function to test if a graph is strongly connected was previously defined in Chapter 13. The first function checks if a node is balanced, the second if a graph is balanced (i.e. all nodes are balanced), and the third if the graph is nearly balanced, according to the definitions above.

```
    def check_balanced_node(self, node):
        return self.in_degree(node) == self.out_degree(node)

    def check_balanced_graph(self):
        for n in self.graph.keys():
            if not self.check_balanced_node(n): return False
        return True

    def check_nearly_balanced_graph(self):
        res = None, None
        for n in self.graph.keys():
            indeg= self.in_degree(n)
            outdeg= self.out_degree(n)
            if indeg - outdeg == 1 and res[1] is None: res = res[0],
 n
            elif indeg - outdeg == -1 and res[0] is None: res = n,
 res[1]
            elif indeg == outdeg: pass
            else: return None, None
        return res
```

Note that to apply the previous functions to DeBruijn graphs, i.e. in the context of the Python class **DeBruijnGraph**, we need to redefine the way the in-degree is calculated, to account for possible multiple edges with the same origin and successor in these graphs.

```
class DeBruijnGraph (MyGraph):

 (...)

    def in_degree(self, v):
        res = 0
        for k in self.graph.keys():
            if v in self.graph[k]:
                res += self.graph[k].count(v)
        return res
```

The next step is to design an algorithm to search for Eulerian cycles, for graphs that contain at least one. This algorithm is quite straightforward, and can be summarized in the following steps:

1. We start by any vertex v in the graph and choose one of its successors; we continue this process always selecting from each node a successor that uses a non-visited edge. As the graph is balanced and connected, we will always reach v after a given number of steps.
2. If the cycle in the previous step is not yet an Eulerian cycle, i.e. it still does not contain all edges, we need to select a vertex w from the path in step 1 with non-visited edges. We repeat step 1 starting in w, until reaching w again.
3. The cycles from steps 1 and 2 can be combined, by going from v to w (part of the first cycle), doing the cycle in the second step and returning to w, and going from w to v (last part of the first cycle).
4. While an Eulerian cycle is not attained, go back to step 2, select a vertex w from the current cycle and perform the steps 2 and 3.

Fig. 15.6 shows an example of the application of the previous algorithm for a given graph. In this case, the first selected node is a, and step 1 gathers the cycle going through nodes *a, d, g, h, a*. Since this is not an Eulerian cycle, we need to select a node with non-visited edges, and select g. In step 2, thus we build the cycle *g, f, e, c, g*, that in step 3 can be combined with the previous to get a cycle doing *a, d, g, f, e, c, g, h, a*. Since this is still not an Eulerian cycle, we return to step 2 select another node with non-visited edges, in this case c, getting the cycle *c, a, b, c*. We, finally, join this cycle with the previous to get *a, d, g, f, e, c, a, b, c, g, h, a*, which is now an Eulerian cycle (Fig. 15.6B).

This algorithm can be implemented within the class **MyGraph** to find Eulerian cycles as shown in the function below.

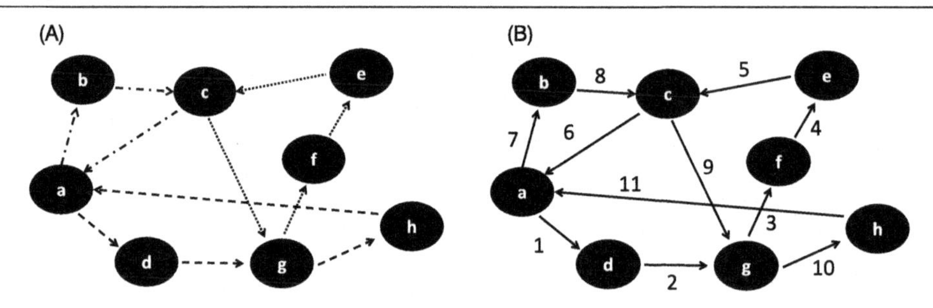

Figure 15.6: Example of the execution of the algorithm to discover Eulerian cycles. (A) shows the three cycles found in steps 1 and 2 of the algorithm with different dashed lines. (B) shows the final Eulerian cycle obtained that goes through all edges labeled with consecutive numbers marking the order of the edges in the circuit.

```
def eulerian_cycle(self):
    if not self.is_connected() or not self.check_balanced_graph()
  : return None
    edges_visit = list(self.get_edges())
    res = []
    while edges_visit:
        pair = edges_visit[0]
        i = 1
        if res != []:
            while pair[0] not in res:
                pair = edges_visit[i]
                i = i + 1
        edges_visit.remove(pair)
        start, nxt = pair
        cycle = [start, nxt]
        while nxt != start:
            for suc in self.graph[nxt]:
                if (nxt, suc) in edges_visit:
                    pair = (nxt,suc)
                    nxt = suc
                    cycle.append(nxt)
                    edges_visit.remove(pair)
        if not res: res = cycle
        else:
            pos = res.index(cycle[0])
            for i in range(len(cycle)-1):
```

```
                res.insert(pos + i +1, cycle[i+1])
        return res
```

To be useful in our context, however, we need to compute Eulerian circuits and not cycles. As we saw above, the difference can be addressed with minor effort. Indeed, we can transform a nearly balanced graph into a balanced one by adding an edge connecting the two semi-balanced nodes to make them balanced. If we calculate an Eulerian cycle for this "extended" graph, removing this extra edge will lead to an Eulerian circuit of the original graph.

This idea is implemented in the next function, again added to the class **MyGraph**.

```
    def eulerian_path(self):
        unb = self.check_nearly_balanced_graph()
        if unb[0] is None or unb[1] is None: return None
        self.graph[unb[1]].append(unb[0])
        cycle = self.eulerian_cycle()
        for i in range(len(cycle)-1):
            if cycle[i] == unb[1] and cycle[i+1] ==  unb[0]:
                break
        path = cycle[i+1:] + cycle[1:i+1]
        return path
```

These functions may be tested within the **DeBruijnGraph** class, considering the previous example, as shown below. Notice that the Eulerian path computed in function **test6** is the one provided in Fig. 15.5.

```
class DeBruijnGraph (MyGraph):

(...) # code above for this class

    def seq_from_path(self, path):
        seq = path[0]
        for i in range(1,len(path)):
            nxt = path[i]
            seq += nxt[-1]
        return seq

def test6():
    frags = [ "ATA", "ACC", "ATG", "ATT", "CAT", "CAT", "CAT", "CCA",
    "GCA", "GGC", "TAA", "TCA", "TGG", "TTC", "TTT"]
```

```
    dbgr = DeBruijnGraph(frags)
    dbgr.print_graph()
    print (dbgr.is_connected())
    print (dbgr.check_nearly_balanced_graph())
    print (dbgr.eulerian_path())

def test7():
    orig_sequence = "ATGCAATGGTCTG"
    frags = composition(3, orig_sequence)
    dbgr = DeBruijnGraph(frags)
    dbgr.print_graph()
    print (dbgr.check_nearly_balanced_graph())
    p= dbgr.eulerian_path()
    print (p)
    print (dbgr.seq_from_path(p))

test6()
test7()
```

The last example shows the closing of the cycle as in the previous section, i.e. we start with a sequence, calculate its composition and try to recover the original sequence.

Fortunately, as you may hint for the above algorithms, this problem has a much lower complexity, when compared to the search for Hamiltonian cycles. Indeed, the above algorithm can be implemented quite efficiently and can be run for large graphs without a problem. Thus, most of the recent genome assembly programs implement this strategy at their core, as we will discuss in the next section.

15.4 Genome Assembly in Practice

In this section, we will cover a number of issues related to genome assembly in the real world, which complicate the life of the software developers in this field, far beyond the idealized versions of the problem considered in previous sections.

One of the main problems in assembling genomes from real reads is the read length. Indeed, the larger this value is the easier it is to assemble genomes, as the DeBruijn graphs tend to become linear (with a single path). However, current technologies are still limited in the read lengths they provide. Another related problem is the assignment of reads to one of the two strands.

Paired-end reads, a commonly used feature in recent technologies help to alleviate these problems. They provide larger read lengths and help to disambiguate the strand. In this case, pairs of reads, with a relatively fixed distance in the genome between themselves, are generated. Thus, each read will be a pair of sequences of length l (being l the read length). The previous algorithms based on DeBruijn graphs can be generalized to accommodate paired end reads, but the details of these algorithms will not be covered here.

Another practical problem, when reconstructing genomes from reads (fragments) is the non-perfect coverage. Depending on the technology and the genome characteristics there will be parts of the original sequence that are not captured by the sequencing reads. A solution is to select a value of k smaller than the original read length l, and replace the original reads by all possible reads of size k contained in the original ones, i.e. all sub-strings of size k of the original string of size l. These new "reads" will then be used as an input to the previous algorithms.

A common problem with real-world assembly is the fact that all technologies produce errors in the reads. In some of the technologies the errors can account to 1 to 5% of the read length. In the DeBruijn graphs, if we insert a read with an error this will cause an alternative path to be created in the graph, in a structure normally called a *bubble*. Modern assembly software tools attempt to find and remove these bubbles from the graph, while performing the assembly. In this case, they need to be able to identify real bubbles, without removing correct paths, which may be a complex task in areas with repeats with some variation.

Depending on the species, the genome may contain many repeated regions, which can be as much as 50% of the entire genome. This represents a very hard problem for assembly programs. Moreover, genomes contain copy-number alterations, i.e. large blocks of sequence that are repeated multiple times. Assembly programs can estimate the number of copies by determining how the coverage in these regions deviate from the expected coverage or by comparing with other regions of the genomes.

Given all these natural and technical problems, even the most sophisticated software tools available have difficulties in generating a single sequence from a set of reads, and in most cases generate a set of *contigs*, the sequences generated from paths in parts (non-branching) of the DeBruijn graph. In many real cases, it is common to end up with hundreds or even thousands of contigs, even when assembling the genomes of simple organisms, as bacteria. In order to assess the quality of an assembly one of the most commonly used measures is the N50. It is a measure of contiguity derived from the length of the assembled contigs, similar to the median of the lengths. The contigs are sorted by their increasing length and the N50 corresponds to the length of the shortest contig at 50% of the summed assembled genome.

The order of the contigs is not known from the result of these algorithms. The process of taking the contigs, ordering them and trying to close possible gaps, to reach a final single genome, is typically quite costly as it involves a lot of manual curation, and may be impossible to achieve. One alternative is to sequence the same genome using different technologies and try to combine their results to make this process easier.

Still, and in spite of all these difficulties, DeBruijn graphs are the basis of most efficient modern assembly software tools able to handle large datasets with many millions of reads. The *Velvet* algorithm [158] was of major importance in the field. Not only it is an efficient algorithm but it was also the starting point of further developments that include: *IDBA* [124], *SPAdes* [20], *SOAPdenovo2* [104] or *Megahit* [99]. Comparison studies show that DeBruijn graph-based assemblers have a very competitive performance when handling very large datasets, with relatively longer read lengths (75bps) requiring shorter runtime and main memory requirements [160]. All the previous alternatives need to face the problems described in this section (and others), and most of the recent advances in the field are related to improvements in the way they handle these challenges.

Bibliographic Notes and Further Reading

Euler's pioneer work on graph theory and Eulerian cycles was described in [60]. DeBruijn graphs were introduced by Nicolaas De Bruijn in [44]. The algorithm for Eulerian cycle searching was developed by Hierholzer in 1873 [76]. The Eulerian approach to genome assembly was first proposed in 1995 by Idury and Waterman [80], and improved in a study by Pevzner et al. in 2001 [127]. The approaches based on DeBruijn graphs for paired end reads were proposed by Medvedev et al. [110], being covered, for instance, in [35] (Chapter 3).

Exercises and Programming Projects

Exercises

1. Consider the algorithm (and its implementation) that allows the determination of Eulerian cycles of a graph. Write a version of this function that can handle non-connected graphs (still balanced). This function should allow to calculate all Eulerian cycles of the different connected components.
2. One of the problems when reconstructing genomes from reads (fragments) is the non-perfect coverage, as we saw above in Section 15.4. Write a function that has as its inputs the original set of fragments, and returns the original sequence that generated the fragments, using DeBruijn graphs, following the strategy explained above. You should test all values of k starting by the read length, and decrement k by 1, until a value of k allows to reach a graph with an Eulerian circuit. If it does not work for $k = 2$, the function should return `None`.

3. Write a method to add to the class **DeBruijnGraph** that given the graph (*self*) identifies the connected components of the graph (notice that in a directed graph, a connected component is a sub-graph that includes a set of nodes that are strongly connected among each other, i.e. there are paths connecting all pairs of nodes), selects the largest one (if it exists) and re-builds the sequence from the graph. If there are no connected components in the graph, the methods should return `None`.

Programming Projects

1. Implement algorithms based on DeBruijn graphs to be able to work with paired end reads.

CHAPTER 16

Matching Reads to Reference Sequences

The problem of pattern finding over target sequences will be revisited in this chapter, looking at scenarios where we have a set of sub-strings to be searched over a large sequence. This has important applications in the processing of DNA sequencing data, when a set of reads/fragments needs to be searched over a full reference genome. We will mainly focus on algorithms and data structures that allow to pre-process the reference sequence, making more efficient to search distinct patterns, also addressing methods to reduce the memory requirements.

16.1 Introduction: Problem Definition and Applications

In this chapter, we will address a common problem in today's Bioinformatics, namely in handling DNA/RNA sequencing data. In many cases, we are faced with the need to find a large set of sequences, typically smaller sequences up to a few hundred nucleotides representing reads from a sequencing project, in a large reference sequence, such as a full genome or a chromosome.

In Chapter 5, we have discussed the problem of finding a fixed pattern (sub-string) in a target sequence and discussed several ways to improve the performance of the respective algorithms, for instance by suitable pre-processing of the pattern to search. This made the search of a single pattern in a set of sequences more efficient. In this case, we want to search a set of different patterns over the same target sequence, a feature that changes the context of the problem.

In the next section, we will start by looking at *tries* a structure that allows to pre-process a set of patterns. We will then move to study a similar structure, *suffix trees*, that allows to pre-process the target sequence, making the search over it more efficient. Finally, we will look at Burrows-Wheeler transforms as a way to further improve the memory needed to keep reference sequences, which are important when the target sequences are very large, as it is the case with full genomes.

16.2 Pre-Processing the Patterns: Tries

16.2.1 Definitions and Algorithms

Tries are n-ary trees that allow to organize a set of string patterns. We have addressed the concept of trees in Chapter 9, while we mostly discussed binary trees. In the case of tries, each internal node can have n branches (with $n \geq 1$).

Bioinformatics Algorithms. DOI: 10.1016/B978-0-12-812520-5.00016-X

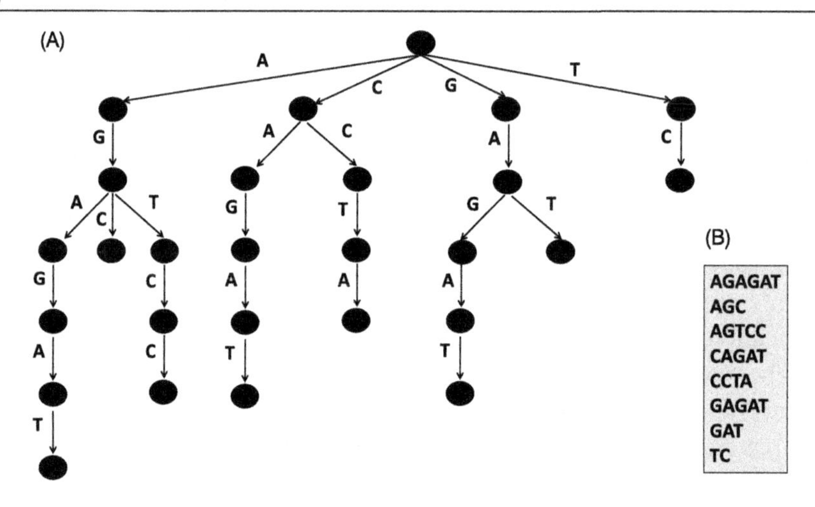

Figure 16.1: Example of a trie for an alphabet with nucleotide symbols: (A) the trie; (B) the patterns represented by the trie in lexicographical order.

In tries, symbols from a given alphabet are associated to edges (arcs) of the tree, and the represented patterns are read walking from the root to a leaf. Thus, the tree will have one leaf for each pattern.

An example is shown in Fig. 16.1, for a set of patterns over a DNA sequence alphabet. Notice that each of the patterns shown in Fig. 16.1B can be obtained following a path from the root to one of the leaves, concatenating the symbols present in the edges belonging to the path.

As it may seem intuitive by looking at the previous figure, a trie can be built from a set of patterns. This task can be achieved following a simple algorithm, which starts with a tree just with the root node, and iteratively considers each of the patterns and adds the necessary nodes to that the tree includes a path from the root to a leaf representing the pattern.

For each pattern p in the input list, we will, in parallel, read the pattern and walk over the tree, starting in the first position of the pattern (p_0), and positioning the current node as the root of the tree. In each iteration i, we will take the symbol p_i in the pattern, and check if there is an edge going out of the current node in the tree which is labeled with p_i. If there is, we descend through that branch and update the current node. If there isn't, we will create a new node, which will become the current one, and connect it from the current node, labeling the branch with p_i. To terminate this iteration, we increment i by 1, and move to the next iteration, repeating the process until we reach the final position in the pattern (i.e. i is the length of p).

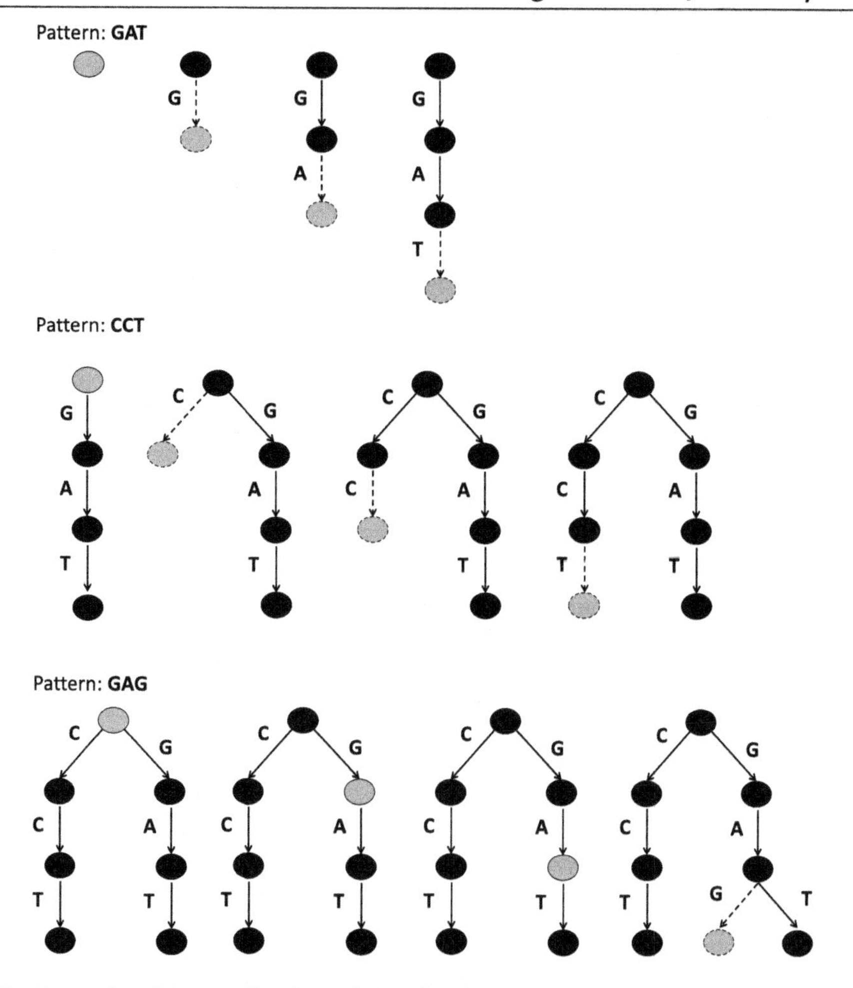

Figure 16.2: Example of the application of the algorithm to build a trie from a set of patterns. In this case, the patterns to be considered are: GAT, CCT, and GAG. The current node in each iteration is marked in gray background. Added nodes and edges have dashed lines in their border and arrow, respectively.

An example of the application of this algorithm is provided in Fig. 16.2, with a trie built from a set of 3 patterns. The addition of each pattern to the current trie is shown in each line, with the different iterations being represented by the vertical sequence of tries, where the current node is highlighted in gray and new nodes/edges have dashed lines.

Given a trie, created from a set of patterns, it can be used to efficiently search the occurrence of any of those patterns in a target sequence. To check how an algorithm for this task can be designed, let us first see how we can check if a pattern in the trie is a prefix of the target sequence.

In this case, we just go through the characters in the sequence and in parallel walk over the trie, starting in the root, following the edges corresponding to the characters in the sequence, until one of the following situations occurs: (i) a leaf of the trie is reached, meaning that a pattern has been identified; (ii) the character in the sequence does not exist as a label to an edge going out from the current node, and thus no pattern in the trie has been identified.

This algorithm can be easily extended to search for pattern occurrences in the whole target sequence, by repeating this process for all suffixes of the sequence, i.e. all sub-strings starting in position i of the sequence and going to its end, for values of i from 0 to the sequence length minus one.

An example of the application of this algorithm is given in Fig. 16.3, over the trie depicted in Fig. 16.1. In this case, we show the application of the algorithm for the first two suffixes, i.e. the sequences starting in the first two positions of the target sequence.

As a limitation of the use of tries for this purpose, notice that when we are searching for patterns in a new suffix, using the previous algorithm, we do not use any information on the previous searches to help in the current one.

Thus, in that aspect, the use of automata put forward in Chapter 5 may be more efficient, while they are more difficult to build to consider multiple patterns. Indeed, automata may be seen as transformations of tries into graphs, adding failure edges that connect some nodes with partial matches avoiding to restart searches from scratch.

16.2.2 Implementing Tries in Python

The first decision to take when implementing data structures as tries is how to represent them. In this case, as it happened with graphs in the previous chapters, taking into account computational and memory efficiency, we will opt to use dictionaries to keep tries.

We will label the nodes of the tree with sequential integer numbers, and these will serve as the keys in our dictionary. The value associated to each node will represent the edges going out of that node, in the form of another dictionary, where there will be an item for each edge: keys will be the symbols labeling the edge and values will be the indexes of the destination nodes. An empty dictionary will identify the leaves of the tree.

An example is shown in Fig. 16.4, where the trie built in the example from Fig. 16.2 is now labeled with numbers in the nodes, being shown the respective representation using dictionaries.

Based on this representation, we will implement tries and related algorithms in Python defining a new class named **Trie**. An object of this class will keep a trie, having as its main attribute (*self.nodes*) a dictionary with the previous structure. Also, there will be an attribute

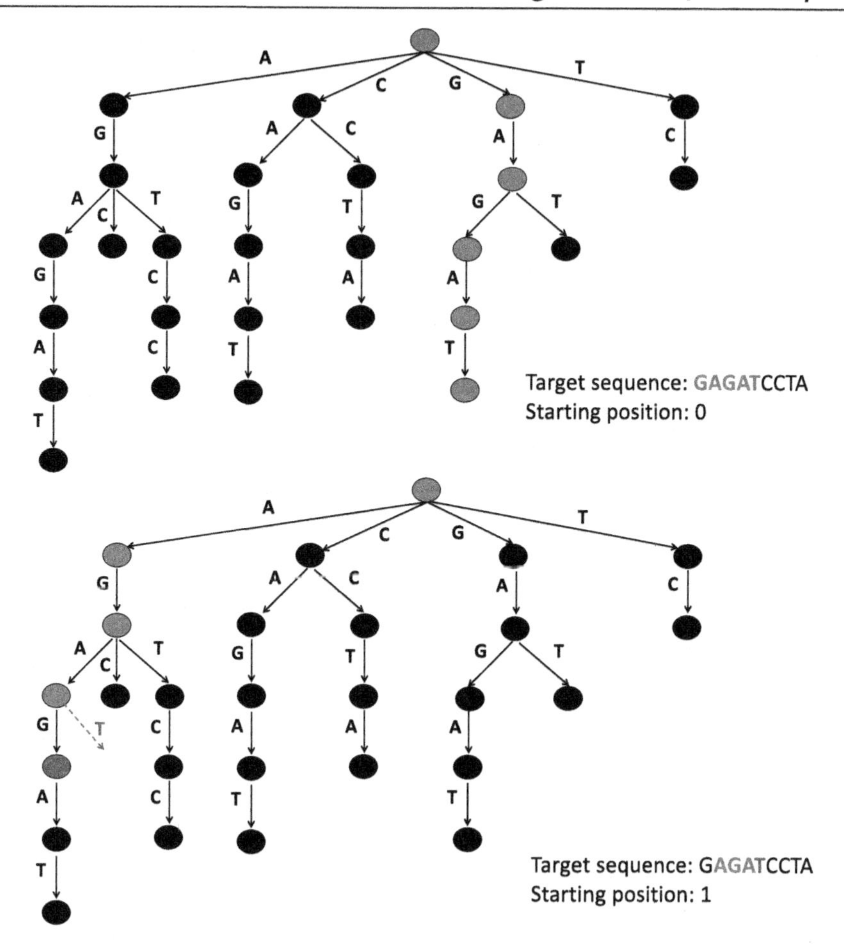

Figure 16.3: Example of the application of the algorithm to search for pattern occurrences in a target sequence. In this case, we show the result of applying the algorithm for the first and second positions of the target sequence. In the first case (top), the search is successful since a leaf is reached (for pattern GAGAT). In the second, the search fails in the last symbol, since an edge labeled with T is not found. Nodes included in the search path are colored in blue (mid gray in print version), while the nodes where it fails and respective edges are colored in red (dark gray in print version).

(*self.num*) to keep the current size of the tree (number of nodes), so that labeling of new nodes is possible.

The following code shows the class definition, its constructor, a method to print tries (in a format similar to the one used in Fig. 16.4), together with methods to add a pattern to a trie and to create a trie from a list of patterns, following the algorithms described in the previous

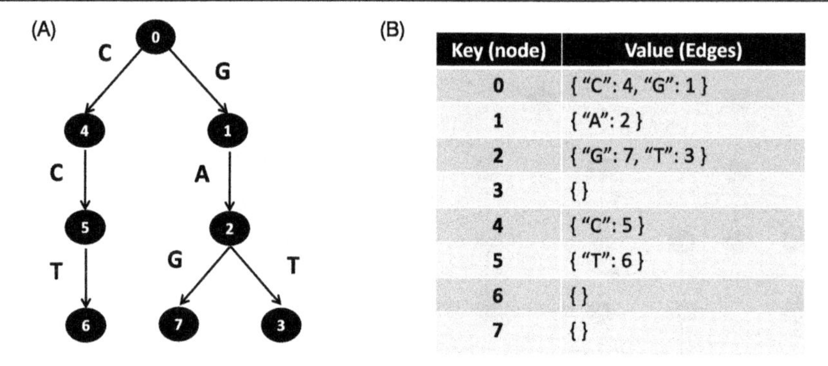

Key (node)	Value (Edges)
0	{ "C": 4, "G": 1 }
1	{ "A": 2 }
2	{ "G": 7, "T": 3 }
3	{ }
4	{ "C": 5 }
5	{ "T": 6 }
6	{ }
7	{ }

Figure 16.4: Representation of an example trie using dictionaries. (A) An example trie with numbered nodes; (B) its representation using a dictionary, being shown the keys and the respective values (other dictionaries with symbols as keys and destination nodes as values; leaves are represented as empty dictionaries).

section (see Fig. 16.2). The example in function **test** builds and prints the simple trie shown in Figs. 16.2 and 16.4.

```
class Trie:

    def __init__(self):
        self.nodes = { 0:{} } # dictionary
        self.num = 0

    def print_trie(self):
        for k in self.nodes.keys():
            print (k, "->" , self.nodes[k])

    def add_node(self, origin, symbol):
        self.num += 1
        self.nodes[origin][symbol] = self.num
        self.nodes[self.num] = {}

    def add_pattern(self, p):
        pos = 0
        node = 0
        while pos < len(p):
            if p[pos] not in self.nodes[node].keys():
                self.add_node(node, p[pos])
```

```
                node = self.nodes[node][p[pos]]
                pos += 1

    def trie_from_patterns(self, pats):
        for p in pats:
            self.add_pattern(p)

def test():
    patterns = ["GAT", "CCT", "GAG"]
    t = Trie()
    t.trie_from_patterns(patterns)
    t.print_trie()

test()
```

Note that the **add_node** method creates a new node and links it to an existing one (specified by the *origin* parameter), labeling the edge with a given symbol, also passed as an argument to the method. This method is used by the **add_pattern** method which follows the algorithm illustrated in Fig. 16.2. The *pos* variable iterates over the symbols in the pattern *p*, while the variable *node* keeps the current node in the tree (starting by the root, which is labeled with number 0). Finally, the **trie_from_patterns** method simply adds each of the patterns in the input list using the previous method.

Next, we will proceed to implement the algorithms to search for patterns in target sequences, using tries. We will first add a method in the previous class to search if a pattern represented in the trie is a prefix of the sequence (method **prefix_trie_match**), and then use it to search for occurrences over the whole sequence (method **trie_matches**), as described in the previous section (see Fig. 16.3).

```
class Trie:

    (...)

    def prefix_trie_match(self, text):
        pos = 0
        match = ""
        node = 0
        while pos < len(text):
            if text[pos] in self.nodes[node].keys():
                node = self.nodes[node][text[pos]]
```

```
                match += text[pos]
                if self.nodes[node] == {}:
                    return match
                else:
                    pos += 1
            else: return None
        return None

    def trie_matches(self, text):
        res = []
        for i in range(len(text)):
            m = self.prefix_trie_match(text[i:])
            if m != None: res.append((i,m))
        return res

def test2():
    patterns = ["AGAGAT", "AGC", "AGTCC", "CAGAT", "CCTA", "GAGAT", "
   GAT", "TC"]
    t = Trie()
    t.trie_from_patterns(patterns)
    print (t.prefix_trie_match("GAGATCCTA"))
    print (t.trie_matches("GAGATCCTA"))

test2()
```

16.3 Pre-Processing the Sequence: Suffix Trees

16.3.1 Definitions and Algorithms

As we saw, tries are an interesting way of processing a set of patterns to make their search over different sequences more efficient. However, in many situations, the most relevant problem is to search a large number of different patterns once (or a few number of times) over a very large sequence. This is the case when seeking to align a large number of reads (fragments) to a known sequence of a large dimension (e.g. a reference genome).

In this case, instead of pre-processing the pattern(s), we may need to process the target sequence, in a way that can lead to the efficient search of occurrences of any given pattern over it. Although the problem is different from the one in the previous section, one of the most popular solutions is quite similar to the tries proposed above.

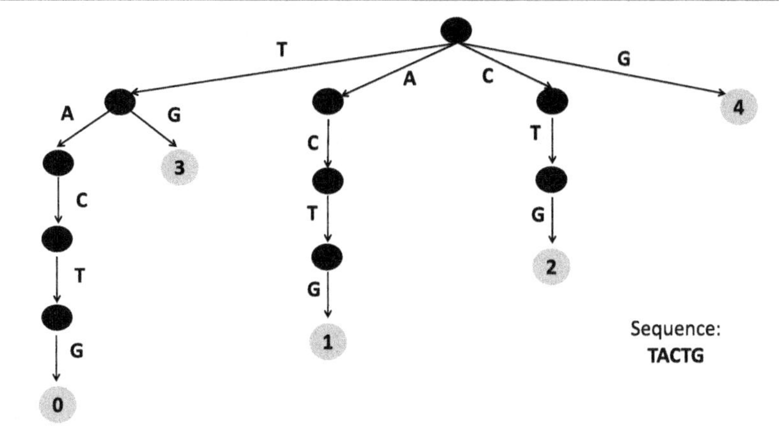

Figure 16.5: Example of a suffix tree for the sequence TACTG. Leaves are represented as nodes with gray background, numbered according to the initial position of the suffix they represent.

Indeed, one solution to pre-process a large sequence is the use of *suffix trees*, which may be seen as a special case of tries, built from the set of all suffixes of the target sequence.

Bringing a more formal definition, a tree $T = (V, E)$ is considered a suffix tree, representing a sequence s, if the following conditions are all met:

- the number of leaves of T is equal to the length of s (leaves are numbered from 0 to the length of $s - 1$);
- each edge in the tree (in E) has as its label a symbol from s;
- every suffix of s is represented by a leaf, and can be reconstructed concatenating the symbols in the path from the root to the leaf;
- all edges going out of a node in V have distinct symbols.

An example is provided in Fig. 16.5, showing the suffix tree for a DNA sequence (TACTG). Note that the leaves (represented with gray background) are numbered according to the starting position of the suffix they represent.

Given the definition above, and the algorithm to build tries proposed in the previous section, building suffix trees for a sequence seems a trivial task. Indeed, we only need to compute suffixes for the sequence and add those to a trie, following the algorithm we have defined by iteratively adding each suffix to the tree in exactly the same way we added patterns in the last section. The only change to be considered here is that we need to number the leaves with the index of the suffix, i.e. its starting position in the overall sequence.

However, the astute reader has, at this point, already realized that there may be situations where we can not create a suffix tree for a given sequence. As an exercise, we recommend that the reader tries to build a tree for the sequence: TACTA.

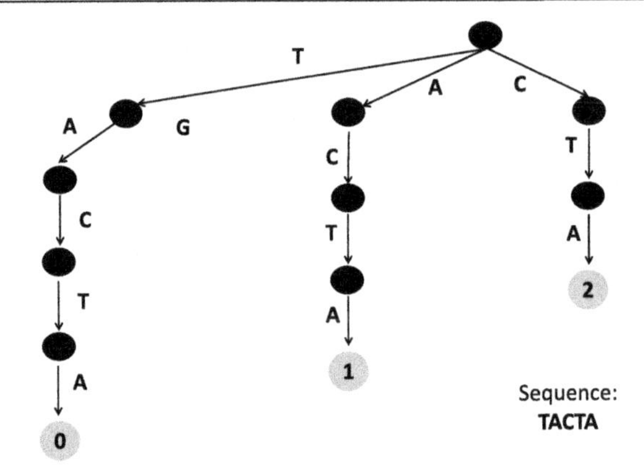

Figure 16.6: Example of a suffix tree for the sequence TACTA. Not all suffixes are represented as leaves, since some are prefixes of other suffixes.

You will quickly realize that the number of leaves of the tree will be smaller than the number of suffixes, since for two of the suffixes we do not reach a leaf. This happens in situations where the suffix is a prefix of another suffix. In this case, TA is prefix of TACTA, and A is a prefix of ACTA. Therefore, in the resulting suffix tree there are no leaves numbered with 3 or 4 (see Fig. 16.6). This means that the suffix tree in the figure does not comply with all conditions set in the above definition.

One way to address the previous problem is to add a symbol in the end of the sequence, which in our examples we will represent as a $. In this way, no suffix will be the prefix of another suffix, as all suffixes terminate with $, which will not appear in any other position in the sequence. Applying this change to the target sequence, we can follow the defined algorithm to build a suffix tree without further problems. We illustrate this strategy by adding this symbol to the previous sequence (TACTA$). The resulting tree is shown in Fig. 16.7.

Another important task is the search for the occurrence of patterns in the sequence, using the suffix tree that represents it. This process is quite simple to achieve in a suffix tree. Indeed, we just need to walk over the suffix tree starting in the root, and following the edges according to the symbols in the pattern.

If those edges exist for the complete set of symbols in the pattern, when we reach the end of the pattern, we will be placed in a given node of the tree. From the definition of a suffix tree, all leaves below that node in the tree correspond to starting positions of the pattern in the sequence.

This process is illustrated in Fig. 16.8A, where the tree from Fig. 16.7 representing the sequence TACTA is used to search for pattern TA. The result shows the pattern occurs in positions 0 and 3 from the original sequence.

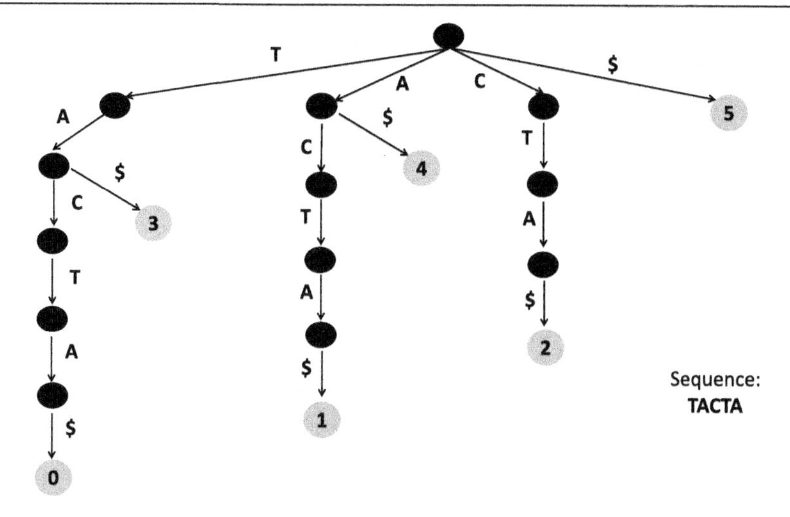

Figure 16.7: Example of a suffix tree for the sequence TACTA, with $ symbol added in the last position.

Note that if the previous process fails in a given node, i.e. if there is no branch leaving the node marked with the symbol in the sequence, this means that the pattern does not occur. This is the case with the search for pattern ACG in Fig. 16.8B.

Since suffix trees can be used in scenarios where the target sequence is very large, one of their main problems is the amount of memory needed, since trees will become very large. Noticing that in many cases there are linear segments, i.e. sequences of nodes only with a single leaving edge, these may be compacted by considering only the first and last node of the segment, and concatenating the symbols in the path.

An example of this process is provided by Fig. 16.9, where the tree from Fig. 16.7 is compacted. Notice that the tree is shown in two versions: the first shows edges with strings of symbols, while the latter shows a tree with ranges of positions. Indeed, since the strings in the edges are always sub-strings of the target sequence, to avoid redundancy it is sufficient to keep the starting and end positions for this string.

Suffix trees can also be used for a number of different tasks when handling strings, and in particular biological sequences. Indeed, from what it was explained above, it seems clear that they are useful to search for repeats of patterns in sequences, thus enabling the identification of many types of repeats in genomes.

Also, suffix trees may be created from more than a sequence, thus enabling to address problems such as discovering which trees contain a given pattern, the longest sub-string shared by a set of sequences or calculating the maximum overlap of a set of sequences.

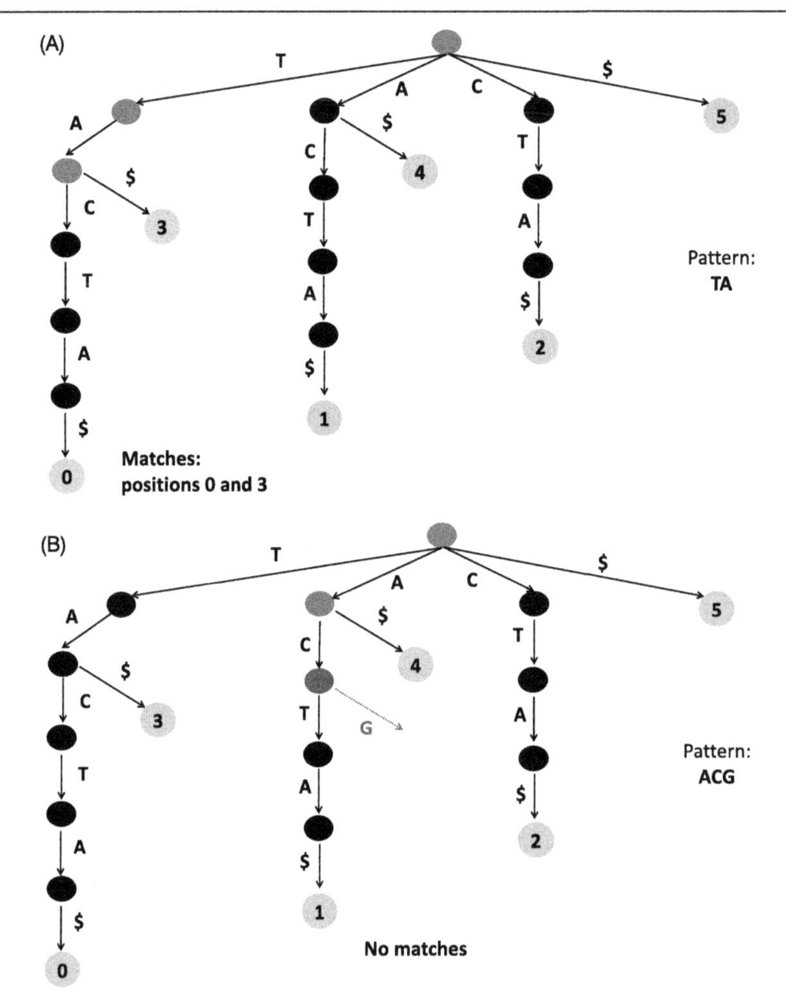

Figure 16.8: Example of the search for the patterns TA, in (A), and ACG, in (B), over a suffix tree (for the sequence TACTA). Nodes in blue (mid gray in print version) represent the walk over the tree, considering the symbols in the pattern, while nodes in red (dark gray in print version) show nodes where the symbol in the sequence does not have a matching edge. Leaves marked in blue (light gray in print version) represent all suffixes matching the pattern.

In Fig. 16.10, we show an example of a suffix tree built from two sequences: TAC and ATA. Two different symbols are used to mark the end of each sequence. The leaves are labeled by joining the sequence index with the starting position of the suffix.

Notice that the pattern search algorithm may be applied over a tree built from multiple sequences in a similar way. In the case of the tree from the figure, for instance, searching for pattern TA would return two matches, one in each sequence.

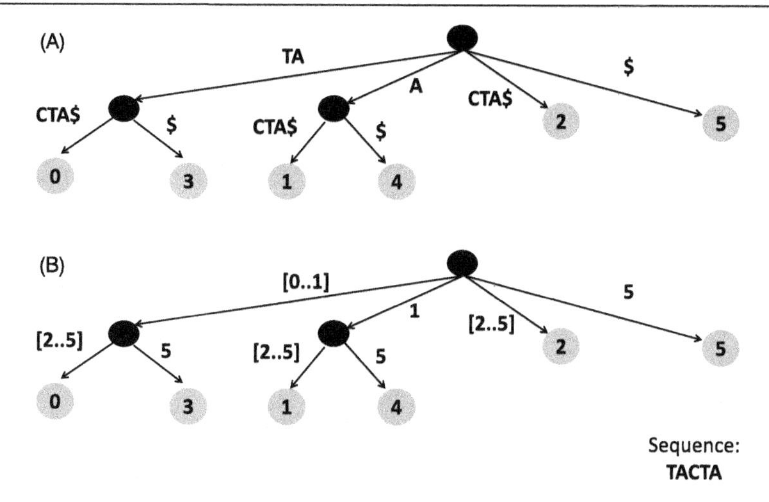

Figure 16.9: Example of a compact suffix tree, representing the same sequence as the one in Fig. 16.7. (A) Edges show sub-strings; (B) Edges show position intervals.

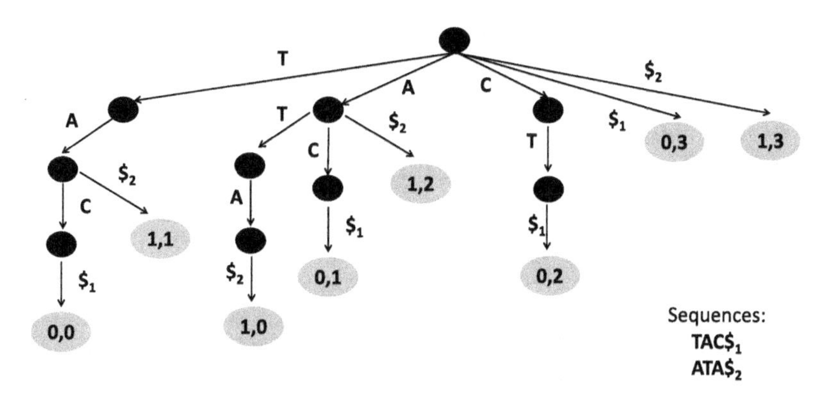

Figure 16.10: Example of a suffix tree, built from two sequences: TAC and ATA. The symbols $\$_1$ and $\$_2$ are added to mark the end of each sequence. Leaves are labeled with a tuple with the sequence and the initial position of the tuple.

16.3.2 Implementing Suffix Trees in Python

To implement suffix trees, we will use a structure very similar to the one used in the previous section for tries. The data structure used to keep tries will need an adaptation to be able to represent suffix trees. So, for each node, we will keep a tuple where the first element is used to represent the position of the suffix (in the case of leaves), being -1 for internal nodes. The second element of the tuple will be a dictionary working similarly to the one used for tries, i.e. keeping the edges (symbols are the keys and destination nodes are the values).

We will define a class **SuffixTree** to keep suffix trees and implement the respective algorithms. In the next code block, we show the definition of the class, along with the constructor and a method to print a tree.

```
class SuffixTree:

    def __init__(self):
        self.nodes = { 0:(-1,{}) } # root node
        self.num = 0

    def print_tree(self):
        for k in self.nodes.keys():
            if self.nodes[k][0] < 0:
                print (k, "->", self.nodes[k][1])
            else:
                print (k, ":", self.nodes[k][0])
```

Next, we can proceed to define methods to implement the algorithm to build suffix trees from a given sequence. This will include the method **add_node** that allows to add a node to the tree, given the origin node, the symbol to label the edge and the leaf number (if it is a leaf; defaults to -1, otherwise). Then, we will define a method **add_suffix** which implements the addition of a suffix to the tree, being an adaptation of the similar method to add a pattern in the previous section (method **add_pattern** of the class **Trie**). Lastly, we will define the method **suffix_tree_from_seq** which will add the $ symbol to the sequence and iteratively call the previous method for each suffix of the sequence. The **test** function builds the suffix tree for an example with the sequence TACTA, building the tree shown in Fig. 16.7.

```
class SuffixTree:

  (...)

    def add_node(self, origin, symbol, leafnum = -1):
        self.num += 1
        self.nodes[origin][1][symbol] = self.num
        self.nodes[self.num] = (leafnum,{})

    def add_suffix(self, p, sufnum):
        pos = 0
        node = 0
        while pos < len(p):
```

```
                if p[pos] not in self.nodes[node][1].keys():
                    if pos == len(p)-1:
                        self.add_node(node, p[pos], sufnum)
                    else:
                        self.add_node(node, p[pos])
                node = self.nodes[node][1][p[pos]]
                pos += 1

    def suffix_tree_from_seq(self, text):
        t = text+"$"
        for i in range(len(t)):
            self.add_suffix(t[i:], i)
def test():
    seq = "TACTA"
    st = SuffixTree()
    st.suffix_tree_from_seq(seq)
    st.print_tree()

test()
```

Finally, we will show the functions that implement the algorithm for pattern search. The method **find_pattern** implements the process of walking the tree until reaching the final node, or failing the search. If the first case occurs, the leaves below the node are returned by calling the auxiliary method **get_leafes_below**, defined using recursiveness (note this might not be a good idea for real world scenarios). In the second, the method returns `None`, for the case where there are no matches. The testing function applies the method to the previous tree, searching for the patterns TA and ACG, mimicking the case depicted in Fig. 16.8.

```
class SuffixTree:

  (...)

    def find_pattern(self, pattern):
        pos = 0
        node = 0
        for pos in range(len(pattern)):
            if pattern[pos] in self.nodes[node][1].keys():
                node = self.nodes[node][1][pattern[pos]]
                pos += 1
```

```
            else: return None
        return self.get_leafes_below(node)

    def get_leafes_below(self, node):
        res = []
        if self.nodes[node][0] >=0:
            res.append(self.nodes[node][0])
        else:
            for k in self.nodes[node][1].keys():
                newnode = self.nodes[node][1][k]
                leafes = self.get_leafes_below(newnode)
                res.extend(leafes)
        return res

def test():
    seq = "TACTA"
    st = SuffixTree()
    st.suffix_tree_from_seq(seq)
    print (st.find_pattern("TA"))
    print (st.find_pattern("ACG"))

test()
```

16.4 Burrows-Wheeler Transforms

16.4.1 Definitions and Algorithms

As we discussed in the previous section, a key aspect in the data structures used to represent the target sequence are the memory requirements. Indeed, even using compact suffix trees, the representation of large sequences can take huge amounts of memory. For instance, the representation of the human genome as a suffix tree takes at least 50 GBytes.

Thus, there is the need to try to compress further the genome to reach a more affordable representation. In this section, we will discuss a method for compressing sequences, named *Burrows-Wheeler Transform* (BWT), which is a fully reversible method, thus allowing to recover the initial sequence without loss of information.

The idea of BWTs is to take advantage of the fact that genomes contain many repeats, of different sizes, along their sequence. BWTs seek to convert these repeats into sequences of the same symbol by an appropriate transformation.

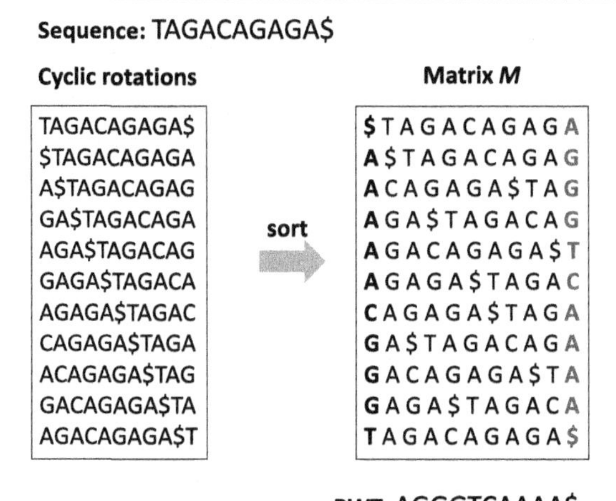

Figure 16.11: Example of the process of construction of a BWT for a sample sequence. The BWT (last column) is highlighted in red (dark gray in print version).

The method is based on cyclic rotations of the original sequence, which are lexicographically ordered, thus creating the Burrows-Wheeler matrix, denoted by M. An example of this process is shown in Fig. 16.11.

In practice, we will not keep the whole matrix, which would obviously not be viable, but only the first and the last columns. The last column is named as the BWT of the original sequence.

Note that the first column is ordered lexicographically and, thus, it is enough to keep the number of occurrences of each symbol in the alphabet, which can be done in a quite compact way (in the example could be: 5AC3GT). Also, the last column will have sets of repeated characters for the repeated patterns in the sequence, which may allow to easily compact its content. Notice, for instance, the three G's in the second, third and fourth positions, due to the repetition of the AGA pattern in the original sequence.

A sequence compression method is only useful if it allows the reverse operation, i.e. if it allows to recover the original sequence given the compressed one, in this case the BWT. We will try to illustrate this process with an example by considering that we know the BWT and want to find the original sequence. In this case, the BWT is ACG$GTAAAAC. The process is depicted in Fig. 16.12.

In the start of the process we know the BWT, thus the last column of M. It is easy to recover the first column of M, since it has the same symbols as the last, but ordered lexicographically. Our purpose is to find the original sequence, in the first row of M. Starting with the first character, we realize that this character follows the $ in every sequence, since these are

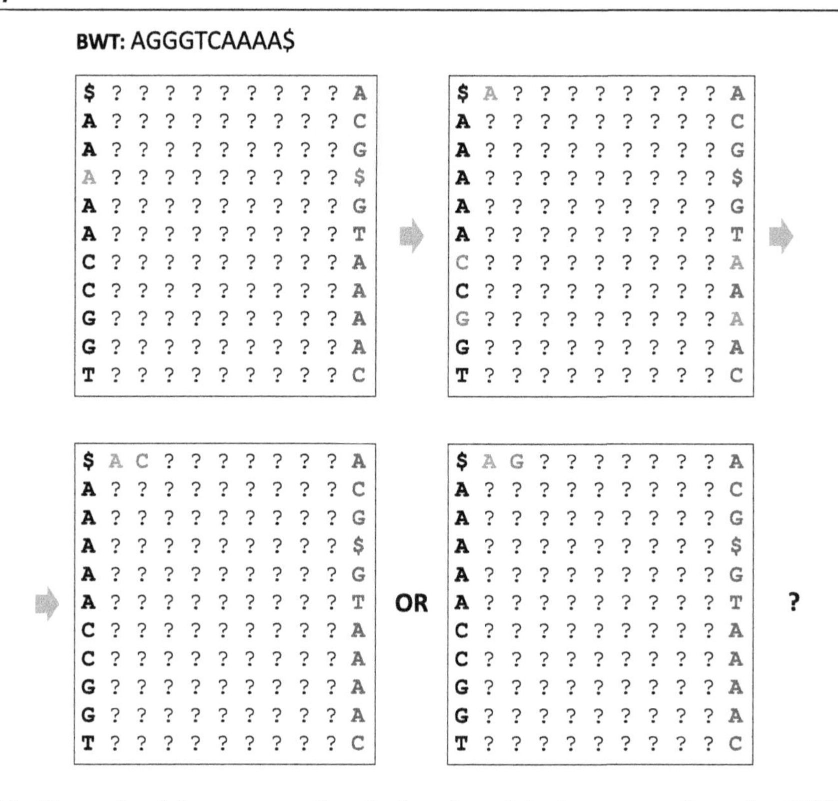

Figure 16.12: Example of the process of retrieving the original sequence from the BWT. The first step fills an A in the first position, but in the second there are several candidates.

cyclic rotations of the sequence. Thus, in the row where the $ is in the last column, the wanted character is in the first (in the example, an A).

We seem to have found a way to easily proceed in finding the whole sequence. However, in the following step of our example we realize that following the same idea brings us to two different hypotheses, since there are different rows with an A in the last column. In those, in some we have a C in the first column and in others we have a G. How to know which one is right?

The solution for this problem is to label each character with an index. It is possible, although not trivial, to show that the order of the different characters is kept from the first to the last column (we will skip the formal proof here).

Applying this change, the process can be implemented following the previous rule. We just identify the character in the last column, find its corresponding in the first, and add this character to the sequence, following this process until the sequence is completely filled. The previous example is shown in Fig. 16.13 now using indexes.

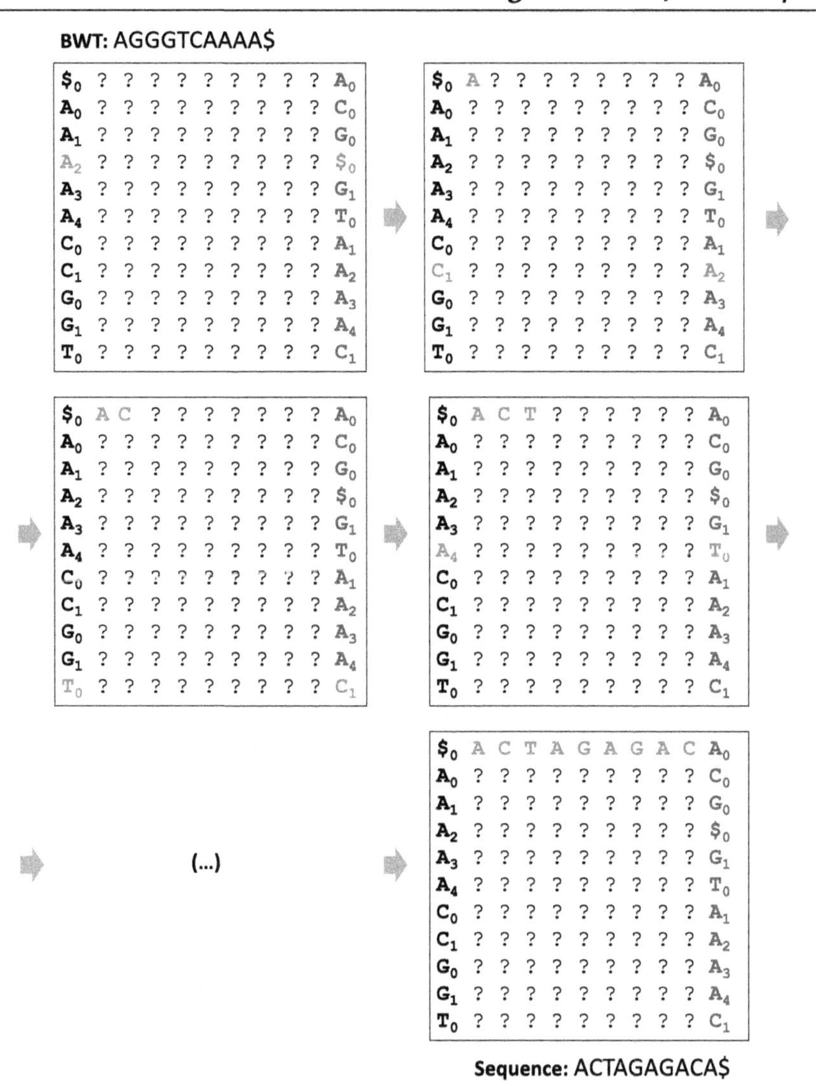

Figure 16.13: Example of the process of retrieving the original sequence from the BWT, indexing the characters. The first four steps are shown, as well as the final result.

We have seen that BWTs can provide an interesting and efficient way to compress sequences. But, along this chapter, we have been worried about pattern search. So, are BWTs good structures to allow pattern search over the original sequence? The answer is positive.

To understand how to efficiently search for patterns in sequences represented as BWTs, the first insight is provided by the fact that, since each row in M starts with a suffix of the sequence, all occurrences of a given pattern may be found in consecutive rows of the matrix M

(these are order in lexicographic order). As an example, in Fig. 16.11, the three occurrences of the pattern AGA can be found in the fourth, fifth, and sixth rows of M.

However, we do not keep the whole matrix M and, as such, we need an algorithm that can search for these patterns only using the first column and the BWT (the last column). The strategy is based on some of the ideas used for recovering the original sequence from the BWT. We will identify the pattern in its reverse order moving within rows of the first and last columns.

The first step is to identify the last symbol of the pattern and check the interval of positions where it matches in the last column, transposing these positions into the respective interval over the first column (notice all equal characters are consecutive in this column). The range of positions is represented by two pointers: top and bottom.

Then, we will move to the next symbol (moving backwards in the pattern) and check where it matches (in the last column), within the sub-set given by the current range. Again, the transposition of these matches to the first column allows to update the range (top and bottom pointers). This process iterates until all symbols in the pattern match, a case where the last range marks the rows in the matrix where there are matches. Also, the process may terminate in failing to find the pattern when the symbol in the pattern does not occur in the last column in the defined range.

The algorithm is depicted in Fig. 16.14, where the BWT from Fig. 16.11 is used to find the pattern AGA.

Notice that the previous algorithm may be used to rapidly find if the pattern occurs and the number of occurrences. Still, to know the positions where the pattern occurs, extra work needs to be done. One option is to recover the first position of each row where the pattern occurs, by following the first column-last column process we saw before until reaching the $ symbol. However, in practice, this is commonly to slow to be useful, given the size of the sequence.

An alternative is to combine BWTs with a structure named as suffix arrays. Basically, a suffix array keeps a list of the starting position of the suffix kept in each row of the matrix M. For instance, for the matrix in Fig. 16.11, the suffix array would be: $[10, 9, 3, 7, 1, 5, 4, 8, 2, 6, 0]$.

Building this structure allows to immediately be able to recover the positions of occurrence of the pattern from the result of the previous algorithm. However, since these are structures that have a size equal to the original sequence, their use may be prohibitive in practice. One alternative is to keep partial suffix arrays for a small proportion ($1/K$) of the positions. Thus, to find a position the last column-first column would only be needed for at most K iterations, while the memory needed would be reduced by a factor of K.

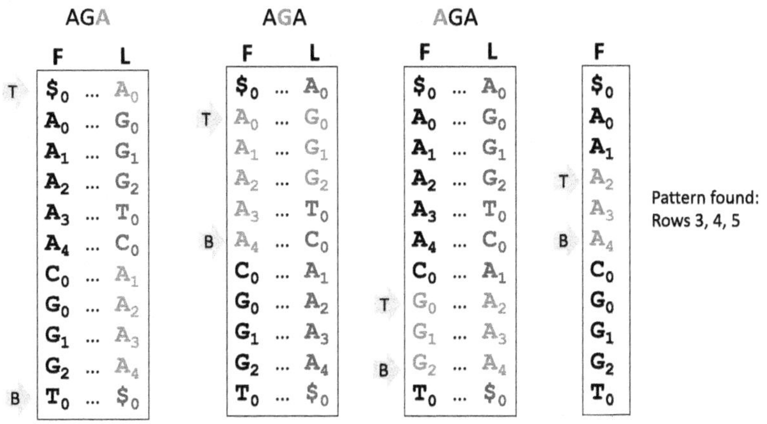

Figure 16.14: Example of the process of searching a pattern using the BWT. We show only the first column (black, labeled with F on top) and the BWT, i.e. the last column (red [dark gray in print version], labeled with L on top) from matrix M. The matching process of the pattern is shown symbol by symbol. The arrows with T and B show the values of the top and bottom pointers marking the range of rows matching the pattern in each stage. Matching characters in each step are marked in green (light gray in print version) in the pattern and in the matrix columns.

16.4.2 Implementation in Python

In this section, we will implement the concept of BWTs by creating a new Python class, named **BWT**. This class will be quite simple in terms of data structures, keeping just the BWT (a string). This can be built by the process depicted in Fig. 16.11, in the constructor, or set by the programmer using an available method.

Although there are efficient algorithms to build the BWT, these are quite complex and will not be explored in this text, and thus we will here explicitly create the matrix M and calculate the BWT as the last column. This is done in the code block below, which will define the class and implement this process of building the BWT. The example in the **test** function is the one provided in Fig. 16.11.

```
class BWT:

    def __init__(self, seq = ""):
        self.bwt = self.build_bwt(seq)

    def set_bwt(self, bw):
        self.bwt = bw
```

```
    def build_bwt(self, text):
        ls = []
        for i in range(len(text)):
            ls.append(text[i:]+text[:i])
        ls.sort()
        res = ""
        for i in range(len(text)):
            res += ls[i][len(text)-1]
        return res

def test():
    seq = "TAGACAGAGA$"
    bw = BWT(seq)
    print (bw.bwt)

test()
```

Next, we will implement the reverse process, i.e. recovering the original sequence from the BWT. This is implemented in the next code block, by the method **inverse_bwt**. The algorithm implemented by this method is the one depicted in Fig. 16.13, being used also the same example in the test function. The auxiliary method **get_first_col** is used to retrieve the first column of the matrix M, while the function **find_ith_occ** is used to find the position of the i-th occurrence of a symbol in a list.

```
class BWT:

(...)

    def inverse_bwt(self):
        firstcol = self.get_first_col()
        res = ""
        c = "$"
        occ = 1
        for i in range(len(self.bwt)):
            pos = find_ith_occ(self.bwt, c, occ)
            c = firstcol[pos]
            occ = 1
            k = pos-1
            while firstcol[k] == c and k >= 0:
```

```
                occ += 1
                k -= 1
            res += c
        return res

    def get_first_col (self):
        firstcol = []
        for c in self.bwt: firstcol.append(c)
        firstcol.sort()
        return firstcol

def find_ith_occ(l, elem, index):
    j,k = 0,0
    while k < index and j < len(l):
        if l[j] == elem:
            k = k +1
            if k == index: return j
        j += 1
    return -1

def test2():
    bw = BWT()
    bw.set_bwt("ACG$GTAAAAC")
    print (bw.inverse_bwt())

test2()
```

The next step will be to implement the pattern searching process, as described in the previous section. This algorithm will be implemented in the method **bw_matching**, which follows the algorithm depicted in Fig. 16.14. The variables *top* and *bottom* keep the pointers of the match, while the auxiliary method **last_to_first** allows to create a lookup table to rapidly convert the position of the same symbol from the last to the first column. The test function builds the same BWT as used in the first code block for this class (Fig. 16.11), and searches for the pattern AGA (the same used in the example from Fig. 16.14).

```
class BWT:

    (...)
```

```
    def last_to_first(self):
        res = []
        firstcol = self.get_first_col()
        for i in range(len(firstcol)):
            c = self.bwt[i]
            ocs = self.bwt[:i].count(c) + 1
            res.append(find_ith_occ(firstcol, c, ocs))
        return res

    def bw_matching(self, patt):
        lf = self.last_to_first()
        res = []
        top = 0
        bottom = len(self.bwt)-1
        flag = True
        while flag and top <= bottom:
            if patt != "":
                symbol = patt[-1]
                patt = patt[:-1]
                lmat = self.bwt[top:(bottom+1)]
                if symbol in lmat:
                    topIndex = lmat.index(symbol) + top
                    bottomIndex = bottom - lmat[::-1].index(symbol)
                    top = lf[topIndex]
                    bottom = lf[bottomIndex]
                else: flag = False
            else:
                for i in range(top, bottom+1): res.append(i)
                flag = False
        return res

def test():
    seq = "TAGACAGAGA$"
    bw = BWT(seq)
    print (bw.last_to_first())
    print (bw.bw_matching("AGA"))

test()
```

Finally, we will implement a method to collect the matching positions of patterns, using suffix arrays, as explained in the previous section. The creation of the suffix array will be done here, optionally, when the BWT is created. Again, the algorithm used will not be usable in practice for large scale sequences, being used here for illustration purposes only.

The suffix array will be kept as an attribute for the class **BWT**, being used in the method **bw_matching_pos** to collect the positions where the pattern occurs, from the results of the method **bw_matching**.

```
class BWT:

    def __init__(self, seq = "", buildsufarray = False):
        self.bwt = self.build_bwt(seq, buildsufarray)

    def build_bwt(self, text, buildsufarray = False):
        ls = []
        for i in range(len(text)):
            ls.append(text[i:]+text[:i])
        ls.sort()
        res = ""
        for i in range(len(text)):
            res += ls[i][len(text)-1]
        if buildsufarray:
            self.sa = []
            for i in range(len(ls)):
                stpos = ls[i].index("$")
                self.sa.append(len(text)-stpos-1)
        return res

    (...)

    def bw_matching_pos(self, patt):
        res = []
        matches = self.bw_matching(patt)
        for m in matches:
            res.append(self.sa[m])
        res.sort()
        return res
```

```
def test():
    seq = "TAGACAGAGA$"
    bw = BWT(seq, True)
    print("Suffix array:", bw.sa)
    print(bw.bw_matching_pos("AGA"))

test()
```

16.4.3 Aligning References to Genomes in Practice

Short read mappers have been developed following sequencing technologies. These tools are optimized to map dozens of millions of short reads (smaller than 100 nucleotides), with very high similarity ($> 98\%$) to reference sequence and with one or fewer matches. When these assumptions are not met, the required changes in the alignment parameters usually result in slower alignment with considerably less accuracy.

BWT-based methods can efficiently handle reads with multiple matches, originating for instance from repeat regions of the genome, since these regions are scanned only once. However, these methods have no established way to handle the variability, either due to sequencing errors or variation, with respect to the reference sequence.

This type of approximate matching can still be achieved by BWTs with some minor changes. Without detailing here the algorithm, we can illustrate these changes by noticing that if our sequence has size k and there is at most 1 mismatch, there is an exact match of at least size $k/2 - 1$. If we look for exact matches of this size, we can then extend the match.

Thus, we can still use a variant of the fast algorithms discussed above for this case. Allowing for a higher number of mismatches will result in slower mapping. Examples of methods based on BWTs, largely used by the Bioinformatics community, include *BWA* [100], *Bowtie2* [93], *SOAP* [102], or *GEM* [107]. As the development of sequencing technologies delivers substantially longer reads, novel methods, or adaptations of existing ones will be required.

Bibliographic Notes and Further Reading

Tries were firstly described by De La Briandais in 1959 [45]. Suffix trees were introduced by Weiner in 1973 [154]. BWTs were proposed in 1994 by Burrows and Wheeler [32], and an efficient implementation of BWTs was proposed by [64].

Exercises and Programming Projects

Exercises

1. Implement a method **repeats(self, k, ocs)**, to add to the class **SuffixTree** that returns a list of patterns of size k, that occur in the tree (or in the sequence used to build it) a number of times of at least ocs.
2. Consider the definition of suffix trees representing two sequences made above, and illustrated in Fig. 16.10.
   a. Write a method to add to class **SuffixTree** that given two sequences S_1 and S_2 (terminating with $\$_1$ and $\$_2$, respectively) builds the suffix tree representing both. Use labels in the leaves in the form of a tuple (sequence, initial position of the suffix).
   b. Adapt the method to find patterns in the suffix tree to account for trees built from the previous method.
   c. Write another method in the same class that, assuming the tree represents two sequences, identifies the largest sub-string that is shared by both sequences.
3. a. Consider the class **BWT** defined in this chapter. Complete the following code so that it prints the sequence: CTTTAAACC$:

```
seq = "..."
bw = BWT(seq)
print(bw.bwt)
```

   b. Write the code in Python to be able to reach the answer to the previous question.

Programming Projects

1. Consider the definition of compact trees provided above and the example in Fig. 16.9.
   a. Implement a method that builds a compact suffix tree from a sequence.
   b. Implement a method to add to the class **SuffixTree** that allows to compact a tree (*self*) that is not compacted.
   c. Implement a method that allows to search patterns over compacted trees.

CHAPTER 17

Further Reading and Resources

In this chapter, we will provide some useful links for the interested reader to deepen the knowledge on Bioinformatics algorithms and tools. We will discuss books with complementary material, educational resources (online courses and sites with relevant resources), journals and conferences in the field, as well as provide a list with a few courses (mainly masters) from universities around the world.

17.1 Complementary Books

There are many published books that can be used to complement the present text in different ways, that encompass books on Bioinformatics, general-purpose algorithms and programming concepts, and more specific Python programming. We will try to mention a few from each group, obviously not attempting to cover the full portfolio of offers, but rather selecting a few examples known by the authors.

Regarding Bioinformatics books, most of the available texts are written by biologists or other biology-related professionals, being written mostly for an audience seeking to improve their skills on using Bioinformatics tools and understanding their results, and not so much for programmers.

From this offer we would select the books by David Mount [117] and Jonathan Pevsner [125]. Indeed, these are two very comprehensive books covering a large set of concepts, algorithms, and tools in Bioinformatics, combining well the definition of the methods and their main features, with the use of tools implementing these methods and the interpretation of their results in a critical perspective. Both these books cover most of the topics we address in this book, while not focusing on the implementation of the methods using a programming language, as it is done here. The latter is also more updated as it has a 2015 edition, while the former had its last edition in 2004.

For a more algorithmic oriented discussion of Bioinformatics, the work of Pavel Pevzner's group is one of the most important efforts. We would emphasize here the first book in collaboration with Neil Jones [84], which covers a number of problems in Bioinformatics and addresses the main algorithmic classes to address these tasks. More recently, in collaboration with Phlippe Compeau, the same author has published a two volume book [35], also focused

Bioinformatics Algorithms. DOI: 10.1016/B978-0-12-812520-5.00017-1

on algorithms to address Bioinformatics problems, motivated in every chapter by a specific biological problem. Although it does not provide implementations of the algorithms, it provides a comprehensive review of the most important algorithmic core topics for Bioinformatics applications.

An important contribution in the same line was the book by Böckenhauer and Bongartz from 2007 [28], which covers a relevant set of algorithms mainly related with sequence/string processing. This is the ideal book for readers who are comfortable with a more formal approach using a Mathematical notation of the main algorithms.

The number of books providing complementary concepts regarding algorithms and programming is very vast, and of course out of the scope here. However, we can mention the book by Dan Gusfield [70] and [39] related to string-based algorithms, while the books by Sedgewick and Wayne [138] and Wirth [155] provide a more general-purpose coverage of algorithms and data structures.

Also, there are a few books with a more practical focus on applications of the Python programming language to Bioinformatics. One of these examples is the book by Sebastian Bassi, recently with its 2nd edition [25], which covers the main concepts in Python programming providing a number of examples for biological data and also addressing the basics of BioPython. The book by Mitchel Model [115] follows a similar approach focusing the explanation on the Python concepts. These are good books to work over Python programming skills, with a flavor of Bioinformatics. Still, they are not focused on the algorithmic aspects of the main Bioinformatics tools and tasks.

17.2 Journals and Conferences

Bioinformatics is a multidisciplinary area, and thus many of the important papers on Bioinformatics tools are spread along journals and conferences of different topics, from computer science to specific biological and biomedical venues. Still, there are a few important journals mainly devoted to this field, or that have specific tracks for Bioinformatics work.

The following list provides an overview of some selected journals and their publishers, and a brief description of their contents:

- *Bioinformatics* (Oxford University Press) – probably the most important journal for this area, focuses mainly on new developments in genome bioinformatics and computational biology. Two sections – Discovery Notes and Application Notes – focus on shorter papers reporting biologically interesting discoveries using computational methods, and the development of bioinformatics applications.

- *Nucleic Acids Research* (Oxford University Press) – its focus are physical, chemical, biochemical and biological aspects of nucleic acids and proteins involved in nucleic acid metabolism and/or interactions; although not a specific Bioinformatics journal, it is known for publishing yearly comprehensive reviews of available databases and web applications in all fields of Bioinformatics; even in the regular issues it publishes many articles on software applications to biological problems.
- *Briefings in Bioinformatics* (Oxford University Press) – it is an international forum for researchers and educators in the life sciences, publishing reviews of databases and tools of genetics, molecular, and systems biology providing practical guidance in Bioinformatics approaches.
- *PLOS Computational Biology* (PLOS) – features studies that focus on the understanding and modeling of living systems in different scales, from cells to populations or ecosystems through the application of computational methods.
- *BMC Bioinformatics* (Biomed Central) – open-access journal that publishes papers on the different aspects of the development, testing and application of bioinformatics methods for the modeling and analysis of biological data, and other areas of computational biology.
- *Computer Methods and Programs in Biomedicine* (Elsevier) – publishes articles related to computing methodology and software in all aspects of biomedical research and medical practice.
- *Computers in Biology and Medicine* (Elsevier) – publishes research results related to the application of computers to the fields of bioscience and medicine.
- *Algorithms for Molecular Biology* (Biomed Central) – publishes articles on algorithms for biological sequence and structure analysis, phylogeny reconstruction, and machine learning.

Regarding international conferences, we list here a few of the most relevant for the Bioinformatics field:

- *Inteligent Systems for Molecular Biology* (ISMB), will have its 26th edition in 2018, being probably the largest and most important venue for Bioinformatics related research, bringing together researchers from computer science, molecular biology, mathematics, statistics, and related fields.
- *European Conference on Computational Biology* (ECCB), will have its 17th edition in 2018, being the main European event in Bioinformatics and Computational Biology.
- *RECOMB*, already having its 22nd edition in 2018, aims to bridge the areas of computational, mathematical, statistical, and biological sciences.
- *IEEE International Conference on Bioinformatics and Biomedicine* (BIBM), is the main IEEE sponsored conference related to Bioinformatics and health informatics.

- *ACM Conference on Bioinformatics, Computational Biology, and Health Informatics* (ACM BCB), the flagship conference of the interest group in Bioinformatics for the ACM, already with 8 editions.
- *Pacific Symposium in Biocomputing* (PSB) – always located in Hawaii, it is a multidisciplinary conference for the discussion of research in the theory and application of computational methods in problems of biological significance, being a conference with a considerable tradition in the field.
- *Workshop on Algorithms in Bioinformatics* (WABI) – already with 17 editions, covers all research in algorithmic work in bioinformatics, computational biology, and systems biology.
- *International Conference on Practical Applications of Computational Biology & Bioinformatics* (PACBB), being a forum already in its 12th edition, mainly for the discussion of practical approaches of Bioinformatics for junior researchers.

17.3 Formal Education

Around the world, there are nowadays many different universities offering degrees in the fields of Bioinformatics and Computational Biology. As we did in previous sections, we review here some examples of this offer, seeking to select degrees spread around the world in institutions known by their quality. Since most of the available degrees are in the masters level, we provide here a number of MSc's in Bioinformatics, which are available for students with a background on Biology (or other biological/biomedical field) or Computer Science (or related degrees in Informatics or Information Technologies).

- *MSc Bioinformatics*, Johns Hopkins Institute, USA – `http://advanced.jhu.edu/academics/graduate-degree-programs/bioinformatics` – available in a fully online version;
- *MSc Bioinformatics and Systems Biology*, Univ. California San Diego, USA – `http://bioinformatics.ucsd.edu/curriculum` – strong component of modeling and Systems Biology, very research oriented;
- *Master Computational Biology and Bioinformatics*, ETH Zurich, Switzerland – `http://www.cbb.ethz.ch/`;
- *MSc Bioinformatics* – Vrije Univ., Amsterdam, Netherlands – `http://masters.vu.nl/en/programmes/bioinformatics-systems-biology`;
- *MSc Bioinformatics and Systems Biology*, University of Manchester, UK – `https://www.bmh.manchester.ac.uk/biology/study/masters/bioinformatics-systems-biology/`;
- *MSc Bioinformatics*, Univ. Minho, Portugal – `http://bioinformatica.di.uminho.pt`;

- *MSc Bioinformatics*, South African National Bioinformatics Institute, South Africa – http://www.sanbi.ac.za/training-2/msc-bioinformatics/;
- *Master of Computational Biology*, University of Melbourne, Australia – http://science-courses.unimelb.edu.au/study/degrees/master-of-computational-biology/;
- *Master Bioinformatics*, Univ. Beijing, China – http://school.cucas.edu.cn/Beijing-Normal-University-27/program/Bioinformatics-4234.html;
- *MSc Bioinformatics*, Nanyang Technological Institute, Singapore – http://sce.ntu.edu.sg/Programmes/CurrentStudents/Graduate/Pages/msc-bioinformatics-intro.aspx.

Apart from these integrated pedagogical offers, most universities, and other research institutions, offer a number of courses in specific subjects related to Bioinformatics. We select next just a few examples:

- University of Cambridge (UK), Bioinformatics Training – http://bioinfotraining.bio.cam.ac.uk;
- Brabaham Institute (UK), Bioinformatics Training – http://www.bioinformatics.babraham.ac.uk/training.html;
- Vienna Biocenter (Austria), BioComp training – http://biocomp.vbcf.ac.at/training;
- Dutch TechCenter for the Life Sciences (Netherlands) – https://www.dtls.nl/courses;
- Gulbenkian Training Programme in Bioinformatics (Portugal) – http://gtpb.igc.gulbenkian.pt/bicourses;
- ECSeq Bioinformatics (Germany; company) – http://ecseq.com/;
- Cold Spring Harbor Lab. (USA) – http://meetings.cshl.edu/courseshome.aspx;
- UC Davis Bioinformatics core (USA) – http://bioinformatics.ucdavis.edu/training;
- Bioinformatics.ca workshops (Canada) – https://bioinformatics.ca/workshops-2017.

These and other resources can be searched using portals as the one of the *Global Organisation for Bioinformatics Learning, Education & Training* (GOBLET), available in http://mygoblet.org/training-portal.

17.4 Online Resources

In the last few years, we have been observing a large growth in the number of available online courses, spread by multiple platforms. The presence of courses related to Bioinformatics and

Table 17.1: Summary of the online courses available in the platform *edX*.

Institution	XSeries/Micromaster	Courses
Harvard University	*Data Analysis for Life Sciences*	*Statistics and R for the Life Sciences, Statistical Inference and Modeling for High-throughput Experiments, High-Dimensional Data Analysis, Introduction to Bioconductor: annotation and analysis of genomes and genomic assays, High-performance computing for reproducible genomics, Case studies in functional genomics*
UC San Diego		*Introduction to Genomic Data Science, Analyze your genome*
Univ. Maryland UC and USM	*Bioinformatics*	*Statistical Analysis in Bioinformatics, DNA Sequences: Alignments and Analysis, Proteins: Alignment, Analysis and Structure*
MIT		*Quantitative Biology Workshop*
Osaka University		*Metabolomics in Life Sciences*
Peking University		*Mathematical Modeling in the Life Sciences*2

Table 17.2: Summary of the online courses available in the platform *Coursera*.

Institution	Specialization	Courses
UC San Diego	*Bioinformatics*	*Finding Hidden Messages in DNA; Genome Sequencing, Comparing Genes, Proteins, and Genomes; Molecular Evolution; Genomic Data Science and Clustering; Finding Mutations in DNA and Proteins, Genome Assembly Programming Challenge*
Johns Hopkins University	*Genomic Data Science*	*Introduction to Genomic Technologies, Genomic Data Science with Galaxy, Python for Genomic Data Science, Algorithms for DNA Sequencing, Command Line Tools for Genomic Data Science, Bioconductor for Genomic Data Science, Statistics for Genomic Data Science*
Mount Sinai	*Systems Biology and Biotechnology*	*Introduction to Systems Biology, Experimental Methods in Systems Biology, Network Analysis in Systems Biology, Dynamical Modeling Methods for Systems Biology, Integrated Analysis in Systems Biology*
University of Toronto		*Bioinformatic Methods I and II*
State University of New York		*Big Data, Genes, and Medicine*
Peking University		*Bioinformatics: Introduction and Methods*
Saint Petersburg State University		*Introduction to Bioinformatics: Metagenomics*

Computational Biology is a constant in these platforms. Given their diversified offer and the quality of the institutions involved, we selected here some available courses in two platforms: *edX* and *Coursera*, which are shown in Tables 17.1 and 17.2, respectively.

Also, there are many sites available that can be excellent resources to learn Bioinformatics and Python programming for the life sciences. In the next list, we selected a few of those that be used by the reader to complement the materials of this book.

- NCBI mini-courses – `https://www.ncbi.nlm.nih.gov/Class/minicourses/`;
- EBI Train Online – `https://www.ebi.ac.uk/training/online/`;
- Biostars – `https://www.biostars.org/`, forum of frequently asked questions in Bioinformatics;
- SeqAnswers – `http://seqanswers.com/`, forum of frequently asked questions when handling next-generation sequencing data;
- Rosalind – `http://rosalind.info/problems/locations/`, a site that allows to learn Bioinformatics by solving programming problems;
- Python for Biologists – `https://pythonforbiologists.com/`, resources for learning to program in Python for people with a background in biology;
- IAP (Introduction to Applied Bioinformatics) – `http://readiab.org/`, open source interactive text that introduces readers to core concepts of Bioinformatics.

Final Words

We have reached the end of our journey through the sea of Bioinformatics algorithms, and we hope it was a pleasant one. Of course, many relevant topics remained untouched, many algorithms are still there to be learned. Our expectation is that we have awakened in our readers the will to learn more about Bioinformatics, and that the resources put together in the last chapter, and in the end of each chapter, are good starting points to proceed this journey into more adventurous algorithms and programs.

Bioinformatics is a multidisciplinary topic and, therefore, the readers of this book probably have very distinct backgrounds. We hope that we were able to address this diversity and explain the topics even for non-experienced readers. By understanding the ideas and algorithms behind the different methods we hope the readers can optimize their usage or be able to develop novel methods that meet their needs. The decision of having simplified algorithms and their Python implementation was to try to accomplish this goal. The readers will be the judges of how this has worked in the end. We will await for their feedback.

Since the field of Bioinformatics is growing, we are expecting the number of books in this area to grow as well, becoming more specialized in the topics to be covered. This is a trend we expect to see also in online and formal courses and degrees. Still, even at this point, we believe that a general purpose overview of the main Bioinformatics algorithms is important, and we tried to fill this gap.

There were many topics we have not considered in this book, but that are important in Bioinformatics. Maybe one of the most relevant is Machine Learning (ML), which makes the core algorithmic support for many tasks in Bioinformatics, from gene prediction, to protein classification and gene expression (or other omics) data analysis. To learn the basic algorithms and Mathematical concepts underlying ML models is, indeed, very relevant for today's Bioinformatics.

One of the main tasks in Bioinformatics is related to protein structure prediction, i.e. the ability to design algorithms able to predict the three-dimensional configuration of the protein based on its sequence of aminoacids. This is also a subject we have not addressed here, which may be approached using ML or other statistical methods, based on features calculated from the sequences.

Genomics for personalized medicine is a research area that already has a large demand of experts. Handling the vast amounts of data that are being generated by all sorts of projects, both in research or clinical laboratories, requires well prepared and trained Bioinformatics specialists. Tasks such as variant calling or gene differential expression were not covered in this book since, beyond the algorithmic details, a solid statistical background is required for a good understanding of these methods. Nevertheless, in this book, we have approached algorithms and methods that form part of the steps of the data analysis pipeline.

In many topics, we also left out some important optimization algorithms, which are used to address complex optimization tasks in Bioinformatics. For instance, we have not covered important meta-heuristics as Evolutionary Algorithms or Simulated Annealing.

In all previous cases, the decision to leave these topics out of this book was related to the complexity of the algorithms and underlying Mathematical and statistical concepts, given our choice to try to explain the main details and implement our algorithms in Python. We hope we can cover these subjects in a future project.

We really hope you have enjoyed this book!

Miguel Rocha

Pedro G. Ferreira

Bibliography

[1] Gwas central, http://www.gwascentral.org/.
[2] Python for beginners, http://www.python.org/doc/Intros.html.
[3] The python language website, http://www.python.org/.
[4] Python tutor, visualization of code execution, http://pythontutor.com/.
[5] The python tutorial, https://docs.python.org/3/tutorial/.
[6] Computational Methods in Molecular Biology, Elsevier Science, 1998.
[7] Biopython: freely available python tools for computational molecular biology and bioinformatics, Bioinformatics 25 (11) (2009) 1422–1423.
[8] Biopython tutorial and cookbook, http://biopython.org/DIST/docs/tutorial/Tutorial.html. (Last update Nov 23, 2016).
[9] Alfred V. Aho, John E. Hopcroft, The Design and Analysis of Computer Algorithms, 1st edition, Addison-Wesley Longman Publishing Co., Inc., Boston, MA, USA, 1974.
[10] B. Alberts, A. Johnson, J. Lewis, M. Raff, K. Roberts, P. Walter, Molecular Biology of the Cell, 4th edition, Garland Science, New York, USA, 2002.
[11] Stephen F. Altschul, Warren Gish, Webb Miller, Eugene W. Myers, David J. Lipman, Basic local alignment search tool, Journal of Molecular Biology 215 (3) (1990) 403–410.
[12] Stephen F. Altschul, Thomas L. Madden, Alejandro A. Schäffer, Jinghui Zhang, Zheng Zhang, Webb Miller, David J. Lipman, Gapped blast and psi-blast: a new generation of protein database search programs, Nucleic Acids Research 25 (17) (1997) 3389–3402.
[13] R. Andersson, et al., An atlas of active enhancers across human cell types and tissues, Nature 507 (7493) (Mar 2014) 455–461.
[14] K. Asai, S. Hayamizu, K. Handa, Prediction of protein secondary structure by the hidden Markov model, Computer Applications in the Biosciences 9 (2) (Apr 1993) 141–146.
[15] A. Auton, et al., A global reference for human genetic variation, Nature 526 (7571) (Oct 2015) 68–74.
[16] Gary D. Bader, Christopher W.V. Hogue, Analyzing yeast protein-protein interaction data obtained from different sources, Nature Biotechnology 20 (10) (2002) 991–997.
[17] T.L. Bailey, Discovering sequence motifs, Methods in Molecular Biology 452 (2008) 231–251.
[18] T.L. Bailey, C. Elkan, Fitting a mixture model by expectation maximization to discover motifs in biopolymers, Proceedings. International Conference on Intelligent Systems for Molecular Biology 2 (1994) 28–36.
[19] Pierre Baldi, Søren Brunak, Bioinformatics: The Machine Learning Approach, 2nd edition, MIT Press, Cambridge, MA, USA, 2001.
[20] Anton Bankevich, Sergey Nurk, Dmitry Antipov, Alexey A. Gurevich, Mikhail Dvorkin, Alexander S. Kulikov, Valery M. Lesin, Sergey I. Nikolenko, Son Pham, Andrey D. Prjibelski, et al., Spades: a new genome assembly algorithm and its applications to single-cell sequencing, Journal of Computational Biology 19 (5) (2012) 455–477.
[21] Albert-László Barabási, Réka Albert, Emergence of scaling in random networks, Science 286 (5439) (1999) 509–512.

[22] Albert-László Barabási, Natali Gulbahce, Joseph Loscalzo, Network medicine: a network-based approach to human disease, Nature Reviews Genetics 12 (1) (2011) 56–68.
[23] Albert-Laszlo Barabási, Zoltan N. Oltvai, Network biology: understanding the cell's functional organization, Nature Reviews Genetics 5 (2) (2004) 101–113.
[24] N.L. Barbosa-Morais, M. Irimia, Q. Pan, H.Y. Xiong, S. Gueroussov, L.J. Lee, V. Slobodeniuc, C. Kutter, S. Watt, R. Colak, T. Kim, C.M. Misquitta-Ali, M.D. Wilson, P.M. Kim, D.T. Odom, B.J. Frey, B.J. Blencowe, The evolutionary landscape of alternative splicing in vertebrate species, Science 338 (6114) (Dec 2012) 1587–1593.
[25] Sebastian Bassi, Python for Bioinformatics, CRC Press, 2016.
[26] T. Beck, R.K. Hastings, S. Gollapudi, R.C. Free, A.J. Brookes, GWAS Central: a comprehensive resource for the comparison and interrogation of genome-wide association studies, European Journal of Human Genetics 22 (7) (Jul 2014) 949–952.
[27] E. Birney, et al., Identification and analysis of functional elements in 1% of the human genome by the ENCODE pilot project, Nature 447 (7146) (Jun 2007) 799–816.
[28] Hans-Joachim Böckenhauer, Dirk Bongartz, Algorithmic Aspects of Bioinformatics, Springer-Verlag New York, Inc., Secaucus, NJ, USA, 2007.
[29] Robert S. Boyer, J. Strother Moore, A fast string searching algorithm, Communications of the ACM 20 (10) (October 1977) 762–772.
[30] J. Buhler, M. Tompa, Finding motifs using random projections, Journal of Computational Biology 9 (2) (2002) 225–242.
[31] C. Burge, S. Karlin, Prediction of complete gene structures in human genomic DNA, Journal of Molecular Biology 268 (1) (Apr 1997) 78–94.
[32] Michael Burrows, David J. Wheeler, A Block-Sorting Lossless Data Compression Algorithm, 1994.
[33] P. Carninci, et al., The transcriptional landscape of the mammalian genome, Science 309 (5740) (Sep 2005) 1559–1563.
[34] Humberto Carrillo, David Lipman, The multiple sequence alignment problem in biology, SIAM Journal on Applied Mathematics 48 (5) (1988) 1073–1082.
[35] Phillip Compeau, Pavel Pevzner, Bioinformatics Algorithms: An Active Learning Approach, Active Learning Publishers, 2015.
[36] Community Content Contributions, Boundless biology, https://www.boundless.com/biology/textbooks/boundless-biology-textbook/, February 2017.
[37] G.M. Cooper, The Cell: A Molecular Approach, 2nd edition, Sinauer Associates, Sunderland, MA, USA, 2000.
[38] Thomas H. Cormen, Clifford Stein, Ronald L. Rivest, Charles E. Leiserson, Introduction to Algorithms, 2nd edition, McGraw-Hill Higher Education, 2001.
[39] Maxime Crochemore, Christophe Hancart, Thierry Lecroq, Algorithms on Strings, Cambridge University Press, 2007.
[40] G.E. Crooks, G. Hon, J.M. Chandonia, S.E. Brenner, WebLogo: a sequence logo generator, Genome Research 14 (6) (Jun 2004) 1188–1190.
[41] M.K. Das, H.K. Dai, A survey of DNA motif finding algorithms, BMC Bioinformatics 8 (Suppl 7) (Nov 2007) S21.
[42] Sanjoy Dasgupta, Christos H. Papadimitriou, Umesh Vazirani, Algorithms, McGraw-Hill, Inc., 2006.
[43] Margaret O. Dayhoff, Atlas of Protein Sequence and Structure, 1965.
[44] N. de Bruijn, A combinatorial problem, Proceedings of the Section of Sciences of the Koninklijke Nederlandse Akademie van Wetenschappen te Amsterdam 49 (7) (1946) 758–764.
[45] Rene De La Briandais, File searching using variable length keys, in: Papers Presented at the March 3–5, 1959, Western Joint Computer Conference, ACM, 1959, pp. 295–298.

[46] T. Derrien, R. Johnson, G. Bussotti, A. Tanzer, S. Djebali, H. Tilgner, G. Guernec, D. Martin, A. Merkel, D.G. Knowles, J. Lagarde, L. Veeravalli, X. Ruan, Y. Ruan, T. Lassmann, P. Carninci, J.B. Brown, L. Lipovich, J.M. Gonzalez, M. Thomas, C.A. Davis, R. Shiekhattar, T.R. Gingeras, T.J. Hubbard, C. Notredame, J. Harrow, R. Guigo, The GENCODE v7 catalog of human long noncoding RNAs: analysis of their gene structure, evolution, and expression, Genome Research 22 (9) (Sep 2012) 1775–1789.
[47] P. D'haeseleer, What are DNA sequence motifs? Nature Biotechnology 24 (4) (Apr 2006) 423–425.
[48] Reinhard Diestel, Graph Theory, 3rd edition, Graduate Texts in Mathematics, vol. 173, Springer, 2005.
[49] I. Dunham, et al., An integrated encyclopedia of DNA elements in the human genome, Nature 489 (7414) (Sep 2012) 57–74.
[50] R. Durbin, S. Eddy, A. Krogh, G. Mitchison, Biological Sequence Analysis: Probabilistic Models of Proteins and Nucleic Acids, Cambridge University Press, 1998.
[51] J.R. Ecker, W.A. Bickmore, I. Barroso, J.K. Pritchard, Y. Gilad, E. Segal, Genomics: ENCODE explained, Nature 489 (7414) (Sep 2012) 52–55.
[52] S.R. Eddy, Multiple alignment using hidden Markov models, Proceedings. International Conference on Intelligent Systems for Molecular Biology 3 (1995) 114–120.
[53] S.R. Eddy, Hidden Markov models, Current Opinion in Structural Biology 6 (3) (Jun 1996) 361–365.
[54] S.R. Eddy, What is a hidden Markov model? Nature Biotechnology 22 (10) (Oct 2004) 1315–1316.
[55] S.R. Eddy, A probabilistic model of local sequence alignment that simplifies statistical significance estimation, PLoS Computational Biology 4 (5) (May 2008) e1000069.
[56] S.R. Eddy, A new generation of homology search tools based on probabilistic inference, Genome Informatics 23 (1) (Oct 2009) 205–211.
[57] S.R. Eddy, Accelerated profile HMM searches, PLoS Computational Biology 7 (10) (Oct 2011) e1002195.
[58] Robert C. Edgar, Muscle: multiple sequence alignment with high accuracy and high throughput, Nucleic Acids Research 32 (5) (2004) 1792–1797.
[59] E. Eskin, P.A. Pevzner, Finding composite regulatory patterns in DNA sequences, Bioinformatics 18 (Suppl 1) (2002) S354–S363.
[60] Leonhard Euler, Solutio problematis ad geometriam situs pertinentis, Commentarii Academiae Scientiarum Petropolitanae 8 (1741) 128–140.
[61] Even Shimon, Graph Algorithms, 2nd edition, Cambridge University Press, New York, NY, USA, 2011.
[62] Joseph Felsenstein, Inferring Phylogenies, vol. 2, Sinauer Associates, Sunderland, MA, 2004.
[63] Da-Fei Feng, Russell F. Doolittle, Progressive sequence alignment as a prerequisitetto correct phylogenetic trees, Journal of Molecular Evolution 25 (4) (1987) 351–360.
[64] Paolo Ferragina, Giovanni Manzini, Opportunistic data structures with applications, in: Foundations of Computer Science, 2000. Proceedings. 41st Annual Symposium on, IEEE, 2000, pp. 390–398.
[65] P.G. Ferreira, P.J. Azevedo, Evaluating deterministic motif significance measures in protein databases, Algorithms for Molecular Biology 2 (Dec 2007) 16.
[66] Jeffrey E.F. Friedl, Mastering Regular Expressions, 2nd edition, O'Reilly & Associates, Inc., Sebastopol, CA, USA, 2002.
[67] Emanuel Gonçalves, Joachim Bucher, Anke Ryll, Jens Niklas, Klaus Mauch, Steffen Klamt, Miguel Rocha, Julio Saez-Rodriguez, Bridging the layers: towards integration of signal transduction, regulation and metabolism into mathematical models, Molecular BioSystems 9 (7) (2013) 1576–1583.
[68] M. Gribskov, S. Veretnik, Identification of sequence pattern with profile analysis, Methods in Enzymology 266 (1996) 198–212.
[69] Stéphane Guindon, Jean-François Dufayard, Vincent Lefort, Maria Anisimova, Wim Hordijk, Olivier Gascuel, New algorithms and methods to estimate maximum-likelihood phylogenies: assessing the performance of PhyML 3.0, Systematic Biology 59 (3) (2010) 307–321.
[70] Dan Gusfield, Algorithms on Strings, Trees and Sequences: Computer Science and Computational Biology, 1st edition, Cambridge University Press, May 1997.
[71] Frank Harary, Graph Theory, Addison-Wesley, 1972.

[72] J. Harrow, A. Frankish, J.M. Gonzalez, E. Tapanari, M. Diekhans, F. Kokocinski, B.L. Aken, D. Barrell, A. Zadissa, S. Searle, I. Barnes, A. Bignell, V. Boychenko, T. Hunt, M. Kay, G. Mukherjee, J. Rajan, G. Despacio-Reyes, G. Saunders, C. Steward, R. Harte, M. Lin, C. Howald, A. Tanzer, T. Derrien, J. Chrast, N. Walters, S. Balasubramanian, B. Pei, M. Tress, J.M. Rodriguez, I. Ezkurdia, J. van Baren, M. Brent, D. Haussler, M. Kellis, A. Valencia, A. Reymond, M. Gerstein, R. Guigo, T.J. Hubbard, GENCODE: the reference human genome annotation for The ENCODE Project, Genome Research 22 (9) (Sep 2012) 1760–1774.

[73] D. Haussler, A. Krogh, I.S. Mian, K. Sjolander, Protein modeling using hidden Markov models: analysis of globins, in: Proceeding of the Twenty-Sixth Hawaii International Conference on System Sciences, IEEE, IEEE, 1993, pp. 792–802.

[74] Steven Henikoff, Jorja G. Henikoff, Amino acid substitution matrices from protein blocks, Proceedings of the National Academy of Sciences 89 (22) (1992) 10915–10919.

[75] G.Z. Hertz, G.W. Hartzell, G.D. Stormo, Identification of consensus patterns in unaligned DNA sequences known to be functionally related, Computer Applications in the Biosciences 6 (2) (Apr 1990) 81–92.

[76] Carl Hierholzer, Über die Möglichkeit, einen Linienzug ohne Wiederholung und ohne Unterbrechung zu umfahren, Mathematische Annalen 6 (1) (1873) 30–32.

[77] Desmond G. Higgins, Paul M. Sharp, Clustal: a package for performing multiple sequence alignment on a microcomputer, Gene 73 (1) (1988) 237–244.

[78] John E. Hopcroft, Rajeev Motwani, Jeffrey D. Ullman, Introduction to Automata Theory, Languages, and Computation, 3rd edition, Addison-Wesley Longman Publishing Co., Inc., Boston, MA, USA, 2006.

[79] Michael Huerta, Gregory Downing, Florence Haseltine, Belinda Seto, Yuan Liu, Nih Working Definition of Bioinformatics and Computational Biology, US National Institute of Health, 2000.

[80] Ramana M. Idury, Michael S. Waterman, A new algorithm for DNA sequence assembly, Journal of Computational Biology 2 (2) (1995) 291–306.

[81] Jan Ihmels, Gilgi Friedlander, Sven Bergmann, Ofer Sarig, Yaniv Ziv, Naama Barkai, Revealing modular organization in the yeast transcriptional network, Nature Genetics 31 (4) (2002) 370.

[82] Ananthararaman Narayana Iyer, pyhmm, https://github.com/ananthpn/pyhmm. (Retrieved October 2017).

[83] Hawoong Jeong, Bálint Tombor, Réka Albert, Zoltan N. Oltvai, A.-L. Barabási, The large-scale organization of metabolic networks, Nature 407 (6804) (2000) 651–654.

[84] N.C. Jones, P. Pevzner, An Introduction to Bioinformatics Algorithms, A Bradford book, London, 2004.

[85] Daniel Jurafsky, James H. Martin, Speech and Language Processing: An Introduction to Natural Language Processing, Computational Linguistics, and Speech Recognition, 1st edition, Prentice Hall PTR, Upper Saddle River, NJ, USA, 2000.

[86] K. Karplus, K. Sjolander, C. Barrett, M. Cline, D. Haussler, R. Hughey, L. Holm, C. Sander, Predicting protein structure using hidden Markov models, Proteins 29 (Suppl 1) (1997) 134–139.

[87] Kazutaka Katoh, Kazuharu Misawa, Kei-ichi Kuma, Takashi Miyata, MAFFT: a novel method for rapid multiple sequence alignment based on fast Fourier transform, Nucleic Acids Research 30 (14) (2002) 3059–3066.

[88] U. Keich, P.A. Pevzner, Subtle motifs: defining the limits of motif finding algorithms, Bioinformatics 18 (10) (Oct 2002) 1382–1390.

[89] H. Keren, G. Lev-Maor, G. Ast, Alternative splicing and evolution: diversification, exon definition and function, Nature Reviews Genetics 11 (5) (May 2010) 345–355.

[90] A. Krogh, Two methods for improving performance of an HMM and their application for gene finding, Proceedings. International Conference on Intelligent Systems for Molecular Biology 5 (1997) 179–186.

[91] A. Krogh, M. Brown, I.S. Mian, K. Sjolander, D. Haussler, Hidden Markov models in computational biology. Applications to protein modeling, Journal of Molecular Biology 235 (5) (Feb 1994) 1501–1531.

[92] Eric S. Lander, Lauren M. Linton, Bruce Birren, Chad Nusbaum, Michael C. Zody, Jennifer Baldwin, Keri Devon, Ken Dewar, Michael Doyle, William FitzHugh, et al., Initial sequencing and analysis of the human genome, Nature 409 (6822) (2001) 860–921.

[93] Langmead Ben, Steven L. Salzberg, Fast gapped-read alignment with Bowtie 2, Nature Methods 9 (4) (2012) 357–359.
[94] C.E. Lawrence, S.F. Altschul, M.S. Boguski, J.S. Liu, A.F. Neuwald, J.C. Wootton, Detecting subtle sequence signals: a Gibbs sampling strategy for multiple alignment, Science 262 (5131) (Oct 1993) 208–214.
[95] M. Lek, et al., Analysis of protein-coding genetic variation in 60,706 humans, Nature 536 (7616) (08, 2016) 285–291.
[96] H.C. Leung, F.Y. Chin, Algorithms for challenging motif problems, Journal of Bioinformatics and Computational Biology 4 (1) (Feb 2006) 43–58.
[97] Vladimir I. Levenshtein, Binary codes capable of correcting deletions, insertions, and reversals, Soviet Physics Doklady 10 (1966) 707–710.
[98] S. Levy, G. Sutton, P.C. Ng, L. Feuk, A.L. Halpern, B.P. Walenz, N. Axelrod, J. Huang, E.F. Kirkness, G. Denisov, Y. Lin, J.R. MacDonald, A.W. Pang, M. Shago, T.B. Stockwell, A. Tsiamouri, V. Bafna, V. Bansal, S.A. Kravitz, D.A. Busam, K.Y. Beeson, T.C. McIntosh, K.A. Remington, J.F. Abril, J. Gill, J. Borman, Y.H. Rogers, M.E. Frazier, S.W. Scherer, R.L. Strausberg, J.C. Venter, The diploid genome sequence of an individual human, PLoS Biology 5 (10) (Sep 2007) e254.
[99] Dinghua Li, Chi-Man Liu, Ruibang Luo, Kunihiko Sadakane, Tak-Wah Lam, MEGAHIT: an ultra-fast single-node solution for large and complex metagenomics assembly via succinct de Bruijn graph, Bioinformatics 31 (10) (2015) 1674–1676.
[100] Heng Li, Richard Durbin, Fast and accurate short read alignment with Burrows-Wheeler transform, Bioinformatics 25 (14) (2009) 1754–1760.
[101] N. Li, M. Tompa, Analysis of computational approaches for motif discovery, Algorithms for Molecular Biology 1 (May 2006) 8.
[102] Ruiqiang Li, Yingrui Li, Karsten Kristiansen, Jun Wang, SOAP: short oligonucleotide alignment program, Bioinformatics 24 (5) (2008) 713–714.
[103] David J. Lipman, Stephen F. Altschul, John D. Kececioglu, A tool for multiple sequence alignment, Proceedings of the National Academy of Sciences 86 (12) (1989) 4412–4415.
[104] Ruibang Luo, Binghang Liu, Yinlong Xie, Zhenyu Li, Weihua Huang, Jianying Yuan, Guangzhu He, Yanxiang Chen, Qi Pan, Yunjie Liu, et al., SOAPdenovo2: an empirically improved memory-efficient short-read de novo assembler, Gigascience 1 (1) (2012) 18.
[105] Nicholas M. Luscombe, Dov Greenbaum, Mark Gerstein, et al., What is bioinformatics? A proposed definition and overview of the field, Methods of Information in Medicine 40 (4) (2001) 346–358.
[106] Daniel Machado, Rafael S. Costa, Miguel Rocha, Eugénio C. Ferreira, Bruce Tidor, Isabel Rocha, Modeling formalisms in systems biology, AMB Express 1 (1) (2011) 45.
[107] S. Marco-Sola, M. Sammeth, R. Guigo, P. Ribeca, The GEM mapper: fast, accurate and versatile alignment by filtration, Nature Methods 9 (12) (Dec 2012) 1185–1188.
[108] Florian Markowetz, All biology is computational biology, PLoS Biology 15 (3) (2017) e2002050.
[109] T. Marschall, S. Rahmann, Efficient exact motif discovery, Bioinformatics 25 (12) (Jun 2009) i356–i364.
[110] Paul Medvedev, Son Pham, Mark Chaisson, Glenn Tesler, Pavel Pevzner, Paired de Bruijn graphs: a novel approach for incorporating mate pair information into genome assemblers, Journal of Computational Biology 18 (11) (2011) 1625–1634.
[111] J. Merkin, C. Russell, P. Chen, C.B. Burge, Evolutionary dynamics of gene and isoform regulation in Mammalian tissues, Science 338 (6114) (Dec 2012) 1593–1599.
[112] F. Mignone, C. Gissi, S. Liuni, G. Pesole, Untranslated regions of mRNAs, Genome Biology 3 (3) (2002), REVIEWS0004.
[113] R.E. Mills, et al., Mapping copy number variation by population-scale genome sequencing, Nature 470 (7332) (Feb 2011) 59–65.
[114] Ron Milo, Shai Shen-Orr, Shalev Itzkovitz, Nadav Kashtan, Dmitri Chklovskii, Uri Alon, Network motifs: simple building blocks of complex networks, Science 298 (5594) (2002) 824–827.

[115] Mitchell L. Model, Bioinformatics Programming Using Python: Practical Programming for Biological Data, 1st edition, OReilly Media, Inc., 2009.
[116] Edward F. Moore, The shortest path through a maze, in: Proc. Int. Symp. Switching Theory, 1959, 1959, pp. 285–292.
[117] David W. Mount, Bioinformatics: Sequence and Genome Analysis, 2nd edition, Cold Spring Harbor Laboratory Press, 2004.
[118] Saul B. Needleman, Christian D. Wunsch, A general method applicable to the search for similarities in the amino acid sequence of two proteins, Journal of Molecular Biology 48 (3) (1970) 443–453.
[119] Cédric Notredame, Desmond G. Higgins, Jaap Heringa, T-coffee: a novel method for fast and accurate multiple sequence alignment, Journal of Molecular Biology 302 (1) (2000) 205–217.
[120] C.M. O'Connor, J.U. Adams, Essentials of Cell Biology, NPG Education, Cambridge, MA, USA, 2010.
[121] Smithsonian's National Museum of Natural History and the National Institutes of Health's National Human Genome Research Institute, Unlocking life's code, https://unlockinglifescode.org/, February 2017.
[122] Q. Pan, O. Shai, L.J. Lee, B.J. Frey, B.J. Blencowe, Deep surveying of alternative splicing complexity in the human transcriptome by high-throughput sequencing, Nature Genetics 40 (12) (Dec 2008) 1413–1415.
[123] William R. Pearson, David J. Lipman, Improved tools for biological sequence comparison, Proceedings of the National Academy of Sciences 85 (8) (1988) 2444–2448.
[124] Yu Peng, Henry C.M. Leung, Siu-Ming Yiu, Francis Y.L. Chin, IDBA—a practical iterative de Bruijn graph de novo assembler, in: Annual International Conference on Research in Computational Molecular Biology, Springer, 2010, pp. 426–440.
[125] Jonathan Pevsner, Bioinformatics and Functional Genomics, 3rd edition, John Wiley & Sons, 2015.
[126] P.A. Pevzner, S.H. Sze, Combinatorial approaches to finding subtle signals in DNA sequences, Proceedings. International Conference on Intelligent Systems for Molecular Biology 8 (2000) 269–278.
[127] Pavel A. Pevzner, Haixu Tang, Michael S. Waterman, An Eulerian path approach to DNA fragment assembly, Proceedings of the National Academy of Sciences 98 (17) (2001) 9748–9753.
[128] Dusty Phillips, Python 3 Object Oriented Programming, Packt Publishing Ltd, 2010.
[129] R. Phillips, R. Milo, Cell Biology by the Numbers, Garland Science, Cambridge, MA, USA, 2015.
[130] Lawrence R. Rabiner, A tutorial on hidden Markov models and selected applications in speech recognition, in: Proceedings of the IEEE, 1989, pp. 257–286.
[131] A. Ralston, Operons and prokaryotic gene regulation, in: Nature Education, vol. 1, Nature Publishing Group, 2008, p. 216.
[132] Erzsébet Ravasz, Anna Lisa Somera, Dale A. Mongru, Zoltán N. Oltvai, A.-L. Barabási, Hierarchical organization of modularity in metabolic networks, Science 297 (5586) (2002) 1551–1555.
[133] Jennifer L. Reed, Thuy D. Vo, Christophe H. Schilling, Bernhard O. Palsson, An expanded genome-scale model of Escherichia coli K-12 (iJR904 GSM/GPR), Genome Biology 4 (9) (2003) R54.
[134] Marie-France Sagot, Spelling approximate repeated or common motifs using a suffix tree, in: Proc. of the 3rd Latin American Symposium on Theoretical Informatics, LATIN'98, Campinas, Brazil, 1998, pp. 374–390.
[135] Naruya Saitou, Masatoshi Nei, The neighbor-joining method: a new method for reconstructing phylogenetic trees, Molecular Biology and Evolution 4 (4) (1987) 406–425.
[136] G.K. Sandve, F. Drabløs, A survey of motif discovery methods in an integrated framework, Biology Direct 1 (Apr 2006) 11.
[137] T.D. Schneider, R.M. Stephens, Sequence logos: a new way to display consensus sequences, Nucleic Acids Research 18 (20) (Oct 1990) 6097–6100.
[138] Robert Sedgewick, Kevin Wayne, Algorithms, Addison-Wesley Professional, 2011.
[139] Fabian Sievers, Andreas Wilm, David Dineen, Toby J. Gibson, Kevin Karplus, Weizhong Li, Rodrigo Lopez, Hamish McWilliam, Michael Remmert, Johannes Söding, et al., Fast, scalable generation of high-quality protein multiple sequence alignments using Clustal Omega, Molecular Systems Biology 7 (1) (2011) 539.
[140] H.O. Smith, K.W. Wilcox, A restriction enzyme from Hemophilus influenzae. I. Purification and general properties, Journal of Molecular Biology 51 (2) (Jul 1970) 379–391.

[141] Temple F. Smith, Michael S. Waterman, Identification of common molecular subsequences, Journal of Molecular Biology 147 (1) (1981) 195–197.
[142] R. Sokal, C. Michener, A statistical method for evaluating systematic relationships, University of Kansas Science Bulletin 28 (1958) 1409–1438.
[143] E.L. Sonnhammer, S.R. Eddy, E. Birney, A. Bateman, R. Durbin, Pfam: multiple sequence alignments and HMM-profiles of protein domains, Nucleic Acids Research 26 (1) (Jan 1998) 320–322.
[144] G.D. Stormo, DNA binding sites: representation and discovery, Bioinformatics 16 (1) (Jan 2000) 16–23.
[145] G.D. Stormo, G.W. Hartzell, Identifying protein-binding sites from unaligned DNA fragments, Proceedings of the National Academy of Sciences of the United States of America 86 (4) (Feb 1989) 1183–1187.
[146] Eric Talevich, Brandon M. Invergo, Peter J.A. Cock, Brad A. Chapman, Bio.Phylo: a unified toolkit for processing, analyzing and visualizing phylogenetic trees in Biopython, BMC Bioinformatics 13 (1) (2012) 209.
[147] Julie D. Thompson, Desmond G. Higgins, Toby J. Gibson, CLUSTAL W: improving the sensitivity of progressive multiple sequence alignment through sequence weighting, position-specific gap penalties and weight matrix choice, Nucleic Acids Research 22 (22) (1994) 4673–4680.
[148] M. Tompa, N. Li, T.L. Bailey, G.M. Church, B. De Moor, E. Eskin, A.V. Favorov, M.C. Frith, Y. Fu, W.J. Kent, V.J. Makeev, A.A. Mironov, W.S. Noble, G. Pavesi, G. Pesole, M. Regnier, N. Simonis, S. Sinha, G. Thijs, J. van Helden, M. Vandenbogaert, Z. Weng, C. Workman, C. Ye, Z. Zhu, Assessing computational tools for the discovery of transcription factor binding sites, Nature Biotechnology 23 (1) (Jan 2005) 137–144.
[149] Guido van Rossum, Personal home page, http://legacy.python.org/~guido/.
[150] J. Craig Venter, Mark D. Adams, Eugene W. Myers, Peter W. Li, Richard J. Mural, Granger G. Sutton, Hamilton O. Smith, Mark Yandell, Cheryl A. Evans, Robert A. Holt, et al., The sequence of the human genome, Science 291 (5507) (2001) 1304–1351.
[151] E.T. Wang, R. Sandberg, S. Luo, I. Khrebtukova, L. Zhang, C. Mayr, S.F. Kingsmore, G.P. Schroth, C.B. Burge, Alternative isoform regulation in human tissue transcriptomes, Nature 456 (7221) (Nov 2008) 470–476.
[152] K. Wang, M. Li, D. Hadley, R. Liu, J. Glessner, S.F. Grant, H. Hakonarson, M. Bucan, PennCNV: an integrated hidden Markov model designed for high-resolution copy number variation detection in whole-genome SNP genotyping data, Genome Research 17 (11) (Nov 2007) 1665–1674.
[153] W. Wei, X.D. Yu, Comparative analysis of regulatory motif discovery tools for transcription factor binding sites, Genomics, Proteomics & Bioinformatics 5 (2) (May 2007) 131–142.
[154] Peter Weiner, Linear pattern matching algorithms, in: Switching and Automata Theory, 1973. SWAT'08. IEEE Conference Record of 14th Annual Symposium on, IEEE, 1973, pp. 1–11.
[155] Niklaus Wirth, Algorithms + Data Structures = Programs, Prentice Hall PTR, Upper Saddle River, NJ, USA, 1978.
[156] D.J. Witherspoon, S. Wooding, A.R. Rogers, E.E. Marchani, W.S. Watkins, M.A. Batzer, L.B. Jorde, Genetic similarities within and between human populations, Genetics 176 (1) (May 2007) 351–359.
[157] Chi-En Wu, PythonHMM, https://github.com/jason2506/PythonHMM. (Retrieved October 2017).
[158] Daniel R. Zerbino, Ewan Birney, Velvet: algorithms for de novo short read assembly using de Bruijn graphs, Genome Research 18 (5) (2008) 821–829.
[159] L. Zhang, S. Kasif, C.R. Cantor, N.E. Broude, GC/AT-content spikes as genomic punctuation marks, Proceedings of the National Academy of Sciences of the United States of America 101 (48) (Nov 2004) 16855–16860.
[160] W. Zhang, J. Chen, Y. Yang, Y. Tang, J. Shang, B. Shen, A practical comparison of de novo genome assembly software tools for next-generation sequencing technologies, PLoS ONE 6 (3) (Mar 2011) e17915.
[161] Konrad Zuse, Der Plankalkül. Number 63, Gesellschaft für Mathematik und Datenverarbeitung, 1972.

Index

Printed in the United States
By Bookmasters